Robust Small Area Estimation

In recent years there has been substantial and growing interest in small area estimation (SAE) that is largely driven by practical demands. Here, the term "small area" typically refers to a subpopulation or domain of interest for which a reliable direct estimate, based only on the domain-specific sample, cannot be produced due to small sample size in the domain.

Keywords in SAE are "borrowing strength". Because there are insufficient samples from the small areas to produce reliable direct estimates, statistical methods are sought to utilize other sources of information to do better than the direct estimates. A typical way of borrowing strength is via statistical modelling. On the other hand, there is no "free lunch". Yes, one can do better by borrowing strength, but there is a cost. This is the main topic discussed in this text.

Features

- A comprehensive account of methods, applications, as well as some open problems related to robust SAE
- Methods illustrated by worked examples and case studies using real data
- Discusses some advanced topics including benchmarking, Bayesian approaches, machine learning methods, missing data, and classified mixed model prediction
- Supplemented with code and data via a website

Robust Small Area Estimation: Methods, Applications, and Open Problems is primarily aimed at researchers and graduate students of statistics and data science and would also be suitable for geography and survey methodology researchers. The practical approach should help persuade practitioners, such as those in government agencies, to more readily adopt robust SAE methods. It could be used to teach a graduate-level course to students with a background in mathematical statistics.

MONOGRAPHS ON STATISTICS AND APPLIED PROBABILITY

Recently Published Titles

Introduction to Time Series Modeling
Genshiro Kitigawa 166

Replication and Evidence Factors in Observational Studies
Paul R. Rosenbaum 167

Introduction to High-Dimensional Statistics, Second Edition
Christophe Giraud 168

Object Oriented Data Analysis
J.S. Marron and Ian L. Dryden 169

Martingale Methods in Statistics
Yoichi Nishiyama 170

The Energy of Data and Distance Correlation
Gabor J. Szekely and Maria L. Rizzo 171

Sparse Graphical Modeling for High Dimensional Data: Sparse Graphical Modeling for High Dimensional Data
Faming Liang and Bochao Jia 172

Bayesian Nonparametric Methods for Missing Data and Causal Inference
Michael J. Daniels, Antonio Linero, and Jason Roy 173

Functional Data Analysis with R
Ciprian M. Crainiceanu, Jeff Goldsmith, Andrew Leroux, and Erjia Cui 174

Mixture Models: Parametric, Semiparametric, and New Directions
Weixin Yao and Sijia Xiang 175

Seminal Ideas and Controversies in Statistics
Roderick J.A. Little 176

Theory of Non-Regular Factorial Designs
Ching-Shui Cheng and Boxin Tang 177

Robust Small Area Estimation: Methods, Applications, and Open Problems
Jiming Jiang and J. Sunil Rao 178

For more information about this series please visit: https://www.crcpress.com/Chapman--HallCRC-Monographs-on-Statistics--Applied-Probability/book-series/CHMONSTAAPP

Robust Small Area Estimation

Methods, Applications, and Open Problems

Jiming Jiang and J. Sunil Rao

CRC Press
Taylor & Francis Group
Boca Raton London New York

CRC Press is an imprint of the
Taylor & Francis Group, an **informa** business
A CHAPMAN & HALL BOOK

First edition published 2026
by CRC Press
2385 NW Executive Center Drive, Suite 320, Boca Raton FL 33431

and by CRC Press
4 Park Square, Milton Park, Abingdon, Oxon, OX14 4RN

CRC Press is an imprint of Taylor & Francis Group, LLC

Library of Congress Cataloging-in-Publication Data

ISBN: 978-1-032-48885-1 (hbk)
ISBN: 978-1-032-49719-8 (pbk)
ISBN: 978-1-003-39517-1 (ebk)

DOI: 10.1201/9781003395171

Typeset in CMR10 font
by KnowledgeWorks Global Ltd.

Publisher's note: This book has been prepared from camera-ready copy provided by the authors.

To Thuan, Vanessa and Emily; Darlene, Nalini and Dev

Contents

Foreword

I am delighted to introduce this monograph by Jiang and Rao on robust Small Area Estimation (SAE). The book is a landmark in SAE theory and practice, more generally in small domain estimation, and much of the content is accessible to masters-level statisticians and data scientists. Coverage is a nice balance between the authors' own and the work of others. In a small way, emphasis on *small*, I motivated the book by inviting Sunil and Jiming to prepare an article [126] for the *Annual Review of Statistics and its Application.*

The SAE context is important in its own right, but also provides an informative platform for identifying fundamental statistical principles and practices. These include trading off variance and bias to produce a relatively small Mean Square Error (MSE), and the ability to confer valuable robustness to model mis-specification with a small sacrifice in efficiency when the model is correct. Robustness is very important in applications because it increases trust in findings. In the SAE and other statistical contexts, there is no free lunch (the trade-off), but there are a wide variety of reduced-price meals (pay a small amount to purchase considerable benefit).

The need to trade-off variance and bias is evident when estimating an area-specific attribute. Unless the area's sample size is large, the direct, unbiased estimate that uses only the information specific to the area will have high variance. Pooling information from nearby areas will stabilize the estimate, but generally increase bias. This hands-on approach generalizes to using a weighted average of the direct estimate and a model-based one; the former unbiased with relatively high variance, the latter, possibly biased, but with relatively low variance.

In a similar vein, methods optimal under specific conditions (probability and sampling distributions, inferential goals) can be substantially sub-optimal when the true situation departs from these (the best of a breed may still be a dog). Happily, there are robust modeling approaches that provide a degree of protection from departures, and in many instances pay a small efficiency price for this protection when the motivating assumptions are correct. Commonly, achieving this robustness is not likelihood-based, and requires the bootstrap or another approach to quantifying uncertainty.

The book features Observed Best Prediction (OBP), one of the authors' signature contributions to SAE with broader, strategic implications. It's an example of going beyond likelihood-based modeling to achieve robustness. OBP addresses the situation wherein the shrinkage target (e.g., a fixed-effects model) is mis-specified and, consequently, so is the putatively best predictor (their EBLUP) which is a weighted average of the direct estimate and the model-based target. To variance-reduce, EBLUP puts higher weight on the model for areas with relatively high-variance direct estimates, and so deficiencies in the model have a larger effect on these estimates. On the other hand, likelihood-based estimation of the target model puts more weight on the relatively low variance areas. So, the areas that "care about" the model have less influence in estimating it. If the target model is properly specified, inferences will still be "optimal", but if it is mis-specified, performance will be relatively poor for these areas.

To address this issue, rather than using the likelihood, the OBP approach assigns additional weight to the relatively high variance areas when estimating the model, so that the shrinkage target is better adapted to them. If the target model is correct, OBP gives up some likelihood-based optimality to purchase this robustness, but the cost is generally low. I leave the details to Chapter 4 and call your attention to their "thurstoning" analogy as a way to think about the approach. Going further, in the linear case, the EBLUP is a weighted average of the direct and the model-based estimates. To produce another effective trade-off, the Compromise Best Predictor (CBP) uses a weighted average (yet another weighted average) of the EBLUP and OBP, with weights based on the relative, empirical goodness of fit of the two candidate procedures.

The foregoing trade-offs are a centerpiece of the book, but are not the full story. Rao and Jiang give serious attention to various threats to validity, including model mis-specification and model checking; accommodating outliers and missing data. They describe and evaluate flexible models such as splines and machine learning approaches, and complement content on linear models with the analysis of discrete data and other non-linear structures. Additional reasons to study this book include the highlighting of finite population models and design-based inference, some content on privacy/confidentiality (one of several open problems), inclusion of real data examples, and an excellent bibliography.

I would have liked to have seen some content on the implications of the approaches to study design. Also, the Bayesian formalism and fully Bayesian analyses have potential as an alternative to using the bootstrap for uncertainty estimation, and to implement non-linear models. These require development and could well appear in the second edition.

Careful study of this book will reward the reader with increased understanding of issues and approaches associated with modern SAE, and the collateral benefit of a tour that touches all bases of a comprehensive statistical assessment.

Thomas A. Louis, PhD
St. Michaels, MD, USA

Preface

In recent years there has been substantial, and growing, interest in small area estimation (SAE) that is largely driven by practical demands. Here the term *small area* typically refers to a subpopulation or domain of interest for which a reliable direct estimate, based only on the domain-specific sample, cannot be produced due to small sample size in the domain. Examples of small areas include geographical regions (e.g., state, county, municipality), demographic groups (e.g., specific age×sex×race groups), demographic groups within geographic regions, etc. Such small areas are often of primary interest, for example, in policymaking regarding allocation of resources to subgroups, or determination of subgroups with specific characteristics (e.g., in health and medical studies) in a population. It is desirable that the decisions regarding such policymaking be made based on reliable estimates. For such reasons, demands and interest in SAE research have increased rapidly in recent years.

For example, SAE is nowadays routinely used for effective planning of health, social, and other services, and for apportioning government funds in the United States, Canada, and many European countries. Between 2013 – 2024, with some gaps due to the pandemic and political turmoil, there has been an Annual International Conference on SAE and related topics in various locations worldwide. Reviews on SAE and related topics can be found, for example, in [40], [77], [116], [192], and [199].

A keyword in SAE is called "borrowing strength" (originally coined by Tukey). Because there are insufficient samples from the small areas to produce reliable direct estimates, statistical methods are sought to utilize other sources of information in order to do better than the direct estimates. This effort of trying to do better by utilizing no additional information is what we call borrowing strength. A typical way of borrowing strength is via statistical modeling.

We are going to discuss various types of statistical models but, before we do this, let us remember another popular quote (originally thought to have been coined by the famed economist Milton Friedman) that says there is no "free lunch". Yes, one can do better by borrowing strength, but there is a cost. The cost is that doing better is not always happening; in fact, in some cases efforts of borrowing strength may make things worse. This is a main topic to be discussed in this monograph. For example, one must understand the advantage, and limitation, of statistical modeling, and one must deal with the consequence of model failure.

In fact, robust SAE has received considerable attention in recent years. See [130] for a recent review on this topic. While significant progress has been made in various directions, there remain open problems related to the topic. In particular, the current SAE methods face new challenges from modern data science, such as big data and privacy-protection. For example, the U.S. Census Bureau has released its 2020 decennial data products with differential privacy (DP). This kind of DP data products is expected to present a growing challenge to SAE. Are the current methods robust to DP-error contamination? If not, what changes need to be made to meet these challenges? Also, as data collections are becoming increasingly more possible in the modern era, so are the definitions of "large" and "small" in terms of sample sizes. What are the impacts of the new definitions to SAE, theoretically and practically? The existing methods may not be "robust" to handle these new challenges.

It is our intention to provide a nearly comprehensive account of methods, theory and applications, as well as some open problems related to robust SAE. It is also our hope that the monograph will help persuade practitioners, such as those in government agencies, to more readily adopt robust SAE methods.

We begin with a brief review of traditional SAE methods including direct and indirect estimation, as well as existing software packages available for implementing SAE methods. We then discuss SAE methods that are developed under weaker assumptions. These methods tend to be more robust to violation of strong assumptions. We next move to discuss SAE methods that are robust in various ways, such as in terms of outliers or model failure. From there, we discuss issues around specific topics such as non-parametric SAE methods, model selection, and diagnostics. The last part of the book discusses further issues such as benchmarking, Bayesian approaches, machine learning methods, missing data, and classified mixed model prediction.

Although extensive references are cited, it should be noted that the text is not a literature review. We intend to cover methods, or developments, that are characteristic within each specific topic. In doing so, a particular piece of work may be chosen to discuss in detail simply because the work is representative of several similar developments. Conversely, a reference may not be cited not because we are not aware of the work (though we might not), but because literature review is not our main purpose.

The book is suitable for a graduate-level (master's or PhD) course in statistics or data science. It may also serve as a reference book for researchers and practitioners. A course in calculus and a first course in mathematical statistics are prerequisites.

Finally, the authors are grateful to their colleagues, in particular, to Professors J. N. K. Rao and P. Lahiri for helpful comments on a draft version of the book and for referring relevant work to the authors to consider. Our special thanks also go to Professor Thomas Louis for writing a delightful Foreword. The authors also wish to thank Dr. Mandi Yu and current PhD student Jiangshan Zhang for their contributions in providing data and discussion regarding a differential privacy algorithm developed by the U.S. Census Bureau.

The authors acknowledge funding support from the National Science Foundation of the United States, Grants DMS-1914465 and DMS-2210569 to Jiming Jiang, and DMS-2210208 to J. Sunil Rao.

Jiming Jiang and J. Sunil Rao
Davis, CA and Twin Cities, MN, USA

Authors

Jiming Jiang is a professor of statistics and chair of the Department of Statistics at the University of California, Davis. His research interests include mixed effects models, small area estimation, model selection, statistical genetics/bioinformatics, and asymptotic theory. He is the author of over 100 peer-reviewed publications and six books/monographs including *Linear and Generalized Linear Mixed Models and Their Applications* (Springer, 2007, 2021), *Large Sample Techniques for Statistics* (Springer 2010, 2022), *The Fence Methods* (World Scientific 2015; with Thuan Nguyen), *Asymptotic Analysis of Mixed Effects Models: Theory, Application, and Open Problems* (Chapman & Hall/CRC, 2017), *Robust Mixed Model Analysis* (World Scientific, 2019), and *Robust Small Area Estimation: Methods, Theory, Applications and Open Problems* (Chapman & Hall/CRC, 2025; with J. Sunil Rao). Dr. Jiang has served on editorial boards of several major statistical journals including *Annals of Statistics* and *Journal of the American Statistical Association.* He is a fellow of the American Association for the Advancement of Science, the American Statistical Association (ASA), and the Institute of Mathematical Statistics, as well as an elected member of the International Statistical Institute. Dr. Jiang is a co-recipient of the Outstanding Statistical Application Award (ASA, 1998), the first co-recipient of the Distinguished Alumni Award (National Institute of Statistical Sciences, 2015), a Yangtze River Scholar (chaired professor, 2017-2020), a primary speaker of the Morris Hansen Lecture (Washington Statistical Society, 2023), and a recipient of the Award for Outstanding Contribution to Small Area Estimation (the SAE Award, 2024).

J. Sunil Rao is a professor in the Division of Biostatistics and Health Data Science at the University of Minnesota, Twin Cities. He is also the director of biostatistics at the Masonic Cancer Center and Founding Chair and professor emeritus in the Division of Biostatistics at the University of Miami. His research interests include mixed modelling, small area estimation, high-dimensional data analysis, modelling of cancer genomic data, and statistical methods for health disparity research. Dr. Rao is author of over 100 peer-reviewed publications and two books/monographs, including *Statistical Methods in Health Disparity Research* (Chapman & Hall/CRC, 2023) and *Robust Small Area Estimation: Methods, Theory, Applications and Open Problems* (Chapman & Hall/CRC, 2025; with Jiming Jiang). He is a fellow of the American Statistical Association and the Institute of Mathematical Statistics, as well as an elected member of the International Statistical Institute. He has served as an associate editor for a number of different statistical journals. Dr. Rao received the V.K. Gupta Endowment Award for Achievements in Statistical Thinking and Practice (2024) and was appointed as an honorary member of the Society for Statistics, Computers and Applications (2024).

1

Small Area Estimation: A Brief Overview

1.1 Introduction

Surveys are traditionally designed to obtain information about large populations. However, for effective planning of health, social, and other services, and for apportioning government funds, there has been growing demand for providing similar information for sub-populations or domains. Such sub-populations or domains may be large in number. For example, in the United States, there are 3,143 counties, which may be viewed as the sub-populations. If the sub-populations are further divided according to demographic variables such as age, gender, and ethnicity, the resulting domains are even much smaller and the total number of such domains is much larger. Typically, the domain sample size is proportional to the size of the domain. Thus, even though the total sample size for the entire U.S. is quite large, say, 370,000,000, when it comes to a small domain formed by sub-dividing a county by age/gender/race groups, the domain sample size can be very small.

Clearly, reliable estimates of characteristics of interest are not possible based on such a small sample. The challenge then is to find a way to produce more reliable estimates than the estimates based on the domain samples only; the latter are called direct estimates. This "business" of finding ways to do better than the direct estimates is called small area estimation, or SAE. Although it is desirable to do better than the direct estimates, to better understand SAE, let us first review some basic elements of the direct estimates.

1.2 Direct estimators

To be specific, let the characteristic of interest be the domain population mean. A standard direct estimator is in the form of

$$\bar{y}_w = \frac{\sum_{i \in S} w_i y_i}{\sum_{i \in S} w_i}, \tag{1.1}$$

DOI: 10.1201/9781003395171-1

where S in the set of indexes corresponding to the sampled units and $w_i = \pi_i^{-1}$ is the sampling weight, with π_i being the inclusion probability for unit i. See, for example, Section 2.4 of [163]. Note that the inclusion probability depends on the sampling scheme. We illustrate with examples.

Example 1.1 (Simple random sampling). If every unit in the domain has the same probability of being sampled, the sampling scheme is called simple random sampling (SRS). Let N be the domain population size, that is, the total number of units in the domain, and n be the sample size, that is, the cardinality of S, $n = |S|$. Then, each unit in the domain has the probability n/N of being sampled. It follows that $\pi_i = n/N$ for each unit i. Thus, (1.1) becomes

$$\bar{y} = \frac{\sum_{i \in S}(N/n)y_i}{(N/n)|S|} = \frac{\sum_{i \in S} y_i}{n}, \tag{1.2}$$

which is simply the sample mean.

Example 1.2 (Stratified random sampling). Suppose that there are H strata and let the population size for the hth stratum be N_h, $h = 1, \dots, H$. The sampling is said to be stratified random sampling (STR) if an SRS is drawn from each stratum, and the samples from different strata are drawn independently. Let S_h be the index set for the sampled units from stratum h, and $n_h = |S_h|$ the corresponding sample size. The stratified sample mean is defined as

$$\bar{y}_{\text{str}} = \sum_{h=1}^{H} \frac{N_h}{N} \bar{y}_h, \tag{1.3}$$

where $N = \sum_{h=1}^{h} N_h$ is the total population size, and $\bar{y}_h$ is the sample mean for the hth stratum, that is, $\bar{y}_h = n_h^{-1} \sum_{j \in S_h} y_{hj}$ with y_{hj} being the jth sampled unit from stratum h. If we define $\pi_{hj} = n_h/N_h$ for any unit j in stratum h, then the right side of (1.1) is equal to

$$\begin{aligned}
\frac{\sum_{h=1}^{H} \sum_{j \in S_h} w_{hj} y_{hj}}{\sum_{h=1}^{H} \sum_{j \in S_h} w_{hj}} &= \frac{\sum_{h=1}^{H} \sum_{j \in S_h} (N_h/n_h) y_{hj}}{\sum_{h=1}^{H} \sum_{j \in S_h} (N_h/n_h)} \\
&= \frac{\sum_{h=1}^{H} N_h \bar{y}_h}{\sum_{h=1}^{H} N_h} \\
&= \sum_{h=1}^{H} \left(\frac{N_h}{N}\right) \bar{y}_h,
\end{aligned}$$

where is the same as the stratified sample mean, $\bar{y}_{\text{str}}$.

Along with the direct estimators are corresponding variance estimators. For example, the variance estimator for (1.2) is

$$\widehat{\text{var}}(\bar{y}) = \left(1 - \frac{n}{N}\right) \frac{s^2}{n}, \tag{1.4}$$

where $s^2 = (n-1)^{-1}\sum_{i\in S}(y_i - \bar{y})^2$; the variance estimator for (1.3) is

$$\widehat{\mathrm{var}}(\bar{y}_{\mathrm{str}}) = \sum_{h=1}^{H}\left(\frac{N_h}{N}\right)^2\left(1-\frac{n_h}{N_h}\right)\frac{s_h^2}{n_h}, \tag{1.5}$$

where $s_h^2 = (n_h-1)^{-1}\sum_{j\in S_h}(y_{hj}-\bar{y}_h)^2$. These variance estimators are constructed such that they are unbiased estimators of $\mathrm{var}(\bar{y})$ and $\mathrm{var}(\bar{y}_{\mathrm{str}})$, respectively, with respect to the corresponding sampling scheme, that is, SRS for $\bar{y}$, and STR for $\bar{y}_{\mathrm{str}}$. See, for example, [163] for details.

The direct estimates do not borrow strength in the sense that the estimator for a given domain does not utilize information from other domains; it also does not make use of information from other sources.

1.3 Design-based indirect estimators

The term design-based means that the method is derived based on sampling from the finite populations or domains without utilizing a model. Throughout this monograph, a model must involve the relationship between variables. So, if only one variable is involved, even if some assumptions are made, the method is design-based rather than model-based according to our definition. Model-based methods will be discussed later.

1.3.1 Synthetic estimators

A simple case of synthetic estimator is to estimate a small area characteristic (e.g., a small area mean), by an estimator of a similar characteristic for a larger population. For example, if the small area is a sub-population of a larger population, the small area mean is estimated by the estimator of the mean of the larger population. The latter may be more stable due to the (much) larger sample size; however, it may be biased if the population mean of the small area is different from the mean of the larger population.

Clearly, for the above synthetic estimator to work well, an implicit assumption is that the small-area and larger population means are equal, or at least approximately equal; otherwise, the estimator is biased. On the other hand, the estimator has a smaller variance due to the larger sample size, compared to the direct estimator of the small area. Thus, there is a bias-variance trade-off.

To illustrate more clearly, let θ be a small area characteristic and $\hat{\theta}$ be its estimator. The mean squared error of the estimator is defined as $\mathrm{MSE}(\hat{\theta}) = \mathrm{E}(\hat{\theta}-\theta)^2$, which has the following decomposition:

$$\mathrm{MSE}(\hat{\theta}) = \{\mathrm{bias}(\hat{\theta})\}^2 + \mathrm{var}(\hat{\theta}), \tag{1.6}$$

where $\text{bias}(\hat{\theta}) = \text{E}(\hat{\theta}) - \theta$. If the variance term dominates the right side of (1.6), the synthetic estimator is likely to do better than the direct estimator. Conversely, if the square of bias dominates the MSE decomposition, the synthetic estimator may not do better than the direct estimator; in fact, it may do worse than the direct estimator in terms of the MSE. In this regard, the synthetic estimator may not be robust if the implicit underlying assumption mentioned in the previous paragraph is not satisfied.

In a way, the above synthetic estimator tries to borrow strength from a larger population that includes the small area as a sub-population. There is another way of borrowing strength through post-stratification. The latter means that the population is divided into strata after the samples are drawn (e.g., [163], sec. 4.4). Suppose there are H post-strata that divide each of the m small areas. The population sizes of the post-stratified small areas are $N_{ih}, 1 \le i \le m, 1 \le h \le H$. Consider estimating the population total of the ith small area. Suppose that an estimator of the hth post-stratum total, $\hat{Y}_{\cdot h}$, is available, $1 \le h \le H$. Let $N_{\cdot h} = \sum_{i=1}^{m} N_{ih}$ be the population size of the hth stratum. Then, a synthetic estimator of the population total of the ith small area is given by

$$\hat{Y}_i = \sum_{h=1}^{H} \left(\frac{N_{ih}}{N_{\cdot h}} \right) \hat{Y}_{\cdot h}. \tag{1.7}$$

It is clear that (1.7) is a weighted average of estimators of the post-stratum totals, where the weights depend on the small-area specific stratum totals. It is also seen that the same post-stratum total estimators are used in all of the small-area total estimators, the only difference being the weights; this way, different small areas can borrow strength from each other.

For more discussion of synthetic estimators, see Chapter 3 of [199]. Note that the latter reference also includes regression-based synthetic estimators, which we consider model-based. The model-based methods are discussed later.

1.3.2 Composite estimator

Suppose that the interest is an estimation of a small area mean, and the small area is a sub-population of a larger population. As noted, the larger population sample mean has smaller variance, but it can be biased. On the other hand, the small area sample mean is unbiased under the standard sampling designs, but its variance can be large due to the small sample size. One way to trade-off the bias with variance is to consider a linear combination of the two:

$$\hat{\theta}_i = w_i \bar{y}_i + (1 - w_i)\bar{y}, \tag{1.8}$$

where $\bar{y}_i$ is the sample mean for the ith small area and $\bar{y}$ is the sample mean of the larger population, and $0 \le w_i \le 1$ is a weight, which may be area-dependent. (1.8) is called a composite estimator. When $w_i = 1$, (1.8) assigns all the weight to $\bar{y}_i$, which is the direct estimator. In this case, there is no borrowing of strength; the bias is zero but the variance is large. When $w_i = 0$,

(1.8) assigns all the weight to $\bar{y}$, which relies entirely on borrowing strength; the variance of this estimator is (much) smaller but it can be seriously biased. It seems that a strategy with either no borrowing of strength or entirely relying on borrowing of strength may not be the optimal one.

At least conceptually, an optimal solution can be derived by finding the optimal weight, w_i, that minimizes the MSE. Let θ_i be the true small area mean. Then, by (1.8), we have $\hat{\theta}_i - \theta_i = w_i(\bar{y}_i - \theta_i) + (1 - w_i)(\bar{y} - \theta_i)$. Thus we have the following expression

$$\begin{aligned} \text{MSE}(\hat{\theta}_i) &= w_i^2 M_{11} + 2w_i(1 - w_i)M_{12} + (1 - w_i)^2 M_{22} \\ &= (M_{11} - 2M_{12} + M_{22})w_i^2 - 2(M_{22} - M_{12})w_i + M_{22}, \quad (1.9) \end{aligned}$$

where $M_{11} = \text{E}(\bar{y}_i - \theta_i)^2$, $M_{12} = \text{E}(\bar{y}_i - \theta_i)(\bar{y} - \theta_i)$, and $M_{22} = \text{E}(\bar{y} - \theta_i)^2$. If one ignores the fact the $0 \le w_i \le 1$, then by differentiating (1.9) with respect to w_i, and setting the derivative equal to zero, one obtains the following solution:

$$\tilde{w}_i = \frac{M_{22} - M_{12}}{M_{11} - 2M_{12} + M_{22}} = \frac{\text{E}(\bar{y} - \bar{y}_i)(\bar{y} - \theta_i)}{\text{E}(\bar{y}_i - \bar{y})^2}. \quad (1.10)$$

Note that, however, (1.10) is not necessarily a solution, even theoretically, because it is not guaranteed to be in $[0, 1]$; regardless, it is not a practical solution anyway, because θ_i is unknown. See [199] (sec. 3.3.1) for a discussion about the approximately optimal weight for the composite estimator.

Later, we will discuss a model-based approach to obtaining the optimal weight for the composite estimator.

1.4 Regression methods

When data of auxiliary variables are available, strength may be borrowed by utilizing a regression relationship. In this approach, the small area characteristic, θ_i, is estimated by a linear combination of the associated auxiliary variables, say, $x_{i1}, \dots, x_{ip}$, that is,

$$\hat{\theta}_i = \hat{\beta}_0 + \hat{\beta}_1 x_{i1} + \cdots + \hat{\beta}_p x_{ip}, \quad (1.11)$$

$i = 1, \dots, m$, where m is the number of small areas for which auxiliary variable data are available. The $\hat{\beta}_i$, $0 \le j \le p$ in (1.11) are estimated regression coefficients, for example, via the least squares (LS) method. Suppose that survey data of the outcome variables, y, and auxiliary variables, $x_1, \dots, x_p$, are available for the sampling units $k \in \mathcal{S}$. One then fits a linear regression model

$$y_k = \beta_0 + \beta_1 x_{1,k} + \cdots + \beta_p x_{p,k} + \epsilon_k, \quad k \in \mathcal{S}, \quad (1.12)$$

where $y_k, x_{1,k}, \dots, x_{p,k}$ are $y, x_1, \dots, x_p$ for the sampling unit k. The LS estimator of $\beta = (\beta_0, \beta_1, \dots, \beta_p)'$ is given by

$$\hat{\beta} = (X'X)^{-1}X'Y, \tag{1.13}$$

where X is the $n \times (p+1)$ matrix $(x_{j,k})_{k \in \mathcal{S}, 0 \le j \le p}$ with $n = |\mathcal{S}|$, the cardinality of $\mathcal{S}$ or overall sample size, and Y is the $n \times 1$ vector $(y_k)_{k \in \mathcal{S}}$. Here it is assumed that X is full-rank in that $X'X$ is non-singular. There is another, more important underlying assumption, that is, the linear regression model (1.12) holds, which means that the errors $\epsilon_k, k \in \mathcal{S}$ must satisfy certain assumptions. See, for example, Chapter 2 of [66]. When such assumptions are violated, an issue of robustness naturally arise regarding the small area estimator, (1.11).

In a way, the estimator (1.11) depends entirely on the regression model (1.12). One may modify the estimator to make it more robust to model failure. One way to do so is to make it less rely on the regression model. Consider the estimation of the small area mean. The small-area sample mean, $\bar{y}_i$, is a direct estimator which is design-unbiased, that is $\mathrm{E}(\bar{y}_i) = \theta_i$, where the expectation is with respect to the finite-population sampling according to the sampling design. However, as noted, $\bar{y}_i$ may have large variance due to the small sample size. A compromise between the direct estimator and the regression estimator is a composite estimator similar to (1.8), with $\bar{y}$ replaced by $x_i'\hat{\beta}$, where $x_i = (1, x_{i1}, \dots, x_{ip})'$, as in (1.11), and $\hat{\beta}$ is given by (1.13). Here, similarly, there is a question on how to choose the weight. We shall discuss this later under a model-based approach.

The regression method is a model-based method. However, it has some shortcomings. First, as noted, the linear regression relationship, (1.12), may not hold. One may try to improve the model by adding more terms to the right side of (1.12), for example, quadratic, and cubic terms; some more sophisticated models may also be considered, such as a spline model (e.g., [188]). Even so, there is no guarantee that one will get a "correct model". Second, there may be heterogeneity among the small areas, and this is not taken into consideration in the regression model.

One way to address these issues is to introduce some small-area specific random effects, say, v_i for the ith small area. The random effect represents the difference between the unknown small area mean and the linear regression predictor, which would be the small area mean when the linear regression model holds. In other words, the random effects "captures" whatever is not captured by the linear regression, something that adding the quadratic, cubic, or spline terms is trying to do, as mentioned above.

Furthermore, the random effect is a natural way to account for the heterogeneity. Roughly speaking, different small areas have different random effects; this difference in addition to that can be explained by the difference in the auxiliary variables.

A model that involves random effects is called a mixed effects model. See, for example, [123]. For the most part, there are three classes of mixed effects models: linear mixed models (LMMs), generalized linear mixed models

(GLMMs), and nonlinear mixed effects models. So far as SAE is concerned, the first two classes are the most relevant. In the next section, we introduce some basic SAE models in these classes.

1.5 Basic small area models

According to [116], there are three basic types of mixed effects models widely used in SAE. We review these three types of models in subsections below.

1.5.1 Fay-Herriot model

It is not uncommon that data are available only at the area level. In fact, even if data are originally collected from sampling units within the small areas, oftentimes only summary statistics, such as the sample mean, is reported for each small area. Thus, area-level models are widely used in SAE, among which the most popular one is the Fay-Herriot model [59]. It was originally proposed by Fay and Herriot for the estimation of per-capita income of small places with population sizes less than 1,000. The model is given by a combination of a sampling model and a linking model. For the sampling model, one assumes the direct estimates are related to the true population values as

$$y_i = \theta_i + e_i, \quad i = 1, \dots, m, \tag{1.14}$$

where e_i is a sampling error. The linking model next formalizes the relationship between the population values and a set of area-level auxiliary variables or covariates x_i as,

$$\theta_i = x_i'\beta + v_i, \quad i = 1, \dots, m, \tag{1.15}$$

where the v_i are area-specific random effects. The model may then be expressed in terms of a linear mixed model (LMM), as

$$y_i = x_i'\beta + v_i + e_i, \quad i = 1, \dots, m, \tag{1.16}$$

It is assumed that v_i's, e_i's are independent with $v_i \sim N(0, A)$ and $e_i \sim N(0, D_i)$. The variance A is unknown, but the sampling variance, D_i, is known, $1 \le i \le m$.

In practice, the sampling variances may not be known exactly; however, they may be estimated with a higher degree of accuracy; thus, the D_i's may be regarded approximately as known. We consider an example.

Example 1.3 (Hospital data). Morris and Christiansen [181] presented a data set involving 23 hospitals (out of a total of 219 hospitals) that had at least 50 kidney transplants during a 27-month period. Specifically, the y_i's are graft failure rates for kidney transplant operations, that is, y_i = number of

TABLE 1.1
Hospital data.

Area	y_i	x_i	$\sqrt{D_i}$
1	0.302	0.112	0.055
2	0.140	0.206	0.053
3	0.203	0.104	0.052
4	0.333	0.168	0.052
5	0.347	0.337	0.047
6	0.216	0.169	0.046
7	0.156	0.211	0.046
8	0.143	0.195	0.046
9	0.220	0.221	0.044
10	0.205	0.077	0.044
11	0.209	0.195	0.042
12	0.266	0.185	0.041
13	0.240	0.202	0.041
14	0.262	0.108	0.036
15	0.144	0.204	0.036
16	0.116	0.072	0.035
17	0.201	0.142	0.033
18	0.212	0.136	0.032
19	0.189	0.172	0.031
20	0.212	0.202	0.029
21	0.166	0.087	0.029
22	0.173	0.177	0.027
23	0.165	0.072	0.025

graft failures $/n_i$, where n_i is the number of kidney transplants at hospital i during the period of interest. The variance for the graft failure rate, D_i, is approximated by $(0.2)(0.8)/n_i$, where 0.2 is the observed failure rate for all hospitals. Thus, the D_i's are approximately known. The severity index, x_i, taken as the average fraction of females, blacks, children, and extremely ill kidney patients at hospital i is considered a covariate. The data are presented in Table 1.1 for convenience. Ref. [67] proposed a Fay-Herriot model, which is (1.16) with $x_i'\beta = \beta_0 + \beta_1 x_i$ and the D_i given above, to fit the data. Here $m = 23$.

1.5.2 Nested-error regression model

When data are available at the unit level, a standard SAE model is the nested error regression (NER) model first proposed by Battese, Harter, and Fuller [8]. The NER model may be expressed as

$$y_{ij} = x_{ij}'\beta + v_i + e_{ij}, \tag{1.17}$$

$i = 1, \ldots, m, j = 1, \ldots, n_i$, where the v_i's are the area-specific random effects and e_{ij}'s are errors so that they are independent and normally distributed with mean zero, $\text{var}(v_i) = \sigma_v^2$ and $\text{var}(e_{ij}) = \sigma_e^2$, where σ_v^2 and σ_e^2 are unknown variances. We use an example to illustrate.

Example 1.4 (Iowa-crops data). Ref. [8] presented data from 12 Iowa counties obtained from the 1978 June Enumerative Survey of the U.S. Department of Agriculture as well as data obtained from land observatory satellites on crop areas involving corn and soybeans. Here, the objective was to estimate the mean hectares of corn and soybeans per segment for those 12 counties using the satellite information. The authors proposed an NER model, expressed as (1.17), in which $x_{ij}'\beta = \beta_0 + \beta_1 x_{ij1} + \beta_2 x_{ij2}$. Here, i represents county and j segment within the county; y_{ij} is the number of hectares of corn (or soybeans); x_{ij1} and x_{ij2} are the number of pixels classified as corn and soybeans, respectively, according to the satellite data. The characteristics of interest are mean hectares of crops, which can be expressed as $\theta_i = \bar{X}_i'\beta + v_i$, where $\bar{X}_{ij1}$ and $\bar{X}_{ij2}$ are the mean numbers of pixels classified as corn and soybeans per segment, respectively, which are available.

1.5.3 Mixed logistic model

Binary responses are often encountered in, for example, health-related surveys. For such type of data, a logistic model is often used. [115] proposed a mixed logistic model for SAE with a binary outcome variable. Let y_{ij} denote the jth binary outcome observation from the ith small area. Let v_i be a random effect associated with the ith small area, $1 \le i \le m$. It is assumed that given $v = (v_i)_{1 \le i \le m}$, y_{ij}, $i = 1, \dots, m$, $j = 1, \dots, n_i$, are conditionally independent binary such that

$$\text{logit}\{\text{P}(y_{ij} = 1 | v_i)\} = x_{ij}'\beta + v_i, \tag{1.18}$$

where $x_{ij} = (x_{ijk})_{1 \le k \le p}$ is a vector of auxiliary variables that are available and $\beta = (\beta_k)_{1 \le k \le p}$ is a vector of unknown parameters. Furthermore, the random effects $v_1, \dots, v_m$ are independent and distributed as $N(0, \sigma^2)$ with σ^2 being an unknown variance. The mixed logistic model is a special case of the generalized linear mixed model (GLMM; e.g., [123]). Again, we use an example to illustrate.

Example 1.5. The Behavioral Risk Factor Surveillance System (BRFSS) is a Centers for Disease Control and Prevention, USA coordinated, state-based random-digit-dialing telephone survey. One survey question of particular interest had to do with the use of mammography among women aged 40 or older, from 1993 to 1995. The survey covered areas from three federal regional offices: Boston (including Maine, Vermont, Massachusetts, Connecticut, Rhode Island, and New Hampshire), New York (including New York and New Jersey), and Philadelphia (including Pennsylvania, Delaware, Washington DC, Maryland, Virginia, and West Virginia). There were a total of 118 health service areas (HSAs) in the region. The HSAs were treated as small areas (see [114]). Initial analysis of the data suggested that mammography rate gradually increases from the age group 40–44 to the age group 50–54, then starts to decrease. To catch this curvature phenomenon, [114] proposed a mixed logistic model for the proportion of women having had mammography. Here, the random effect corresponding to the HSA. Given the HSA effect, the proportion of women having had mammography, p, satisfies

$$\begin{aligned}\text{logit}(p) &= \beta_0 + \beta_1 * \text{age} + \beta_2 * \text{age}^2 + \beta_3 * \text{Race} + \beta_4 * \text{Edu} \\ &\quad + \text{HSA effect},\end{aligned}$$

where Age is grouped as 40–44, 45–49, ..., 75–79, and 80 and above; Race is categorized as white (1) and others (0); and Edu is the percentage of people in the HSA aged 25 or older with at least high school education.

1.6 Other types of SAE models

There are other types of models, mostly mixed effects models, that have been proposed in SAE. In this subsection we provide a brief summary.

1. Poisson models for small-area counts. In the context of small-area counts, that is, when the outcome variable is a count (e.g., the number of deaths due to heart disease; number of cigarettes consumed within a month), a Poisson model may be appropriate. Ref. [34] considered two models for area-level count data. The two models are the same at the data-level. They are different at the mean-level, one being a gamma model without random effects and the other being a GLMM.

At the data-level, the outcome data, denoted by y_i, and a vector of auxiliary variables, x_i, are available. It assumes that, given the area means, $\mu_i, 1 \leq i \leq m$, where m is the number of small areas, $y_i, 1 \leq i \leq m$ are independent such that $y_i \sim \text{Poisson}(\mu_i)$.

At the mean-level, the first model assumes that the $\mu_i, 1 \leq i \leq$ are independent such that $\mu_i \sim \text{Gamma}(x_i'\beta/\phi, \phi)$, where $\text{Gamma}(\alpha, \phi)$ is the Gamma distribution with shape parameter α, scale parameter ϕ, and probability density function (pdf) given by

$$f(u|\alpha, \phi) = \frac{1}{\Gamma(\alpha)\phi^\alpha} u^{\alpha-1} e^{-u/\phi}, \quad u > 0.$$

The second model is a mixed-effect log-normal model, a special case of GLMM (e.g., [123]), expressed as $\log(\mu_i) = x_i'\beta + v_i$, where β is a vector of fixed effects, and v_i is an area-specific random effect that is assumed to be independent, for different i's, and distributed as $N(0, \sigma^2)$ for some unknown variance σ^2.

Example 1.6. Ghosh *et al.* [71] used Poisson area-level models to estimate county-level COVID-19 deaths in 2023 in Florida. Specifically, they used case-tracking data from the state of Florida which we have updated as of February 24, 2023 in Table 1.2. Direct death counts are listed as the two covariates they found to be important – confirmed cases in a county and adjusted gross

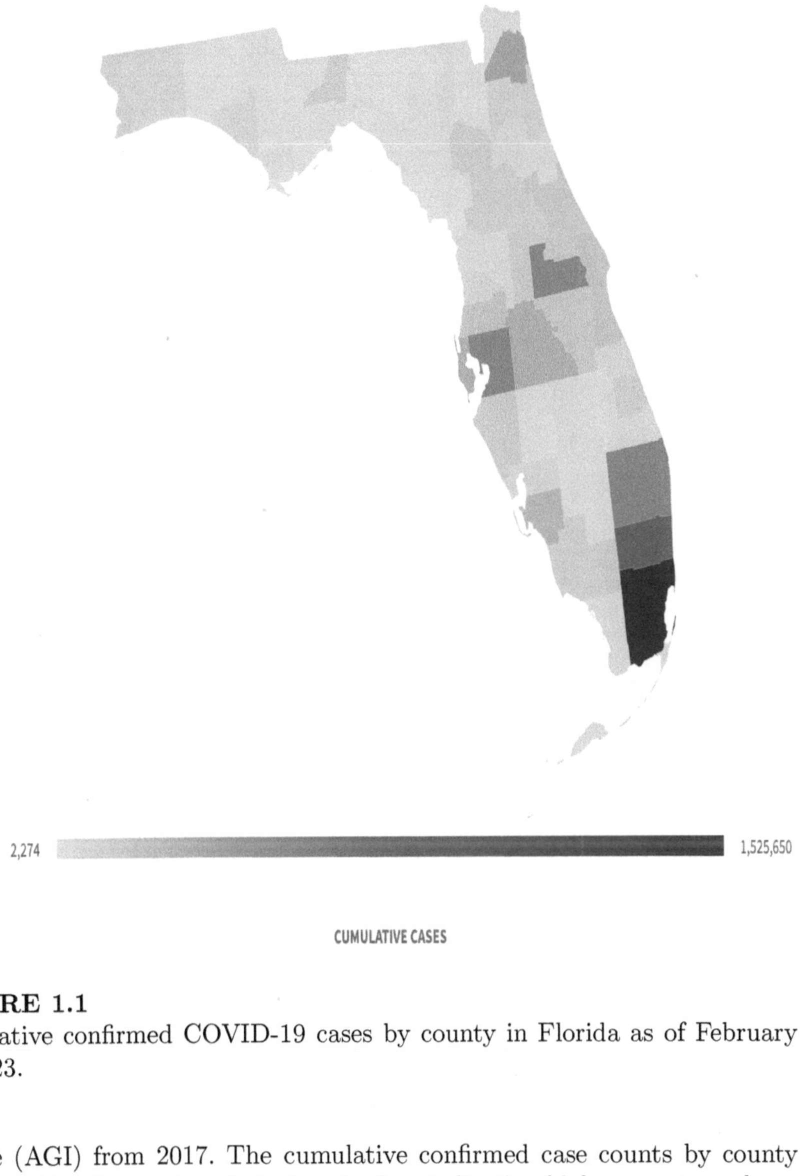

FIGURE 1.1
Cumulative confirmed COVID-19 cases by county in Florida as of February 24, 2023.

income (AGI) from 2017. The cumulative confirmed case counts by county are plotted in Figure1.1 with darker shading indicating higher count numbers. The data consist of $m = 57$ counties and [71] transformed their direct counts using a square root transformation thus allowing them to assume that $\sqrt{y}_i \sim N(x_i'\beta, A + 0.25)$ which is a FH model but with homoscedastic known error variance.

2. Spatial LMM for SAE. Quite often, the small areas are associated with geographical regions. It is possible that data from nearby regions are more correlated than from regions far away. To take such spatial correlations into

TABLE 1.2
COVID-19 cases and deaths in Florida as of February 24, 2023.

County	Confirmed Cases	AGI (2017) in millions	Deaths
Alachua	87980	7.294066	285
Baker	10595	0.554345	62
Bay	54736	4.571565	394
Bradford	9092	0.478624	58
Brevard	172577	17.959964	914
Broward	763410	68.207212	3079
Calhoun	4864	0.207429	45
Charlotte	47809	5.233347	437
Citrus	38510	3.447436	463
Clay	66314	6.127044	352
Collier	103860	27.616972	551
Columbia	25323	1.304030	170
DeSoto	11160	0.527165	97
Dixie	4864	0.540883	24
Duval	313854	28.523854	1483
Escambia	93405	8.068327	706
Flagler	28610	3.411798	114
Franklin	3514	0.283821	20
Gadsden	17448	0.824111	99
Gilchrist	5556	0.328444	44
Glades	2274	0.232870	19
Gulf	4975	0.321757	45
Hamilton	4498	0.190548	25
Hardee	10237	0.417308	45
Hendry	13641	0.709752	87
Hernando	53134	3.958244	482
Highlands	30347	1.882105	366
Hillsborough	472896	46.284944	1832
Holmes	6285	0.280332	49
Indian River	42065	7.607307	306
Jackson	17933	0.819253	158
Jefferson	4941	0.309690	28
Lafayette	3110	0.106457	26
Lake	108592	9.534123	663
Lee	237793	27.089733	1009
Leon	111744	7.918907	332
Levy	12595	0.721669	56
Liberty	2964	0.113382	16
Madison	6950	0.324513	45
Manatee	121207	13.804729	689
Marion	104146	7.971880	996
Martin	39875	9.243197	335
Miami-Dade	1525650	92.544722	6472
Monroe	22278	5.227991	52
Nassau	26923	3.389901	127
Okaloosa	61193	6.812245	367
Okeechobee	12320	0.720267	89
Orange	469296	41.729653	1310
Osceola	142222	7.340821	529
Palm Beach	472649	84.441622	2883
Pasco	157416	13.333365	802
Pinellas	271522	35.959433	1671
Polk	256399	15.491829	1387
Putnam	20970	1.217151	160
Santa Rosa	58396	5.192317	291
Sarasota	119052	20.149336	844
Seminole	130489	16.199289	516
St. Johns	77617	12.966328	221
St. Lucie	94666	7.683491	657
Sumter	29653	4.709867	284
Suwannee	15722	0.701966	140
Taylor	8022	0.362501	48
Union	5072	0.217377	74
Volusia	148608	14.096377	846
Wakulla	12031	0.710564	56
Walton	19521	3.317614	89
Washington	7639	0.397700	52

Note: Adapted from [71].

consideration, [227] proposed a spatial LMM for SAE. It is an area-level model, assuming that the area-level data satisfy

$$y_i = x_i'\beta + z_i'v_i + e_i, \quad i = 1, \dots, m, \tag{1.19}$$

where y_i, x_i, β have the same meanings as in (1.16); and z_i is a known vector. As for the area-level random effect, v_i, and error, e_i, it is better described when (1.19) is written in a matrix form as

$$y = X\beta + Zv + e, \tag{1.20}$$

where $y = (y_i)_{1\le i\le m}$, $X = (x_i')_{1\le i\le m}$, $Z = \text{diag}(z_i', 1 \le i \le m)$, $v = (v_i)_{1\le i\le m}$, and $e = (e_i)_{1\le i\le m}$. It is assumed that v, e are independent with $v \sim N(0, G)$ and $e \sim N(0, R)$. Here the covariance matrices G and R are assumed to have some structures. In particular, the structure of G may incorporate certain spatial correlations, for example,

$$G = \sigma_1(I_m - \sigma_2 D)^{-1}C^{-1},$$

where σ_1, σ_2 are parameters accounting for spatial dispersion and spatial correlation, respectively; D is an $m \times m$ matrix with elements $d_{ii} = 0$ and $d_{ij} = w_{ij}/w_{i+}$ if areas i and j are adjacent, denoted by $i \sim j$, and $d_{ij} = 0$ otherwise. Here, w_{ij} are known weights, and $w_{i+} = \sum_{j\sim i} w_{ij}$. Furthermore, $C = \text{diag}(w_{i+}, 1 \le i \le m)$. The simplest model for R is $R = \tau^2 I_m$, where τ^2 is an unknown variance.

Example 1.7. May *et al.* [172] reported on work from the USDA Forest Service's Forest Inventory and Analysis (FIA) program in which they constructed estimates of forest attributes using spatial FH models based on a quasi-systematic sample of field plots distributed across the contiguous US. In particular, they made small-area estimates of above-ground biomass density (ABGD) within 64000-hectare hexagons across the US using data from the Global Ecosystem Dynamics Investigation (GEDI) and the National Land Cover Database 2016 Tree Canopy Cover (TCC) as auxiliary data in the model. Figure 1.2 shows the potential informativeness of each auxiliary variable in the spatial FH model. In fact, [172] discovered that the model with both auxiliary variables provided the most accurate estimates.

3. Semi-parametric/non-parametric mixed models. There have been uses of semi-parametric or non-parametric mixed effects models in SAE. For example, [164] considered the following semi-parametric mixed model:

$$g(\text{E}(y_{ij}|v_i, t_{ij}, x_{ij})) = \lambda(t_{ij}) + x_{ij}'\beta + z_{ij}'v_i, \tag{1.21}$$

where $g(\cdot)$ is a known link function; v_i is a vector-valued random effect; x_{ij}, t_{ij} are observed vectors of auxiliary variables or covariates, and z_{ij} is a subvector of $(1, x_{ij}')'$. Furthermore, $\lambda(\cdot)$ is an unknown function. Note that (1.21) is different from a GLMM because the function $\lambda(\cdot)$ is unknown.

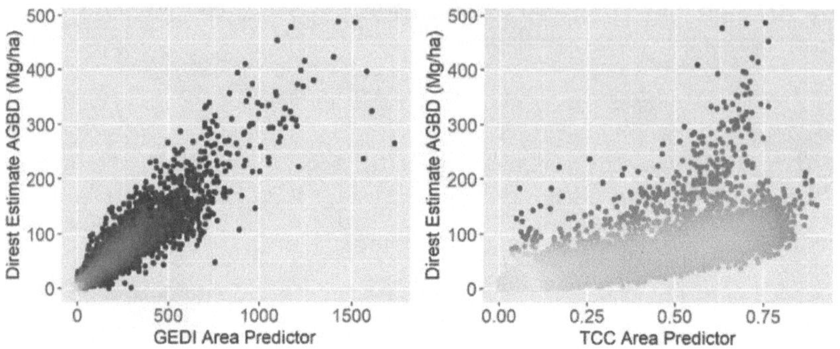

FIGURE 1.2
Plots of direct estimates of ABGD versus the GEDI and TCC auxiliary variables for all 64000 hexagons in the US sample. (Taken from [172].)

In terms of the non-parametric models, [188] proposed a spline-based non-parametric mixed effects model in the context of SAE. In the case of area-level model, this is an extension of the Fay-Herriot model, expressed as

$$y_i = f(x_i) + v_i + e_i, \quad i = 1, \dots, m,$$

where the assumptions about v_i and e_i are the same as in (1.16), but $f(\cdot)$ is an unknown function like the $\lambda(\cdot)$ in (1.21). In order to make inference about $f(\cdot)$, [188] used a P-spline approximation to $f(\cdot)$. Here P-spline refers to penalized spline, which uses penalization methods to ensure smoothness of the fitting function and avoid over-fitting. Here, the P-spline can be expressed as

$$\tilde{f}(x) = \beta_0 + \beta_1 x + \cdots + \beta_p x^p + \gamma_1 (x - \kappa_1)_+^p + \cdots + \gamma_q (x - \kappa_q)_+^p, \quad (1.22)$$

where p is the degree of the spline, q is the number of knots, κ_j, $1 \le j \le q$ are the knots, and $x_+ = x 1_{(x>0)}$.

Example 1.8. In [188], they analyzed data from the Environmental Monitoring Assessment Program (EMAP) between 1991 and 1996 where a survey was conducted of 344 lakes out of a population of 21,026 lakes in the northeastern U.S. Some of these lakes were repeatedly sampled resulting in a total of 551 measurements. The authors were interested in estimating acid neutralizing capacity (ANC) for each of 113 small areas that were defined by Hydrologic Unit Codes (HUC). Auxiliary covariates of interest included geographical coordinates (UTM style) and elevation. They fit a non-parametric FH small area model using a bivariate radial spline function for the two-dimensional geographical coordinate covariates and a linear term for the elevation covariates. Figure 1.3 shows the knot locations used to model the bivariate spline for the geographical coordinate covariates.

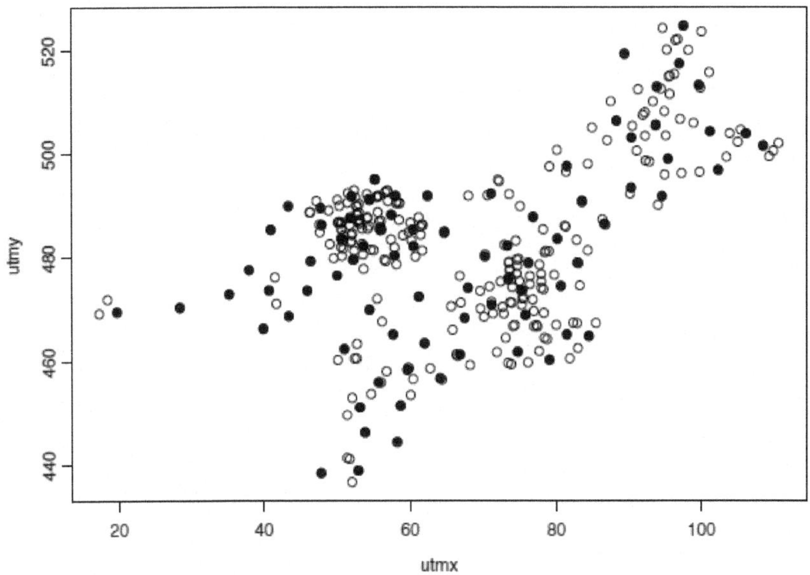

FIGURE 1.3
Knot locations for the geographical coordinates in the estimation of ANC in the northeastern US. (Taken from [188].)

1.7 Mixed model prediction

Under a mixed effects model, a small area characteristic, such as a small area mean, can be expressed as a mixed effect. For example, under the Fay-Herriot model (1.16), the small area mean is $\theta_i = x_i'\beta + v_i$, which is a linear mixed effect. There has been a well-developed field, called mixed model prediction (MMP), with well-developed methods of estimating, or predicting, the mixed effects. See, for example, sec. 2.3 of [123]. The term prediction is used because a mixed effect is a random variable, as opposed to a fixed parameter (see [206]). Therefore, methods of MMP naturally apply to SAE problems.

A general prediction can be formulated as follows. Let ζ be an unobserved random variable, and y a vector of observations. A predictor of ζ, $\hat{\zeta}$, is a function of y. The prediction accuracy is typically measured by the mean squared prediction error (MSPE), defined as $\mathrm{E}(\hat{\zeta} - \zeta)^2$. Here, y is also considered a (vector-valued) random variable, although, unlike ζ, y is observable; thus, $\hat{\zeta}$ is also a random variable. The expectation in the MSPE is with respect to the joint distribution of ζ and y. Theoretically, it is known that the best predictor

(BP) of ζ, in the sense of minimizing the MSPE, is

$$\tilde{\zeta} = \mathrm{E}(\zeta|y). \tag{1.23}$$

Although (1.23) is known to have the minimum MSPE, it may not be computable because the underlying distribution may be unknown. For example, it may involve an unknown vector of parameters, say, ψ. In other words, the BP can be expressed as

$$\tilde{\zeta} = g(y, \psi) \tag{1.24}$$

for some function $g(\cdot, \cdot)$. We use an example to illustrate.

Example 1.9. Consider the Fay-Herriot model of sec. 1.5.1. It can be shown that the BP of $\theta_i = x_i'\beta + v_i$ is

$$\tilde{\theta}_i = \frac{A}{A + D_i} y_i + \frac{D_i}{A + D_i} x_i'\beta. \tag{1.25}$$

It is clear that the BP is a function of y (through y_i only) and the vector of unknown parameters $\psi = (\beta', A)'$. Note that this function also depends on i through D_i, which is known.

When the BP involves unknown parameters, the parameters ψ are replaced by a (consistent) estimator, $\hat{\psi}$, if it is available. This leads to the empirical best predictor (EBP), given by (1.24) with ψ replaced by $\hat{\psi}$, that is,

$$\hat{\zeta} = g(y, \hat{\psi}). \tag{1.26}$$

In the case of LMM, the EBP is typically done in two steps. Consider the LMM (1.20). Here there are two types of unknown parameters, one is β, the vector of fixed effects, and one is γ, the vector of dispersion parameters, or variance components, involved in G and R. Now consider a linear mixed effect, expressed as $\theta = a'\beta + b'v$, where β, v are the vectors of fixed and random effects in (1.20). Under the normality assumption, the BP of θ is given by

$$\mathrm{E}(\theta|y) = a'\beta + b'\mathrm{E}(v|y), \quad \mathrm{E}(v|y) = GZ'V^{-1}(y - X\beta), \tag{1.27}$$

assuming that β and γ are known. To obtain the EBP, first assume γ is known. Then, β is replaced by its maximum likelihood estimator (MLE), given by

$$\tilde{\beta} = (X'V^{-1}X)^{-1}X'V^{-1}y, \tag{1.28}$$

where $V = R + ZGZ'$. (1.28) is also known as the best linear unbiased estimator (BLUE) of β (e.g., [123], sec. 2.3.2). With β in (1.27) replaced by the BLUE, one obtains the best linear unbiased predictor (BLUP) of θ,

$$\tilde{\theta} = a'\tilde{\beta} + b'\tilde{v}, \quad \tilde{v} = GZ'V^{-1}(y - X\tilde{\beta}). \tag{1.29}$$

Note that the BLUP is still not computable unless γ is known. One then replaces the γ in (1.29) by a consistent estimator, $\hat{\gamma}$, leading to

$$\hat{\theta} = a'\hat{\beta} + b'\hat{v}, \quad \hat{v} = \hat{G}Z'\hat{V}^{-1}(y - X\hat{\beta}), \tag{1.30}$$

where $\hat{G}, \hat{V}$ are G, V, respectively, with γ replaced by $\hat{\gamma}$, and $\hat{\beta} = (X'\hat{V}^{-1}X)^{-1}X'\hat{V}^{-1}y$. Methods to obtain a consistent estimator of γ include maximum likelihood and restricted maximum likelihood (REML). See, for example, sec. 1.3 of [123]. Eq. (1.30) is referred to as the empirical BLUP (EBLUP) It is clear that EBLUP is a special case of EBP.

The BP typically does not have an analytic expression when the underlying mixed model is not linear. For example, under the mixed logistic model of sec. 1.5.3, suppose that one is interested in predicting a mixed effect in the form of $\theta = g(x'\beta + v_i)$ for a known vector x and function $g(\cdot)$ (not necessarily linear). The BP of θ is given by

$$\begin{aligned} \tilde{\theta} &= \mathrm{E}(\theta|y) \\ &= \frac{\mathrm{E}[g(x'\beta + \sigma\xi)\exp\{y_{i\cdot}\sigma\xi - \sum_{j=1}^{n_i}\log(1 + e^{x'_{ij}\beta+\sigma\xi})\}]}{\mathrm{E}[\exp\{y_{i\cdot}\sigma\xi - \sum_{j=1}^{n_i}\log(1 + e^{x'_{ij}\beta+\sigma\xi})\}]}, \end{aligned} \tag{1.31}$$

where $y_{i\cdot} = \sum_{j=1}^{n_i} y_{ij}$, and the expectations are taken with respect to $\xi \sim N(0,1)$. We consider some special cases of (1.31).

Example 1.10. If the auxiliary variables are at the area level, that is, $x_{ij} = x_i$, and $g(u) = \mathrm{logit}^{-1}(u) = e^u/(1 + e^u)$, then, (1.31) reduces to

$$\tilde{p} = \frac{\mathrm{E}[\mathrm{logit}^{-1}(x'\beta + \sigma\xi)\exp\{y_{i\cdot}\sigma\xi - n_i\log(1 + e^{x'_i\beta+\sigma\xi})\}]}{\mathrm{E}[\exp\{y_{i\cdot}\sigma\xi - n_i\log(1 + e^{x'_i\beta+\sigma\xi})\}]}, \tag{1.32}$$

which is the BP of $p = \mathrm{logit}^{-1}(x'\beta + v_i)$. In this case, the mixed effect is an area-specific (conditional) probability, such as the proportion of women in an HSA (see Example 1.5) having had mammography in the past year.

Example 1.11. If $x = 0$, and $g(u) = u$, (1.31) reduces to

$$\tilde{v}_i = \sigma\frac{\mathrm{E}[\xi\exp\{y_{i\cdot}\sigma\xi - \sum_{j=1}^{n_i}\log(1 + e^{x'_{ij}\beta+\sigma\xi})\}]}{\mathrm{E}[\exp\{y_{i\cdot}\sigma\xi - \sum_{j=1}^{n_i}\log(1 + e^{x'_{ij}\beta+\sigma\xi})\}]}, \tag{1.33}$$

which is the BP of v_i, the area-specific (e.g., HAS) random effect.

To obtain the EBP, the unknown parameters β, σ in (1.31) are replaced by their consistent estimators, such as the MLE or method of moments (MoM) estimators; see, for example, ch. 4 of [123]. In particular, computation of MLE under a GLMM can be numerically challenging; see Sec. 4.1 of [123].

1.8 Assessing uncertainty in SAE

The most popular uncertainty measure in SAE is the MSPE. Prediction intervals, which combine prediction with uncertainty measures, are also widely used. Below we focus on these two types of measures.

1.8.1 Estimation of MSPE

Although the application of MMP to SAE is relatively straightforward, especially in the linear case, an important issue in SAE is measures of uncertainty. In practice, not only the numerical values of the SAE estimates are needed, it is also desirable to provide uncertainty measures, such as the square roots of the MSPEs or prediction intervals. In fact, such topics have been more extensively studied in SAE than in any other field.

One of the challenging problems in this context is estimation of the MSPE of the EBLUP. Although computation of the EBLUP is relatively simple, its MSPE is complex due to the fact that the EBLUP is not a linear function of y. On the other hand, the MSPE of BLUP, which is a linear function of y, is much simpler. Let $\tilde{\theta}$ denote the BLUP of θ [see above (1.27)], assuming that γ is known. Then, we have

$$\mathrm{MSPE}(\tilde{\theta}) = g_1(\gamma) + g_2(\gamma), \tag{1.34}$$

where $g_1(\gamma)$ is the MSPE of BP; $g_2(\gamma)$ is the mean squared difference between BLUP and BP, that is, $g_1(\gamma) = \mathrm{MSPE}(\tilde{\theta}_{\mathrm{BP}})$, where $\tilde{\theta}_{\mathrm{BP}} = \mathrm{E}(\theta|y)$, and $g_2(\gamma) = \mathrm{E}(\tilde{\theta} - \tilde{\theta}_{\mathrm{BP}})^2$. Analytically, we have the following expressions:

$$\begin{aligned} g_1(\gamma) &= a'(G - GZ'V^{-1}ZG)a, \\ g_2(\gamma) &= (b - X'V^{-1}ZGa)'(X'V^{-1}X)^{-1}(b - X'V^{-1}Za) \end{aligned}$$

(note that γ is involved in G and V).

However, the MSPE of BLUP underestimates the MSPE of EBLUP. In fact, under the normality assumption, we have another decomposition:

$$\begin{aligned} \mathrm{MSPE}(\hat{\theta}) &= \mathrm{MSPE}(\tilde{\theta}) + \mathrm{E}(\hat{\theta} - \tilde{\theta})^2 \\ &= \mathrm{MSPE}(\tilde{\theta}) + g_3(\gamma) + o(m^{-1}). \end{aligned} \tag{1.35}$$

See [137]. Thus, Combining (1.34) and (1.35), we have

$$\mathrm{MSPE}(\hat{\theta}) = g_1(\gamma) + g_2(\gamma) + g_3(\gamma) + o(m^{-1}). \tag{1.36}$$

Note that, typically, $g_3(\gamma)$ is of the same order as $g_2(\gamma)$; namely, one has $g_1(\gamma) = O(1)$, $g_2(\gamma) = O(m^{-1})$ and $g_3(\gamma) = O(m^{-1})$. Therefore, the g_3 term cannot be simply ignored.

It should also be noted that the actual expression of $g_3(\gamma)$ depends on how the variance components, γ, are estimated in the EBLUP. In contrast, $g_1(\gamma)$, $g_2(\gamma)$ have nothing to do with how one estimates γ; in other words, they remain the same regardless of how one estimates γ. We use an example to illustrate.

Example 1.9 (continued). Note that $\gamma = A$ in this case. The EBLUP, $\hat{\theta}_i$, is given by the right side of (1.25) with A replaced by $\hat{A}$ and β replaced by

$$\hat{\beta} = \left(\sum_{i=1}^{m} \frac{x_i x_i'}{\hat{A} + D_i}\right)^{-1} \sum_{i=1}^{m} \frac{x_i y_i}{\hat{A} + D_i}. \tag{1.37}$$

It can be shown that, in this case, one has

$$\begin{aligned} g_1(A) &= \frac{AD_i}{A+D_i}, \\ g_2(A) &= \left(\frac{D_i}{A+D_i}\right)^2 x_i' \left(\sum_{j=1}^{m} \frac{x_j x_j'}{A+D_j}\right)^{-1} x_i. \end{aligned}$$

As for $g_3(A)$, [45] showed that, if $\hat{A}$ is the REML estimator, then, we have

$$g_3(A) = \frac{2D_i^2}{(A+D_i)^3} \left\{\sum_{j=1}^{m}(A+D_j)^{-2}\right\}^{-1}.$$

Regarding estimation of the MSPE, it is desirable to obtain a second-order unbiased estimator, that is, an MSPE estimator whose bias is $o(m^{-1})$, which is of the same order as the lower-order term on the right side of (1.36). For the most part, there are three main approaches, which we summarize below.

(i) The Prasad-Rao linearization methods. This approach was pioneered by the seminal paper of Prasad and Rao [196]. It is a two-step procedure. In the first step, a second-order approximation of the MSPE is obtained, which is the right side of (1.36) with specified g terms. A naive MSPE estimator is the plug-in estimator, that is, replacing the γ by $\hat{\gamma}$ in those g terms. This way, the bias is $o(m^{-1})$ for estimating g_2 and g_3; however, the bias is $O(m^{-1})$ for estimating g_1, which is not good enough. Thus, in the secondstep, one bias-corrects the plug-in estimator of g_1 to obtain an estimator whose bias is $o(m^{-1})$. Again, we consider an example.

Example 1.9 (continued). If the REML estimator of A is used in the EBLUP, the Prasad-Rao MSPE estimator is given by

$$\widehat{\text{MSPE}}(\hat{\theta}_i) = g_1(\hat{A}) + g_2(\hat{A}) + 2g_3(\hat{A}). \tag{1.38}$$

In other words, the bias-corrected estimator of $g_1(A)$ is $g_1(\hat{A}) + g_3(\hat{A})$. The estimator (1.38) has a bias of $o(m^{-1})$. Another nice feature of this estimator is that it is guaranteed positive, a desirable property of an MSPE estimator. Very few MSPE estimators are known to be both second-order unbiased and guaranteed positive. See [117] for discussion in this regard.

(ii) Re-sampling methods. Ref. [118] proposed a jackknife MSPE estimator. The method is not restricted to LMM or EBLUP. Note the EBLUP may be viewed as an empirical BP (EBP). In general, suppose that the BP of ζ, a mixed effect (not necessarily linear), can be expressed as

$$\tilde{\zeta} = \text{E}(\zeta|y_1, \dots, y_m) = \pi(y_S, \psi), \tag{1.39}$$

where $y_S = (y_i)_{i \in S}$ and $S \subset \{1, \dots, m\}$, and ψ is a vector of parameters. The EBP of ζ, $\hat{\zeta}$, is $\tilde{\zeta}$ with ψ replaced by $\hat{\psi}$, a consistent estimator. Similar to (1.35), we have the following decomposition:

$$\text{MSPE}(\hat{\zeta}) = \text{MSPE}(\tilde{\zeta}) + \text{E}(\hat{\zeta} - \tilde{\zeta})^2. \tag{1.40}$$

The idea is, therefore, to jackknife the two terms on the right side of (1.40) separately, and then combine the results. Suppose that MSPE($\tilde{\zeta}$) $= b(\psi)$, a known function of ψ. Then, MSPE($\tilde{\zeta}$) can be jackknifed by

$$b(\hat{\psi}) - \frac{m-1}{m}\sum_{i=1}^{m}\{b(\hat{\psi}_{-i}) - b(\hat{\psi})\}, \tag{1.41}$$

where $\hat{\psi}$ is an M estimator of ψ (see [118]) and $\hat{\psi}_{-i}$ is the jackknife replication of $\hat{\psi}$. On the other hand, $\mathrm{E}(\hat{\zeta} - \tilde{\zeta})^2$ can be jackknifed by

$$\frac{m-1}{m}\sum_{i=1}^{m}(\hat{\zeta}_{-i} - \hat{\zeta})^2, \tag{1.42}$$

where $\hat{\zeta}_{-i} = \pi(y_S, \hat{\psi}_{-i})$. Note that, in $\hat{\zeta}_{-i}$, one cannot delete the ith group of observations from y_S; in other words, the "naive" jackknife does not work here. The combined jackknife estimator of MSPE($\hat{\zeta}$) is obtained by summing the right sides of (1.41) and (1.42). Ref. [117] developed a Monte-Carlo version of the jackknife, called McJack, for estimating the log-MSPE. The method does not involve a term like (1.42). The jackknife estimator is known to be second-order unbiased for estimating the MSPE. The McJack is second-order unbiased for estimating the log-MSPE.

An alternative approach to re-sampling methods is the bootstrap. Parametric or model-based bootstrap is relatively simpler, at least conceptually. For example, under an LMM, let $\hat{\psi}$ be a consistent estimator of the model parameters, including the fixed effects and variance components. Then, one can bootstrap data, treating $\hat{\psi}$ as the true parameters. For the bootstrapped data, one knows the true mixed effect (because this is how the data are generated) as well as the predictor based on the bootstrapped data. One can therefore compute the empirical MSPE based on the bootstrap and use it as an estimator of the MSPE. See, for example, [19]. Bootstrap without bias-correction typically produces a first-order unbiased MSPE estimator, that is, the bias is $O(m^{-1})$. Ref. [89] proposed a double-bootstrap procedure to bias-correct the bootstrap estimator, resulting in a second-order unbiased MSPE estimator.

Non-parametric bootstrap method is more difficult to develop in the context of mixed effects models. Ref. [88] proposed a version of the bootstrap, which may be considered non-parametric, for SAE under a nested-error regression model [8]. The basic idea is to bootstrap the random effects and errors, whose distributions are assumed unknown, under some "working" distributions whose third and fourth moments match the estimated moments. Again, double-bootstrap is used to achieve the second-order unbiasedness.

(iii) The Sumca method. More recently, [133] proposed an approach to second-order unbiased MSPE estimation, which may be viewed as a hybrid of Prasad-Rao method and re-sampling method. The authors noted that any $O(m^{-1})$ terms do not need bias-correction—a plug-in estimator will have a bias of order $o(m^{-1})$. Thus, they proposed to use an analytic method similar to Prasad-Rao to estimate a leading term of the MSPE, and a Monte-Carlo (bootstrap) method to evaluate a lower-order term. The combined estimator is second-order unbiased, as shown by the authors. The name *Sumca* is an abbreviation for "simple, unified, Monte-Carlo assisted". To illustrate the method, note that the MSPE of $\hat{\zeta}$ can be expressed as

$$\text{MSPE} = \text{E}(\hat{\zeta} - \zeta)^2 = \text{E}\left[\text{E}\{(\hat{\zeta} - \zeta)^2 | y\}\right]. \tag{1.43}$$

Suppose that the underlying distribution of y depends on a vector of unknown parameters, ψ. Then, the conditional expectation inside the expectation on the right side of (1.43) is a function of y and ψ, which can be written as

$$\begin{aligned} a(y, \psi) &= \text{E}\{(\hat{\eta} - \eta)^2 | y\} \\ &= \hat{\eta}^2 - 2\hat{\eta}\text{E}(\eta|y) + \text{E}(\eta^2|y) \\ &= \{\hat{\eta} - \text{E}(\eta|y)\}^2 + \text{var}(\eta|y) \\ &= \{\hat{\eta} - a_1(y, \psi)\}^2 + a_2(y, \psi), \end{aligned} \tag{1.44}$$

where $a_1(y, \psi) = \text{E}(\eta|y)$ and $a_2(y, \psi) = \text{var}(\eta|y)$. For simplicity, let us focus on a special class of predictors, the EBP, that is, $\hat{\eta} = a_1(y, \hat{\psi})$, where $\hat{\psi}$ is a consistent estimator of ψ. Then, by (1.44), we have the simplified expression

$$a(y, \hat{\psi}) = a_2(y, \hat{\psi}) = \text{var}(\eta|y)|_{\psi=\hat{\psi}}. \tag{1.45}$$

If we replace the ψ in $a(y, \psi)$ by $\hat{\psi}$, the result is a first-order unbiased estimator, that is, we have

$$\text{E}\{a(y, \hat{\psi}) - a(y, \psi)\} = O(m^{-1}). \tag{1.46}$$

On the other hand, both MSPE $= \text{E}\{a(y, \psi)\}$ [by (1.43), (1.44)] and $\text{E}\{a(y, \hat{\psi})\}$ are functions of ψ, denoted by $b(\psi)$ and $c(\psi)$, respectively. By (1.46), we have $d(\psi) = b(\psi) - c(\psi) = O(m^{-1})$; thus, if we replace, again, ψ by $\hat{\psi}$ in $d(\psi)$, the difference is a lower-order term, that is, we have

$$d(\hat{\psi}) - d(\psi) = o_{\text{P}}(m^{-1}) \tag{1.47}$$

Now consider the estimator

$$\widehat{\text{MSPE}} = a(y, \hat{\psi}) + d(\hat{\psi}) = a_2(y, \hat{\psi}) + b(\hat{\psi}) - c(\hat{\psi}), \tag{1.48}$$

using (1.45). We have, by (1.43), (1.44), (1.46)–(1.48),

$$\begin{aligned} \text{E}(\widehat{\text{MSPE}}) &= \text{E}\{a(y, \psi)\} + \text{E}\{a(y, \hat{\psi}) - a(y, \psi)\} + \text{E}\{d(\hat{\psi})\} \\ &= \text{MSPE} + \text{E}\{d(\hat{\psi}) - d(\psi)\} \\ &= \text{MSPE} + o(m^{-1}). \end{aligned}$$

Essentially, this one-line, heuristic derivation shows the second-order unbiasedness of the proposed MSPE estimator, (1.48), provided that the terms involved can be evaluated.

Note that, in (1.48), $a(y, \hat{\psi})$ is the leading term. These terms may be evaluated analytically. The lower-order term corresponds to a bias correction. This term is typically much more difficult to evaluate than the leading term. Ref. [133] propose to approximate this term using a Monte-Carlo method. Let P_ψ denote the distribution of y with ψ being the true parameter vector. Given ψ, one can generate y under P_ψ. Let $y_{[k]}$ denote y generated under the kth Monte-Carlo sample, $k = 1, \dots, K$. Then, we have

$$b(\psi) - c(\psi) \approx \frac{1}{K} \sum_{k=1}^{K} \left\{ a(y_{[k]}, \psi) - a(y_{[k]}, \hat{\psi}_{[k]}) \right\}, \qquad (1.49)$$

where $\hat{\psi}_{[k]}$ denotes $\hat{\psi}$ based on $y_{[k]}$. If K is sufficiently large, which one has control over during the Monte-Carlo simulation, the difference between the two sides of (1.49) is $o(m^{-1})$. Write the right side of (1.49) as $d_K(\psi)$. Then, the Sumca estimator is given by

$$\begin{aligned} \widehat{\text{MSPE}}_K &= a(y, \hat{\psi}) + d_K(\hat{\psi}) \\ &= a_2(y, \hat{\psi}) + \frac{1}{K} \sum_{k=1}^{K} \left\{ a(y_{[k]}, \hat{\psi}) - a(y_{[k]}, \hat{\psi}_{[k]}) \right\}, \qquad (1.50) \end{aligned}$$

where $y_{[k]}, k = 1, \dots, K$ are generated as above with $\psi = \hat{\psi}$, and $\hat{\psi}_{[k]}$ is, again, the estimator of ψ based on $y_{[k]}$. Regarding the choice of the Monte-Carlo sample size, K, [133] suggests $K = m$.

1.8.2 Prediction intervals

Under the LMM (1.20), a small-area characteristic, such as the small area mean, can often be expressed in the form of $\zeta = c'(X\beta + Zu)$ for some known constant vector c. Under the normality assumption—that is, $v \sim N(0, G)$, $e \sim N(0, R)$, and v and e are independent—the conditional distribution of ζ given the data y is $N(\tilde{\mu}, \tilde{\sigma}^2)$, where

$$\tilde{\mu} = c'\{X\beta + ZGZ'V^{-1}(y - X\beta)\}, \qquad (1.51)$$
$$\tilde{\sigma}^2 = c'Z(G - GZ'V^{-1}ZG)Z'c, \qquad (1.52)$$

and $V = \text{Var}(y) = R + ZGZ'$. Therefore, for any $\alpha \in (0, 1)$, we have

$$\text{P}\left(\left| \frac{\zeta - \tilde{\mu}}{\tilde{\sigma}} \right| \le z_{\alpha/2} \middle| y \right) = 1 - \alpha,$$

where z_α is the α-critical value of $N(0, 1)$ [i.e., $\Phi(\alpha) = 1 - \alpha$, where Φ is the cdf of $N(0, 1)$]. It follows that the probability is $1 - \alpha$ that the interval

$$I = [\tilde{\mu} - z_{\alpha/2}\tilde{\sigma}, \tilde{\mu} + z_{\alpha/2}\tilde{\sigma}]$$

covers ζ.

Of course, the latter is not a practical solution for the prediction interval, because there are unknown parameters involved. In addition to β, the covariance matrices G and R typically also depend on some unknown dispersion parameters or variance components. Suppose that $G = G(\gamma)$ and $R = R(\gamma)$, where ψ is a q-dimensional vector of variance components, and $\psi = (\beta', \psi')'$. It is customary to replace ψ by an estimator, $\hat{\psi} = (\hat{\beta}', \hat{\gamma}')'$. However, with the replacement of ψ by $\hat{\psi}$, the resulting prediction interval no longer has the coverage probability $1 - \alpha$. We illustrate with an example.

Example 1.9 (continued). Note that the Fay-Herriot model may be thought of equivalently as a two-level model, as follows.
Level 1: Conditional on $\zeta_1, \dots, \zeta_n$, that $y_1, \dots, y_n$ are independent such that $y_i \sim N(\zeta_i, D_i)$.
Level 2: $\zeta_i = x_i'\beta + u_i$, $1 \le i \le n$, where the x_i's are observed vectors of covariates, β is a $p \times 1$ vector of unknown regression coefficients, the u_i's are small-area-specific random effects that are independent and distributed as $N(0, A)$, and $A > 0$ is an unknown variance.
A prediction interval for ζ_i based on the Level 1 model only is given by

$$I_{1,i} = [y_i - z_{\alpha/2}\sqrt{D_i}, y_i + z_{\alpha/2}\sqrt{D_i}].$$

This interval has the right coverage probability $1 - \alpha$ but is hardly useful because its length is too large due to the high variability of the predictor based on a single observation y_i. Alternatively, one may construct a prediction interval based on the Level 2 model only. We consider a special case.

Let $D_i = 1$ and $x_i'\beta = \beta$. At Level 2, the ζ_i's are i.i.d. with the distribution $N(\beta, A)$. If β and A were known, a Level 2 prediction interval that has the coverage probability $1-\alpha$ would be $I_2 = [\beta - z_{\alpha/2}\sqrt{A}, \beta + z_{\alpha/2}\sqrt{A}]$. Note that the interval is not area-specific (i.e., not dependent on i). When β and A are unknown, they are replaced by estimators. The standard estimators are $\hat{\beta} = \bar{y} = n^{-1}\sum_{i=1}^n y_i$ and $\hat{A} = \max(0, s^2 - 1)$, where $s^2 = (n-1)^{-1}\sum_{i=1}^n (y_i - \bar{y})^2$. This leads to the interval

$$\tilde{I}_2 = [\hat{\beta} - z_{\alpha/2}\sqrt{\hat{A}}, \hat{\beta} + z_{\alpha/2}\sqrt{\hat{A}}].$$

But, again, the coverage probability is no longer $1 - \alpha$. However, because $\hat{\beta}$ and $\hat{A}$ are consistent estimators, the coverage probability is $1 - \alpha + o(1)$. One can do better with a simple Level 2 bootstrap, as follows.
(a) Draw independent samples $\zeta_1^*, \dots, \zeta_n^*$ from $N(\hat{\beta}, \hat{A})$.
(b) Draw $y_i^*, i = 1, \dots, n$, independently such that $y_i^* \sim N(\zeta_i^*, 1)$; compute $\hat{\beta}^*$ and $\hat{A}^*$ the same way as $\hat{\beta}$ and $\hat{A}$ except using the y_i^*'.
(c) Determine the cutoff points t_1 and t_2 such that

$$\mathrm{P}^*\left(\hat{\beta}^* - t_1\sqrt{\hat{A}^*} \le \zeta_i^* \le \hat{\beta}^* + t_2\sqrt{\hat{A}^*}\right) = 1 - \alpha,$$

where P^* denotes the bootstrap probability, evaluated using a large number of bootstrap replications.

A (Level 2) bootstrap prediction interval is then given by

$$\hat{I}_2 = [\hat{\beta} - t_1\sqrt{\hat{A}}, \hat{\beta} + t_2\sqrt{\hat{A}}].$$

It can be shown that $\hat{I}_2$ has a coverage probability of $1 - \alpha + o(n^{-1/2})$, which is an improvement but still not accurate enough.

Intuitively, one should do better by combining both levels of the model. In fact, [37] proposed the following empirical Bayes interval for ζ_i using information of both levels:

$$I_i^{\mathrm{C}}: \quad (1 - \hat{B}_i)y_i + \hat{B}_i x_i'\hat{\beta} \pm z_{\alpha/2}\sqrt{D_i(1 - \hat{B}_i)}, \tag{1.53}$$

where $\hat{B}_i$ and $\hat{\beta}$ are (consistent) estimators of $B_i = D_i/(A + D_i)$ and β, respectively. Note that the center of the interval (1.53) is a weighted average of the centers of the Level 1 and Level 2 intervals. If D_i is much larger than A, $1 - \hat{B}_i$ is expected to be close to zero. In this case, the center of I_i^{C} is close to the Level 2 center, but the length will be close to zero and, in particular, much smaller than that of the Level 1 interval (which is $\sqrt{D_i}$). On the other hand, if A is much larger than D_i, then the center of I_i^{C} is expected to be close to the Level 1 center, and the length is close to $\sqrt{D_i}$, much smaller than that of the Level 2 interval (which would be $\sqrt{A}$ if A were known). Therefore, the empirical Bayes interval combines strength of both levels. In fact, under regularity conditions, the coverage probability of I_i^{C} is $1 - \alpha + O(n^{-1})$.

The interval I_i^{C} can be further improved. [32] proposed a parametric bootstrap prediction interval, whose coverage probability is $1 - \alpha + O(n^{-3/2})$.

1.9 A survey of reviews, monographs, and software

1.9.1 Reviews and monographs

One of the earliest reviews on SAE was given in 1994 by Ghosh and Rao [77]. This was the first time that the problems of SAE, as well as the idea of "borrowing strength", were brought to the general attention of the statistical community. The review covered main SAE methods such as synthetic, EBLUP, empirical Bayes (EB), and hierarchical Bayes (HB). Nearly thirty years later, these remain as the main approaches to SAE.

The 1994 paper has stimulated research in SAE as well as applications. The field had expanded when Rao [198] published the first edition of his monograph on SAE in 2003. Since then, a number of review papers have appeared, with relatively more specialized focuses.

The 2006 review of Jiang and Lahiri [116]; with discussion focuses on the use of MMP techniques in SAE. The first part of the review surveyed this unique field of statistics where the prediction of mixed effects is of primary interest. The second part of the review naturally linked MMP to the estimation of small area characteristics, which are viewed as prediction of mixed effects under a mixed effects model. One particular aspect of MMP has to do with measures of uncertainty. This part of the development is mainly done in the context of SAE.

The 2009 review by Datta [40] focuses on model-based SAE methods. It covers both frequentist and Bayesian methods, ranging from basic SAE models to estimation of MSPE, and to multivariate and cross-sectional/time-series SAE models. One feature of the review is that it has significant content on Bayesian methods in SAE.

With a very different focus, the review of Lehtonen and Veijanen [155], published in the same year as the review of [40], targets design-based methods. In contrast to model-based methods, design-based methods are largely model-free, that is, they are based on the mechanism of finite-population sampling. The review covers direct and indirect estimators for domain estimation. At the time, it served as a complement to [40].

Pfeffermann wrote two reviews on developments in SAE, first in 2002 and second in 2013. The second review [192], in particular, surveys new developments after the first review [191], especially since Rao's 2003 monograph [198]. In particular, the survey discusses pros and cons of design-based methods, new developments in robust SAE, benchmarking, robustification, informative sampling and non-response, and model selection and checking in the context of SAE.

The second edition of Rao's 2003 monograph, authored by Rao and Molina and published in 2015 [199], provides comprehensive coverage of SAE, including methods, theory, applications and software code. Compared to the first edition, the second edition provides an up-to-date account of the new developments. In model-based methods, in particular, the new developments include a discussion of the advantage and limitation of various methods, estimators (predictors) that are more robust to model misspecification, measures of uncertainty, model diagnostics and model selection in the context of SAE. In design-based methods, the new developments include information on survey designs, synthetic estimation, and further refinements. Numerous examples are provided, including applications in U.S. Federal programs. Some data sets with R code are also provided.

Most recently, Sugasawa and Kubokawa published a new monograph in 2023 [220]. The authors' approach has some similarity to the 2006 review by Jiang and Lahiri [116], that is, it highlights the connection between mixed effects models and SAE. Natually, the monograph provides a more comprehensive coverage of the topic than the review. The style of the monograph is also very different from [198] and [199] in that it provides a (much) more rigorous theoretical treatment of SAE topics, including theorems and proofs. Additional covered topics include adjusted likelihood methods, observed best prediction (OBP), and SAE models for non-normal responses such as generalized linear mixed models (e.g., [123], ch. 3).

1.9.2 Software

It is informative to know what software is currently in the public domain for implementing some of the SAE methods. While this information is certainly going to change over time, to date, this is what is currently available in R:

The sae package [178] produces the P-R MSPE estimator for the F-H model. The package will also return re-sampling-based (parametric and non-parametric bootstrap) MSPE estimates for various forms of the F-H model as well as unit-level models.

The robustsae package [75] does fully non-subjective Bayesian analysis for general area-level models. Ref. [6] developed an R package, OBPSAE, which implement OBP for both the area level and unit level models. Also implemented in the package is benchmarked OBP methods, which the author developed.

The BIAS project provides an R package called ecoreg along with Win-BUGS resources for fitting a variety of area level and unit level models.

The emdi R package [145] estimates and maps regional disaggregated indicators. Either direct estimation or the EBP approach proposed by Molina and Rao [179], who developed the approach for estimating nonlinear small area population parameters but the approach is applicable to general nonlinear parameters. Estimates of the MSPE for the EBLUPs are done using the parametric bootstrap technique developed in [78].

2

SAE Methods Built on Weaker Assumptions

2.1 Assumption or no assumption?

This might sound like a philosophical question but are there methods that require no assumptions? For example, the sample mean, used as an estimator of the population mean, seems to require no assumption. Really?

To engage in a more serious conversation, we need to be clear about what an assumption means. Here are some points: (i) It should not depend on whether or not it was assumed when the method was proposed, because it might not be assumed in the first place but was later found necessary to ensure some properties of the proposed method. (ii) It is needed only when one cares about the behavior, or performance, of the method. In other words, if one does not care about the performance of the method, no assumption is needed. (iii) There must be solid evidence that the assumption is needed. In other words, one should not impose an assumption that is not needed, or almost not needed. The latter means that the assumption is satisfied in virtually all practical situations.

Having this in mind, let us look again at the sample mean. The method seems to require no assumption because taking an average seems to be a natural thing to do—people have learned to do this ever since they learned how to count, and no one said anything back then. But, when it comes to the performance, the theoretical guarantee behind the sample mean is the law of large numbers (LLN). The latter is not without assumptions; in fact, there are examples, practical ones, in which the sample mean may not perform well. For example, the sample mean may not be robust to outliers. Note that there are finite-population extensions of the limit theorems, including the LNN. See, for example, [66].

It is therefore reasonable to link the assumptions to the performance of the method. Whether or not a method requires assumptions, and what assumptions, are determined by what it takes for the method to have a good performance. The standard measures of statistical performance include unbiasedness, variability, mean squared error (MSE), and large-sample properties such as consistency and asymptotic distribution. In today's data science, there

DOI: 10.1201/9781003395171-2

are also measures of computational performance that need to be taken into consideration. Sometimes, the assumptions are laid out when the method is proposed, so one knows it from the beginning. Sometimes, however, the assumptions are not clearly made when the method is proposed; it may take some arguments, or future studies, to find out what it takes for the method to have a good performance.

Furthermore, some performance measures are considered weaker than others. For example, unbiasedness may be weaker than optimality in terms of the MSE, and consistency is weaker than asymptotic efficiency (e.g., [153], sec. 6.2). Naturally, weaker assumptions are needed for weaker performance guarantee, and stronger assumptions are needed for stronger one.

Generally speaking, design-based methods (see Chapter 1) require weaker assumptions. For the most part, such an assumption requires that the samples are drawn in some probabilistic way so that the standard design-based small-area estimators are unbiased. However, the variance of the estimator may be large if the sample size associated with the small area is relatively small. On the other hand, model-based methods (also see Chapter 1) typically impose stronger assumptions so that the estimator has smaller variance than the design-based estimator when the assumptions hold. However, as is often said, "there is no free lunch". We shall come back to this point, multiple times, in the sequel.

Two classical methods of developing weaker-assumption estimators are the least squares (LS) and the generalized estimating equations (GEE) . The latter may be viewed as an extension of the former. See, for example, Chapter 2 of [112]. The basic assumptions of GEE are in terms of the first two moments of the observations. Sometimes, the assumptions are too weak to obtain estimators with well-established behaviors. We consider some special cases below.

2.2 Robust empirical Bayes estimator

Ghosh and Lahiri [72] considered estimation of stratum means in finite population. Suppose that there are m strata, denoted by $i = 1, \dots, m$. The corresponding stratum population means are $\gamma_i = N_i^{-1}\sum_{j=1}^{N_i} y_{ij}$, where N_i is the population size for stratum i. The assumptions state that there are some "small area means", $\theta_i, 1 \le i \le m$, at a super-population level, that is, an infinite population from which the finite population is viewed as samples. The super-population model is a standard framework used to study theoretical behaviors of finite-population estimators (e.g., [66]). Furthermore, it is assumed that

(i) conditional on $\theta = (\theta_i)_{1\le i\le m}$, $y_{ij}, 1 \le j \le N_i$, $1 \le i \le m$ are independent; $y_{ij}, 1 \le j \le N_i$ have the same conditional distribution such that $\mathrm{E}(y_{ij}|\theta) = \theta_i$, $\mathrm{var}(y_{ij}|\theta) = \mu_2(\theta_i)$, where $\mu_2(\cdot)$ is a known function.

(ii) $\theta_i, 1 \leq i \leq m$ are i.i.d. with mean μ and variance σ^2.
(iii) $0 < \tau^2 = \mathrm{E}\{\mu_2(\theta_1)\} < \infty$, where $\mu_2(\theta_i) = \mathrm{var}(y_{ij}|\theta)$.
Note that assumptions (i)—(iii) do not specify the form of the (conditional) distribution of y_{ij}, such as a normal distribution. Thus, the assumptions are weaker than assuming normality or any specific distribution. But, there is another assumption, called posterior linearity. Let $y = \{y_{ij}, 1 \leq j \leq n_i, 1 \leq i \leq m\}$ be the sampled values of y, and $\bar{y}_i = n_i^{-1} \sum_{j=1}^{n_i} y_{ij}$. The additional assumption states that
(iv) $\mathrm{E}(\theta_i|y) = a_i\bar{y}_i + b_i$, $i = 1, \dots, m$, a_i, b_i are constants.
Under assumptions (i)–(iv), the Bayes estimator of $\gamma = (\gamma_i)_{1 \leq i \leq m}$ can be shown to have simple expressions. The empirical Bayes (EB) estimator is obtained by replacing some hyper-parameters involved in the Bayes estimator by their estimator.

It is not immediately clear how much weaker assumption (iv) is than the normality assumption, which implies (iv). We illustrate with an example.

Example 2.1. Suppose that, (a) conditional on θ_i and τ, $y_{ij}, 1 \leq j \leq n_i$ are i.i.d. and distributed as $N(\theta_i, 1/\tau)$; (b) conditional on τ, $\theta_i, 1 \leq i \leq m$ are i.i.d. and distributed as $N(\mu, (1/\lambda\tau)$, where λ is a positive constant; and (c) τ has a proper prior distribution. Then, it can be shown that (iv) holds with

$$a_i = \frac{n_i}{\lambda + n_i}, \quad b_i = \frac{\lambda\mu}{\lambda + n_i}.$$

Note that, under (a), (b), (c), the distribution of y_{ij} is not normal. Such a distribution may occur when there is heterogeneity in the population. The model may be better explained under a mixed effects model (see Section 1.5.2) except for the prior.

Under the model defined via (i)–(iv), the EB estimator can be developed in very much the same way as the EBLUP; see Section 1.7. In particular, the authors showed that, under suitable conditions, the Bayes risk of the EB estimator approaches that of the Bayes estimator as $m \to \infty$. Because the Bayes estimator has the smallest Bayes risk, the EB estimator is thus in a sense asymptotically optimal. See [72] for detail.

Simulation Study. Ghosh and Lahiri [72] conducted a simulation study to compare the Bayes risk (BR) of the EB estimator of Ghosh and Meeden [74] and the Bayes risk of $e_0 = \bar{Y}$ when the loss is given by $L(e, \gamma) = m^1 \sum_{j=1}^{m}(e_j - \gamma_j)^2$ for some estimators e and the prior described by the above model. They considered a case with $m = 5$ and $n_1 = 1$, $n_2 = 2$, $n_3 = 3$, $n_4 = 5$, and $n_5 = 10$ when the underlying population is normal. The true stratum means were set to $\gamma_1 = 0$, $\gamma_2 = 2$, $\gamma_3 = 4$, $\gamma_4 = 6$, and $\gamma_5 = 3$ and $\mu = 0$. Simulated risks were reported as the average losses after 1000 repetitions of the simulation. Table 2.1 displays their findings.

In the simulation design, exchangeability does not hold among the strata. However, e_{EB} performed better than e_0, indicating a robustness of the EB estimator. The same authors also conducted more detailed comparisons under

TABLE 2.1
Empirical risks over 1000 repetitions of the simulation.

σ	BR for e_0	BR for e_{EB}	Perc. imp. over e_0	BR for e_B
0.3	0.3840	0.3757	2.1614	5.3434
0.5	0.3840	0.3760	2.0833	2.5559
0.8	0.3840	0.3794	1.1979	1.0072
1.0	0.3840	0.3816	0.6250	0.6637
1.5	0.3840	0.3837	0.0781	0.4152
2.0	0.3840	0.3831	0.2344	0.3739
2.5	0.3840	0.3827	0.3385	0.3678

Note: Adapted from Ghosh and Lahiri (1987) [72].

the assumption that the underlying population was binomial, from different beta distributions or Poisson. Overall, they were able to show that the EB estimator was similarly robust as compared to e_0.

2.3 Regression average

Recall the EBP method in the case of a mixed logistic model for binary outcomes (see Section 1.7). The expression of the EBP is not simple, which involves integrals that can only be evaluated numerically; see (1.32), for example. The MLE under the mixed logistic model, which is a special case of GLMM, is also computationally not straightforward to obtain (e.g., [123], sec. 4.1). Furthermore, the EBP relies on assumptions about the distribution of the random effects, and the assumption that the mixed logistic model holds not only for the sampled units but also for the non-sampled units as well. These assumption are often not satisfied in practice.

In an earlier work, Matloff [171] proposed a regression average (RA) method. It assumes a fixed effects logistic regression model, instead of a mixed logistic model. Let y_{ij} be a binary outcome associated with a vector of covariates, x_{ij}. Suppose that the interest is to estimate the unconditional probability, $p_i = \mathrm{P}(y_{ij} = 1)$, which does not depend on j. If (y_{ij}, x_{ij}) are independent, by fitting the logistic regression, $\mathrm{logit}\{\mathrm{P}(y_{ij} = 1|x_{ij})\} = x_{ij}'\gamma$, one obtains the MLE of γ, say, $\hat{\gamma}$. The procedure is computationally straightforward. One then computes the AR estimator of p_i by

$$\hat{p}_i = \frac{1}{n_i}\sum_{j=1}^{n_i} \mathrm{logit}^{-1}(x_{ij}'\hat{\gamma}). \tag{2.1}$$

Comparing (1.32) and (2.1), it is clear that the RA predictor is much simpler, both analytically and computationally.

In general, the RA predictor for the ith small area mean is given by

$$\hat{\mu}_i = \frac{1}{n_i}\sum_{j=1}^{n_i} g(x_{ij}, \hat{\beta}), \tag{2.2}$$

where $g(\cdot,\cdot)$ is a known mean function, that is, $\mathrm{E}(y_{ij}) = g(x_{ij}, \beta)$, and $\hat{\beta}$ is an estimator of the parameter vector β involved in the mean function. Thus, (2.1) is just a special case of (2.2).

It was shown [171] that, for estimating the mean of a single population, the RA predictor is strictly more efficient than the direct estimator, $\bar{y}_{i\cdot}$, unless the mean function is linear. The above result is regarding one population while, in SAE, there are many (sub)populations corresponding to the small areas. The above result of [171] holds if $x_{ij}, y_{ij}, j = 1, \dots, n_i$, for a fixed i, are all of the data we have, hence $\hat{\beta}$ is the least squares (LS) estimator based on these data only. However, if β is estimated by combining all of the data, including $x_{i'j}, y_{i'j}, j = 1, \dots, n_{i'}$, $i' \neq i$ (hence borrowing strength from other small areas), the RA predictor is expected to be more efficient than the direct estimator even when the mean function is linear. We illustrate with a numerical example.

Example 2.2. The R package **lme4** includes a dataset **InstEval**, consisting of 73,421 student evaluations of instructors, with various covariates such as student ID, professor/lecturer ID, student's age, lecturer's teaching experience (measured by the number of semesters), whether the course is a service course, department ID. The main outcome variable is students' rating, with a discrete scale of 1–5. Here, each instructor is considered as a "small area". The average number of evaluations per instructor is 65.09; some instructors have 500 or more evaluations. We took a random sample of 1,000 evaluations; the mean number of the sampled evaluations per instructor was 1.98. A linear regression model was fitted using all of the samples, leading to the RA predictor (2.2), whose mean absolute prediction error (MAPE; over different small areas) was 0.46. Note that the population means for the small areas are known; therefore, the MAPE can be computed. On the other hand, we computed the sample means, $\bar{y}_{i\cdot}$, for the small areas, whose MAPE was 0.81. Thus, in this example, the RA method has a reduction of 43% of MAPE compared to the sample mean (in other words, the MAPE of the sample mean is 76% larger than that of the RA predictor).

Although we use the binary outcome variable and linear regression as examples to illustrate the RA method, the latter, in fact, applies to a more general class of models covering continuous, discrete or categorical responses. As mentioned, the RA method relies on fairly weak assumptions. The basic assumptions are that (y_{ij}, x_{ij}) are i.i.d. within each small area i and have finite second moments and that the mean function, $g(\cdot,\cdot)$ [see (2.2)], is correctly specified. Also, note that the RA predictor is a mean of model-based predictors only. It is possible to consider a weighted average of a direct predictor (e.g., the sample mean) and an RA predictor, which may further improve the RA predictor. Note that the EBP typically can be expressed as a weighted average.

In some cases there are missing outcome data. The missing outcome could occur in some of the small areas due to limited sample size; in such a case the outcome data are still available for other small areas. The missing outcome may also be due to certain privacy concerns, in which case the outcome data may be available for the un-grouped data, $y_k, x_k, k = 1, \dots, N$, but not for the grouped data, $y_{ij}, x_{ij}, i = 1, \dots, m, j = 1, \dots, n_i$, where $N = \sum_{i=1}^m n_i$. In this case, one can still fit the model with the un-grouped data to obtain the parameter estimator, $\hat{\beta}$, which is then used in the RA predictor with the partial grouped data, x_{ij}. In other words, the RA predictor for the ith small area mean is again given by (2.2), where $\hat{\beta}$ is obtained by fitting the model $\mathrm{E}(y_k) = g(x_k, \beta)$ with the un-grouped data. Note that (2.2) also applies to small area i when no outcome data is available from that area but available for some other areas. In such a case, $\hat{\beta}$ may be obtained by fitting the model $\mathrm{E}(y_{ij}) = g(x_{ij}, \beta)$ with the grouped data for which the outcome data are available. As for the areas for which $\bar{y}_{i\cdot}$ is available, one may still use (2.2) as a predictor, which is more efficient than $\bar{y}_{i\cdot}$; alternatively, one may consider a weighted average as mentioned above, which combines $\bar{y}_{i\cdot}$ with the RA predictor.

Although the RA approach is described based on a fixed effects model, it may be extended to a mixed effects model. For example, the method may be used for prediction about future small area means when the data are collected at different time points. In this case, some components of x_{ij} may be time varying. Consider, for example, a mixed effects model, such that
(i) conditional on α_i, $(y_{ijt}, x_{ijt}), 1 \le j \le n_i, 1 \le t \le T$ are i.i.d.;
(ii) $\mathrm{E}(y_{ijt}|x_{ijt}, \alpha_i) = h_i(x_{ijt}, \beta, \alpha_i)$; and
(iii) $\mathrm{var}(y_{ijt}|x_{ijt}, \alpha_i) = v_i(x_{ijt}, \psi)$
for some known functions $h_i(\cdot,\cdot,\cdot)$ and $v_i(\cdot,\cdot)$, where β, ψ are unknown vectors of parameters with possible overlap. Note that the above model assumptions are weaker than the GLMM in that no specific distributional assumption is made for the random effects, α_i. Now suppose that one wishes to make a prediction about the ith small area mean at a future time, say, t_{f}, for which $x_{ijt_{\mathrm{f}}}$ are available. By fitting the mixed effects model with the current data, for example, using the GEE method (e.g., [112], ch. 2), one can obtain estimators of β and ψ, say, $\hat{\beta}$ and $\hat{\psi}$, as well as predictor of α_i, say, $\hat{\alpha}_i$, for example, using the penalized generalized WLS approach (e.g., [123], sec. 3.6.1.3), which again does not require specification of the random effect distribution. One then predicts the future small area mean by (2.2) with $g(x_{ij}, \hat{\beta})$ replaced by $h_i(x_{ijt_{\mathrm{f}}}, \hat{\beta}, \hat{\alpha}_i)$, and n_i replaced by $n_{i,t_{\mathrm{f}}}$, which is the sample size for $x_{ijt_{\mathrm{f}}}, 1 \le j \le n_{i,t_{\mathrm{f}}}$ that are available.

It remains unclear, at this point, whether the result of [171] regarding the superiority of the RA predictor over the direct estimator also holds under the mixed model assumption. An investigation on this would be of practical interest because, if the superiority holds, then even if the outcome data are available, there is a potential advantage of the RA estimator over the direct estimator.

2.4 LMM with non-Gaussian distribution

The standard Fay-Herriot model (see Section 1.5.1) assumes that the random effects and errors are normally distributed. There are concerns that the normality-based methods are not robust to outliers, because the normal distribution has a very "light" tail. One way to obtain an estimator (or predictor) that is more robust to outliers is to develop the estimation procedure based on a heavy-tailed distribution. Such distributions include the t-distribution with a low degree of freedom and the Laplace (or double-exponential) distribution.

Bell and Huang (2006) [11] considered two scenarios in this direction under the Fay-Herriot model (see Section 1.5.1). In the first scenario, the random effect, v_i, has a t distribution with k degree of freedom, suitably normalized to have mean 0 and variance A; the distribution of the sampling error, e_i, remains as $N(0, D_i)$. In the second scenario, the distribution of v_i remains as $N(0, A)$ but e_i is assumed to have a t distribution with k degrees of freedom, suitably normalized to have mean 0 and variance D_i.

It should be noted that the t-distribution assumptions can be expressed in terms of a gamma/normal mixture. Namely, in the first scenario, it is assumed that $v_i|\gamma_i \sim N(0, \gamma A)$ and $1/\gamma_i \sim \text{Gamma}(k/2, (k-2)/2)$. Similarly, under the second scenario, it is assumed that $e_i|\gamma_i \sim N(0, \gamma_i D_i)$ and $1/\gamma_i$ has the same Gamma distribution. In both cases, it implies the (marginal) t-distribution (e.g., [69]). Also note that, because when k is large, the t-distribution approximates the normal distribution, the class of t-distributions, in a way, is wider than the normal distribution. In this sense, the normality assumption is weakened under this setting.

Bell and Huang (2006) [11] develop methods of Bayesian analysis under a Bayesian version of the Fay-Herriot model (see Section 5.2). The implementation involves posterior simulation via Markov-Chain Monte-Carlo (MCMC).

Example 2.3. Bell and Huang (2006) [11] explored univariate t-distribution Fay-Herriot models for the 5-17 year old school age children state poverty ratio model for 1989 and focused their attention on the state of Connecticut (CT), hereafter CT 1989 5-17. Direct poverty estimates from the U.S Current Population Survey (CPS) Annual Social and Economic Supplement (ASEC) were available. The models incorporated covariates related to poverty from administrative data and group-wise poverty estimates from the prior census.

The authors examined standardized residuals from the usual (normal) model and found that for the three younger age groups (0-4, 5-17 and 18-64), the CT in 1989 could be regarded as an outlier which seemed to be due to very low CPS direct estimates. They concluded that for CT 1989 5-17 that the outlier was in the survey error rather than in the model error.

Table 2.2 shows the posterior means of the true poverty ratios for CT when the survey errors are assumed to follow a t-distribution with varying degrees of freedom. As can be seen, under the t-distribution model, the posterior means

TABLE 2.2
Posterior means of true poverty ratio from state poverty models for CT for 1989.

	$df = 3$	$df = 4$	$df = 5$	$df = 8$	$df = \infty$ (normal)
Posterior means	9.7%	10.0%	10.0%	9.8%	8.6%

Note: Adapted from Bell and Huang (2006) [11].

are slightly higher than in the normal model and actually in the direction of the regression fit (10.5%). Therefore, one can conclude that the t-distribution models are giving less weight to the direct estimate.

Bell and Huang (2006) [11] also fitted similar t-distribution Fay-Herriot models for the model errors and found that the posterior means decreased as the df decreased moving away from the regression fit and actually towards the direct estimate for CT (2.2%).

In a recent work, [240] consider a Fay-Herriot model, where the sampling error e_i is assumed to have an asymmetric Laplace (AL) distribution. The AL distribution is an extension of the Laplace distribution, which depends on three parameters, μ, σ and p, where $\mu \in R$, $\sigma > 0$, and $p \in (0,1)$, such that its pdf has the following expression:

$$f(x|\mu,\sigma,p) = \frac{p(1-p)}{\sigma} \exp\left[-\frac{x-\mu}{\sigma}\{p - 1_{(x \le \mu)}\}\right], \quad x \in (-\infty, \infty). \tag{2.3}$$

The distribution (2.3) is denoted by AL(μ,σ,p). Here, σ and p are called scale and kurtosis parameters, respectively. Like the Laplace distribution, the AL distribution has a heavier tail than the normal distribution. There is, however, another purpose of introducing the AL distribution—for quantile mixed-effects model; see [240] for detail. Here, in the context of Fay-Herriot model, the e_i is assume to be distributed as AL$(0, \sqrt{D_i}, p)$; the rest of the model assumption is the same as that of the Fay-Herriot model.

There is a convenience expression of the AL distribution in terms of a normal-variance-mean mixture (e.g., [141]). If $X \sim$ AL(μ,σ,p), X can be expressed via a latent random variable, Z, such that $X|Z \sim N(\mu + \eta Z, \tau^2\sigma Z)$ and $Z \sim$ Exp(σ), where $\eta = (1-2p)/p(1-p)$, $\tau^2 = 2/p(1-p)$, and the exponential distribution, Exp(σ), has the pdf $f(x|\sigma) = \sigma^{-1}e^{x/\sigma}$, $x > 0$. It follows that this variation of the Fay-Herriot model can be expressed as

$$y_i = x_i'\beta + v_i + \eta z_i + \tau D_i^{1/4}\sqrt{z_i}\epsilon_i, \tag{2.4}$$

where $z_i \sim$ Exp$(\sqrt{D_i})$ and $\epsilon_i \sim N(0,1)$. Alternatively, with $\theta_i = x_i'\beta + v_i$, the model can be expressed as

$$y_i|\theta_i, z_i \sim N(\theta_i + \eta z_i, \tau^2\sqrt{D_i}z_i),\ \theta_i \sim N(x_i'\beta, A),\ z_i \sim \text{Exp}(\sqrt{D_i}),$$

and θ_i, z_i are independent.

It can be shown that

$$\theta_i|y_i, z_i \sim N\left[\frac{A(y_i - \eta z_i) + \tau^2\sqrt{D_i}z_i(x_i'\beta)}{A + \tau^2\sqrt{D_i}z_i}, \frac{\tau^2 A\sqrt{D_i}z_i}{A + \tau^2\sqrt{D_i}z_i}\right].$$

Thus, in particular, the conditional mean, $\mathrm{E}(\theta_i|y_i, z_i)$, is a weighted average of $y_i - \eta z_i$ and $x_i'\beta$. It follows that

$$\begin{aligned} \mathrm{E}(\theta_i|y_i) &= \mathrm{E}\left\{\frac{A(y_i - \eta z) + \tau^2\sqrt{D_i}z(x_i'\beta)}{A + \tau^2\sqrt{D_i}z}\right\}, &(2.5)\\ \mathrm{var}(\theta_i|y_i) &= \mathrm{E}\left(\frac{\tau^2 A\sqrt{D_i}z}{A + \tau^2\sqrt{D_i}z}\right)\\ &\quad +\mathrm{E}\left\{\frac{A(y_i - \eta z) + \tau^2\sqrt{D_i}z(x_i'\beta)}{A + \tau^2\sqrt{D_i}z}\right\}, &(2.6)\end{aligned}$$

where the expectations on the right sides are with respect to $z \sim \mathrm{Exp}(\sqrt{D_i})$. Expressions (2.5) and (2.6) can be used to derive the BP (EBP) of θ_i as well as a measure of uncertainty. The expectations involved do not have analytic expressions; however, they can be evaluated numerically via numerical integration or Monte-Carlo simulation. Note that η and τ^2 are functions of p, and unknown model parameters are β, A and p. See [240] for further detail.

The mixture-model approach is actually quite commonly used in introducing models that have robust features. As another application of the mixture-model approach, [70] considered mixture of normal distribution in a NER model (see Section 1.5.2) for estimating the finite population means. Here, the random effects v_i are assumed to be normal as in the NER model, but the errors e_{ij} are assumed to have a mixture of normal distribution, or normal-mixture distribution. To motivate such a consideration, suppose that the error distribution has two modes, one corresponding a "typical" observation and the other to an outlying observation. The outlying observation is more likely to come from the distribution with the larger variance; it has a small probability to occur, but it does occur. A mixture of two normal distributions, one with a relatively small variance and the other with a large variance may be able to describe such a scenario. Thus, a method developed under such a mixture-distribution may better address the outlier issue than one based on the normality assumption.

Similar to the AL distribution discussed above, when the mixture has two components, it can be conveniently described using a latent random variable. Here, the latent variable, z_{ij}, is assumed to have a Bernoulli(p) distribution. Furthermore, there are two normal random variables, $\epsilon_{ijr} \sim N(0, \sigma_r^2)$, $r = 1, 2$, corresponding to the mixture components. Here σ_1^2 is the smaller variance and σ_2^2 the larger one. It is assumed that $z_{ij}, \epsilon_{ij1}, \epsilon_{ij2}$ are independent, and

$$e_{ij} = (1 - z_{ij})\epsilon_{ij1} + z_{ij}\epsilon_{ij2}.$$

It follows that the pdf of ϵ_{ij} is a weighted average of the pdf of $N(0, \sigma_1^2)$, with weight $1 - p$, and the pdf of $N(0, \sigma_2^2)$, with weight p.

When the number of mixture components is larger than 2, the latent variable expression gets more complicated. However, the weighted-average expression of the pdf can be straightforwardly generalized. Suppose that e_{ij} has a normal-mixture distribution with K components. Then, the pdf of e_{ij} can be expressed as $\sum_{k=1}^{K} p_k f_k(y|X)$, where $f_k(y|X)$ is the n-dimensional multivariate normal distribution (n is the total sample size) with mean vector $X\beta$ and covariance matrix Σ_k, $1 \leq k \leq K$. The covariance matrix Σ_k may depends on a (much) lower-dimensional parameter vector, say, ψ_k.

Example 2.4. Gershunskaya and Lahiri [70] used a two-component normal mixture for the error distribution in the NER model when estimating monthly employment changes in metropolitan statistical areas (MSAs) using data from the Current Employment Statistics Survey conducted by the U.S. Bureau of Labor Statistics. The framework being put forward here was not so much about robustness to outliers but rather trying to capture real finite population structure that may not be described by an assumption based model. The component normal distributions were assumed to have a common mean and different variances. To generate unit level data, the author expanded around a hypothetical true super-population parameter at the state level.

Inference about the normal-mixture model is known to be challenging computationally. On the theoretical side, the identifiability of the model parameters is among the most researched topics in this area. See, for example, [33]. A standard approach for computing the MLE is the E-M algorithm [48]. Below we focus on the two-component mixture.

Multiple further robustification techniques were also proposed but we will focus on the core proposed estimator. Results were compared against estimators from the standard area-level Fay-Herriot model, and estimators from different unit-level models.

Focus of attention was on four states (Alabama, California, Florida and Pennsylvania) where separate models were fit to each state's industrial supersector (i.e., a set of MSAs within these super-sectors defining the set of small areas). It was found that the new estimator was often more efficient than other estimators.

In the finite-population setting, the NER model is assumed to hold for a super-population, and units of the finite population are considered samples from the super-population. The real-life samples are selected from the finite population via a certain sampling scheme, such as simple random sampling (SRS; e.g., [163]). The non-sample part of the finite-population mean can be estimated by $\bar{X}_i'\beta + \tilde{v}_i$ for the ith population, where $\bar{X}_i$ is the mean of the non-sampled auxiliary variables, which is assumed to be available, and $\tilde{v}_i$ is a predictor of v_i [see (1.17)]. The BP of v_i is

$$\tilde{v}_i = \mathrm{E}(v_i|y) = \mathrm{E}(v_i|y_i) = \mathrm{E}\{\mathrm{E}(v_i|y_i, z_i)|y_i\},$$

where $y_i = (y_{ij})_{1 \leq j \leq n_i}$ and z_i is defined similarly. Under the assumed model, it can be derived (e.g., [123], sec. B.1) that

$$\mathrm{E}(v_i|y_i, z_i) = \frac{\sigma_v^2}{\sigma_v^2 + \tau_i^2}(\bar{y}_i - \bar{x}_i'\beta),$$

where $\tau_i^2 = (\sum_{j=1}^{n_i} \tau_{ij}^{-2})^{-1}$ with $\tau_{ij}^{-2} = (1 - z_{ij})\sigma_1^{-2} + z_{ij}\sigma_2^{-2}$, $\bar{y}_i = \tau_i^2 \sum_{j=1}^{n_i} \tau_{ij}^{-2} y_{ij}$ and $\bar{x}_i$ is defined similarly. Thus, we have

$$\tilde{v}_i = \mathrm{E}\left\{ \left. \frac{\sigma_v^2}{\sigma_v^2 + \tau_i^2}(\bar{y}_i - \bar{x}_i'\beta) \right| y_i \right\},$$

where the conditional expectation is with respect to $z_i|y_i$. This is where a complication arises because although z_{ij} has a Bernoulli distribution, $z_{ij}, 1 \leq j \leq n_i$ are not independent conditional on y_i. In [70], the evaluation of the conditional expectation is combined with the E-M algorithm for obtaining the MLE. As for measures of uncertainty, the authors proposed a parametric bootstrap method to construct prediction intervals.

2.5 Heteroscedastic NER models

On a different aspect, [121] relaxed the NER model assumption regarding the error variance. Recall under the NER model (see Section 1.5.2), the variance of the errors, e_{ij}, is assumed to be a constant, σ_e^2. Ref. [121] argues that this assumption is sometimes violated. To relax the constant error-variance assumption, first note that the BP of v_i can be expressed as

$$\mathrm{E}(v_i|y) = \frac{n_i\sigma_v^2}{\sigma_e^2 + n_i\sigma_v^2}(\bar{y}_{i\cdot} - \bar{x}_{i\cdot}'\beta) = \frac{n_i\gamma}{1 + n_i\gamma}(\bar{y}_{i\cdot} - \bar{x}_{i\cdot}'\beta), \tag{2.7}$$

where $\gamma = \sigma_v^2/\sigma_e^2$. It follows that the BP depends on the variance components, σ_v^2, σ_e^2, only through their ratio γ. Alternatively, because under the NER model, the within-class correlation coefficient, $\rho = \mathrm{cor}(y_{ij}, y_{ik}) = \gamma/(1+\gamma)$ for $j \neq k$, the dependence of the BP on the variance components can also be expressed in terms of ρ.

Ref. [121] relaxed the constant error-variance assumption by allowing σ_e^2 to vary with the area, that is, $\mathrm{var}(e_{ij}) = \sigma_i^2$, where $\sigma_i^2, 1 \leq i \leq m$ are unknown variances. On the other hand, the variance ratio, $\mathrm{var}(v_i)/\mathrm{var}(e_{ij})$, is assumed to be a constant, say, $\gamma \geq 0$. The authors called the extension a heteroscedatic NER (HNER) model.

There is an alternative and more intuitive interpretation of the HNER model. Suppose that, for the jth individual from the ith small area, the mean of y_{ij} is $x_{ij}'\beta$, and the variance is proportional to σ_i^2, that is, $\mathrm{var}(y_{ij}) \propto \sigma_i^2$. Now consider the standardized observation, $z_{ij} = (y_{ij} - x_{ij}'\beta)/\sigma_i$, which has zero mean and constant variance. Suppose that the standardized observation satisfies a NER model, that is, $z_{ij} = \xi_i + \epsilon_{ij}$, where ξ_i and ϵ_{ij} are independent with $\xi_i \sim N(0, \sigma_\xi^2)$ and $\epsilon_{ij} \sim N(0, \sigma_\epsilon^2)$. Then, the original observation can be expressed as $y_{ij} = x_{ij}'\beta + v_i + e_{ij}$, where $v_i = \sigma_i \xi_i$ and $e_{ij} = \sigma_i \epsilon_{ij}$. It follows that $\mathrm{E}(v_i) = \mathrm{E}(e_{ij}) = 0$, $\mathrm{var}(v_i) = \sigma_i^2 \sigma_\xi^2$, and $\mathrm{var}(e_{ij}) = \sigma_i^2 \sigma_\epsilon^2$. Thus, we have

$$\frac{\mathrm{var}(v_i)}{\mathrm{var}(e_{ij})} = \frac{\sigma_i^2 \sigma_\xi^2}{\sigma_i^2 \sigma_\epsilon^2} = \frac{\sigma_\xi^2}{\sigma_\epsilon^2} \equiv \gamma,$$

a constant. In other words, the original observation satisfies a HNER model. Thus, the HNER model may be interpreted as assuming the standardized observations satisfying the NER model. Because the standardized observation has constant variance, it is more likely to satisfy a NER model than the non-standardized observation. Therefore, the HNER assumption is, indeed, less restrictive than the NER assumption. On the other hand, the (unknown) standardizing constant is allowed to vary with the small area. Thus, the HNER model assumption is more flexible. We illustrate with a real-data example.

Example 2.3. In the original paper that introduced the NER model [8], the authors discussed a crops data set from 12 Iowa counties obtained from the 1978 June Enumerative Survey of the U.S. Department of Agriculture. Also available are data obtained from land observatory satellites on crop areas. The objective was to predict the mean hectares of crops per segment for the 12 counties using the satellite information. Some of the counties have very few observations so it is hard to assess the within-county variation empirically. Therefore, some counties are pooled together to form a larger "small area". Specifically, the first three counties have only 1 sampled segment from each county; the fourth county has 2 sampled segments, and the rest of the counties have the numbers of sampled segments ranging from 3 to 6. We combine the first three counties to form the first small area. The rest of the small areas consist of counties 4–12. This way, there are a total of $m = 10$ small areas.

Following [8], we consider the following model:

$$y_{ij} = \beta_0 + \beta_1 x_{ij1} + \beta_2 x_{ij2} + v_i + e_{ij}, \tag{2.8}$$

$i = 1, \ldots, 10, j = 1, \ldots, n_i$, where y_{ij} is the jth sampled hectare in area i; x_{ij1} and x_{ij2} are the corresponding numbers of pixels classified by the satellite as corn and soybeans, respectively. Eq. (2.8) is a special case of the NER model. Under this model, the within-area variance is a constant, which is $\tau^2 = \sigma^2(1 + \gamma)$. However, initial exploration of the data suggests that the within-area variation changes dramatically from small area to small area. For example, Figure 2.1 shows the boxplots for different areas, which indicates significant changes in the within-area variation.

It is known [121] that, under the HNER model, the MLE of β and γ are consistent. On the other hand, the authors showed that, under a real HNER model, that is, the σ_i^2's are indeed different, and the data are unbalanced (that is, the n_i's are not equal), the MLE of γ obtained under the NER model

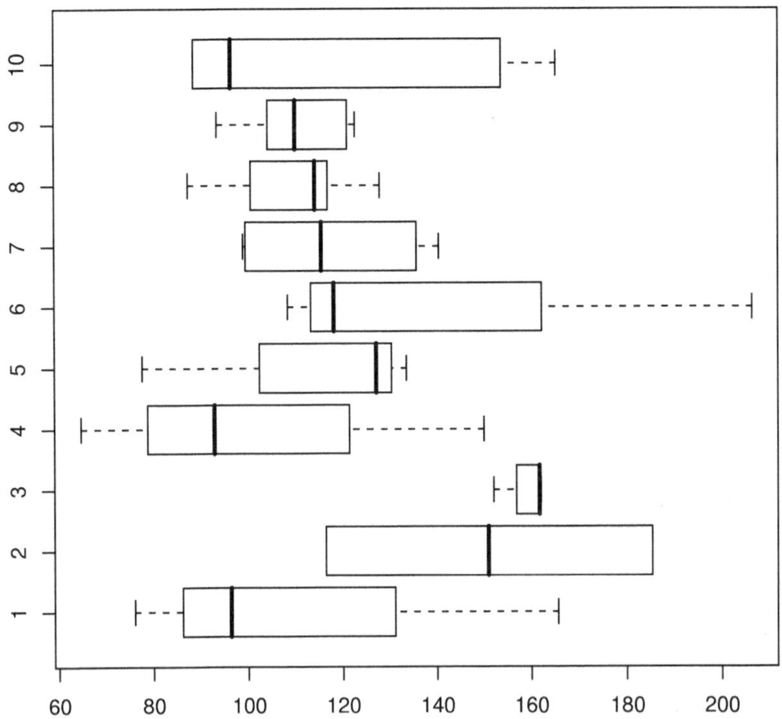

FIGURE 2.1
Boxplots for the Iowa crops data.

is generally inconsistent. Furthermore, simulation results showed that, under a real HNER model, the NER-based EBLUP may perform poorly compared to the HNER-based EBLUP; on the other hand, under an NER model, the performance of the NER-based EBLUP and that of the HNER-based EBLUP are similar. Thus, the HNER-based EBLUP is, indeed, robust regardless of the underlying situation. Of course, there is a computational cost because fitting a HNER model is not as straightforward as fitting an NER one. A computational procedure was provided in [121] as well as measures of uncertainty.

There has been further work related to the HNER model. It was noted [121] that the error variance, $\sigma_i^2, 1 \leq i \leq m$ are not consistently estimable due to the boundedness of the n_i's, which is typically the case in SAE. Due to such a concern, some researchers have considered random dispersion models as opposed to fixed and unknown variances. In [166], the authors considered an extension of the Fay-Herriot model, where the sampling variances are unknown and dependent on the area. The sampling variances are then modeled as having an inverse gamma distribution. The resulting EB estimator shrinks both the mean and the variance; however, the EB does not have an analytic expression. [146] considered the HNER model but proposed a random dispersion model for the σ_i^2's so that the resulting EB has an analytic form.

More recently, [219] proposed a variance function approach for the within-area dispersion. The variance function is in the form of $\sigma_{ij}^2 = \sigma^2(z_{ij}'\gamma)$ for $\sigma_{ij}^2 = \text{var}(e_{ij})$, where $\sigma^2(\cdot)$ is a known function, z_{ij} is a vector of known covariates, and γ is a vector of unknown parameters. The variance function offers some flexibility over the NER model, which may be viewed as having $\sigma^2(x) = x$ and $z_{ij}'\gamma = \sigma_e^2$. On the other hand, it reduces the number of unknown parameters associated with the heteroscedastic variances in the HNER model assuming that the σ_i^2's are completely unknown. Note that, in the latter case, there are m unknown variances, $\sigma_i^2, 1 \le i \le m$, associated with the variances of e_{ij}'s while, with the variance function, the unknown parameters are in the vector γ, which has a fixed dimension. Thus, if m is large, there is a significant reduction in the number of variance parameters. Other parts of the model proposed by [219], which we call HNER-VF model (VF stands for variance function), are the same as in the NER model except without the normality assumption. In particular, one has $\text{var}(v_i) = \sigma_v^2$. As for the normality assumption, it is replaced by the assumption that the v_i's and e_{ij}'s are independent, with means zero and variances specified above. Clearly, the HNER-VF model has much weaker assumptions than the NER model.

The unknown parameters under the HNER-VF model are β, σ_v^2, and γ. Due to the weaker assumptions, maximum likelihood is not possible. The authors proposed a three-step procedure for estimating the unknown parameters. Namely, if σ_v^2, γ are known, β can be estimated via the BLUE method as in (1.28). If γ is known, one can estimate σ_v^2 via an empirical method of moments (EMM; e.g., [108]). Specifically, from the equations $\text{E}(y_{ij} - x_{ij}'\beta)^2 = \sigma_v^2 + \sigma^2(z_{ij}'\gamma), 1 \le i \le m, 1 \le j \le n_i$, one obtains the moment estimator

$$\tilde{\sigma}_v^2 = \frac{1}{N}\sum_{i=1}^{m}\sum_{j=1}^{n_i}\{(y_{ij} - x_{ij}'\beta)^2 - \sigma^2(z_{ij}'\gamma)\}, \tag{2.9}$$

assuming that β is known. Because β is unknown, one replace it in (2.9) by its ordinary least squares (OLS) estimator,

$$\hat{\beta}_{\text{OLS}} = (X'X)^{-1}X'y. \tag{2.10}$$

This leads to the EMM estimator of σ_v^2. Finally, γ can be estimator via a similar EMM method based on the equations

$$\text{E}\{y_{ij} - \bar{y}_i - (x_{ij} - \bar{x}_i)'\beta\}^2 = \left(1 - \frac{2}{n_i}\right)\sigma^2(z_{ij}'\gamma) + \frac{1}{n_i^2}\sum_{k=1}^{n_i}\sigma^2(z_{ik}'\gamma), \tag{2.11}$$

where $\bar{y}_i = n_i^{-1}\sum_{j=1}^{n_i} y_{ij}$ and $\bar{x}_i$ is defined similarly. This leads to the estimator of γ as the solution to the following equation:

$$\sum_{i=1}^{m}\sum_{j=1}^{n_i}\left[\{y_{ij} - \bar{y}_i - (x_{ij} - \bar{x}_i)'\beta\}^2 - \left(1 - \frac{2}{n_i}\right)\sigma^2(z_{ij}'\gamma)\right.$$
$$\left. - \frac{1}{n_i^2}\sum_{k=1}^{n_i}\sigma^2(z_{ik}'\gamma)\right] = 0, \tag{2.12}$$

assuming that β is known. Because β is unknown, we again replaced it in (2.12) by its OLS estimator given by (2.10). This leads to the EMM estimator of γ. In summary, one begins with the OLS estimator of β, then use (2.12) to obtain the EMM estimator of γ; then (2.9) to obtain the EMM estimator of σ_v^2; and finally (1.28) to update the OLS estimator of β.

Once the parameter estimates are obtained, the estimates of small area characteristics of interest as well as the associated MSPE estimates are obtained using the EBLUP and Prasad-Rao methods. Here, a complication for implementing the Prasad-Rao method is that, because normality is not assumed, some higher moments, namely, the kurtoses, appear in the expression of the MSPE that cannot be ignored up to an $o(m^{-1})$ term. Once again, the EMM is called in to obtain consistent estimators of the kurtoses. See [219] for more details.

2.6 MSPE estimation under weaker assumptions

An important part of SAE is obtaining measures of uncertainty for the small area estimates. This is reasonable because, by nature, data collected from many of the small areas are limited; in other words, the sample sizes for many of the small areas are small. As a result, there is significant uncertainty associated with those small area estimates that one needs to "get right". In fact, no other field in statistics or, more generally, data science has given more attention to the topic of uncertainty measures than in SAE (e.g., [199]).

A standard measure of uncertainty is the MSPE. In this regard, the seminal paper of Prasad and Rao [196], known as the Prasad-Rao method, is highly influential. The method uses a linearization method to approximate the MSPE of the EBLUP to the second order, that is, the approximation error is $o(m^{-1})$, where m is the total number of small areas, from which data are available. The next step of the Prasad-Rao method is to bias correct a plug-in estimator, obtained by replacing the unknown parameters involved in the MSPE approximation by their (consistent) estimators. The bias-correction also achieves a second-order unbiasedness, that is, the bias of the resulting MSPE estimator is also $o(m^{-1})$. One assumption underlying the original Prasad-Rao method is that the random effects and errors are both normally distributed.

Ref. [150] considered an area-level model similar to the Fay-Herriot model except that the normality assumption about the area-specific random effects, v_i, is relaxed. It is assumed that the v_i's are i.i.d. with $\mathrm{E}(v_i) = 0$, $\mathrm{var}(v_i) = A$, and $\mathrm{E}(|v_i|^{8+\delta}) < \infty$ for some $0 < \delta < 1$. The normality assumption about the sampling errors, e_i, remains the same as in the Fay-Herriot model. The Prasad-Rao (P-R) estimator of A is given by

$$\hat{A} = \frac{\sum_{i=1}^{m}(y_i - x_i'\hat{\beta})^2 - \mathrm{tr}(DP_{X^\perp})}{m - p}, \tag{2.13}$$

where $p = \dim(\beta)$, $D = \text{diag}(D_1, \ldots, D_m)$, $P_{X^\perp} = I_m - P_X$ with $P_X = X(X'X)^{-1}X'$, the projection matrix onto the linear space spanned by the columns of $X = (x_i')_{1 \le i \le m}$. Note that the P-R estimator of A is a method of moments (MoM) estimator; it does not require the normality assumption, neither on the random effects nor on the sampling errors. The original Prasad-Rao MSPE estimator [196] is derived based on the P-R estimator of A Ref. [150] showed that, under the relaxed assumption of the random effects, the Prasad-Rao estimator of the MSPE of the EBLUP, with the P-R estimator of A, remains second-order unbiased. Note that the Prasad-Rao MSPE estimator was derived under the normality assumption [196]; nevertheless, the estimator remains second-order unbiased even if the normality assumption about the random effects is relaxed. It should be noted that the normality assumption about the sampling errors is still needed.

A "trick" used in the authors' arguments is worth noting. As shown in [150], without the normality assumption on v_i, the MSPE of the EBLUP involves additional unknown parameters other than the variance components, namely, the kurtosis of v_i defined as $\kappa_4 = \text{E}(v_i^4)/A^2 - 3$. Under the normality assumption, $\kappa_4 = 0$, so no additional parameter is involved in the MSPE. However, when the random effects are non-normal, κ appears in the expression of the MSPE among the $O(m^{-1})$ terms; therefore, to obtain a second-order unbiased MSPE estimator, the unknown kurtosis cannot be simply ignored. It turns out that, in the end, the kurtosis can indeed be ignored, after all. The reason is that the same expression involving the kurtosis also appears in the expectation of the g_1 term [see (1.38)] so that, when taking the expectation of the Prasad-Rao MSPE estimator, given by (1.38), the kurtosis cancels; as a result, the second-order unbiasedness of the Prasad-Rao MSPE estimator is unaffected.

As another note, this nice result of a robust MSPE estimator is only known to hold for EBLUP with the P-R estimator of A. It would be interesting to explore the extension of the result to other estimator of A, such as the REML and ML estimators, under the area-level model. It would also be interesting to explore the extension of the result to unit-level models.

Ref. [27] developed a bias-robust mean squared error estimator for pseudo-linear small area estimators, that is, estimators that can be expressed as weighted sums, but where the weights can depend on the sample values of the variable of interest. Specifically, they considered MSE estimation for estimators of area means when different linear models apply within different areas. For notation simplicity, let y_j denote the value of the variable of interest for unit j of the population and assuming this unit is in area i, then an area-specific linear model for y_j can be expressed in the form $y_j = x_j'\beta_i + e_j$, where x_j is a $p \times 1$ vector of unit level auxiliary variables for unit j, β_i is a $p \times 1$ vector of regression coefficients and e_j is a unit level random effect with mean zero and variance σ_j^2 that is uncorrelated between different population units. Usually, estimates $\hat{\sigma}_j^2$ are derived under a 'working model' (e.g. $\sigma_j^2 = \sigma^2$). Ref. [27] developed squared bias and prediction variance expressions for a small area mean. Since any assumptions regarding σ_j^2 in the working model only affected second order terms of the prediction variance, their estimator of prediction variance is bias-robust (i.e., it remains approximately unbiased under misspecification of the variance components of the working model). In their work, [27] discussed how this approach to MSPE estimation actually can underestimates the MSPE because most small area estimators like the EBLUP are pseudo-linear in that they are composed of weights that depend on the sampled values of y. For large values of the overall sample size n, we can treat these weights as fixed if the linear mixed model assumption is true. Still, the potential underestimation of MSPE has to be weighed against the robustness of the MSPE estimator under misspecification of second-order moments of y.

Example 2.4 (Synthetic data on grape production). Here we illustrate the robust MSPE estimator of [27] using synthetic data on grape production from 274 municipalities in Tuscany available in the R package saeRobust. We have direct estimators of the mean agrarian surface area used for production of grape (in hectares) for each Tuscany municipality and auxiliary variables that include area (agrarian surface area use for production in hectares) and workdays (average number of working days in the year 2000). Sampling variances of the direct estimators are also available. Also produced are bias-corrected robust EBLUPs (REBLUPBC) for each area using the method [26]. Figure 2.2

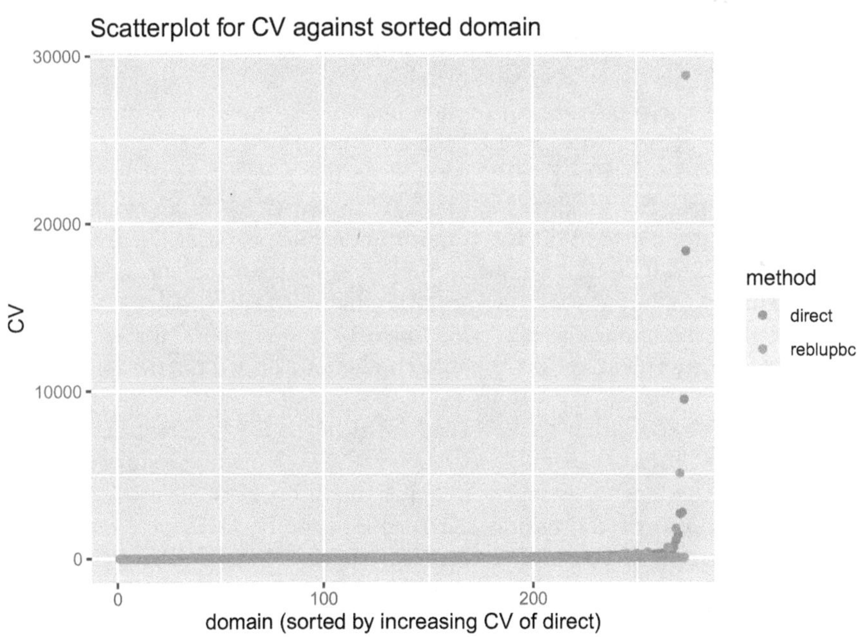

FIGURE 2.2
Coefficients of variation for direct and REBLUPBC for each domain sorted by CV for the synthetic data grape production example. (Taken from [27].)

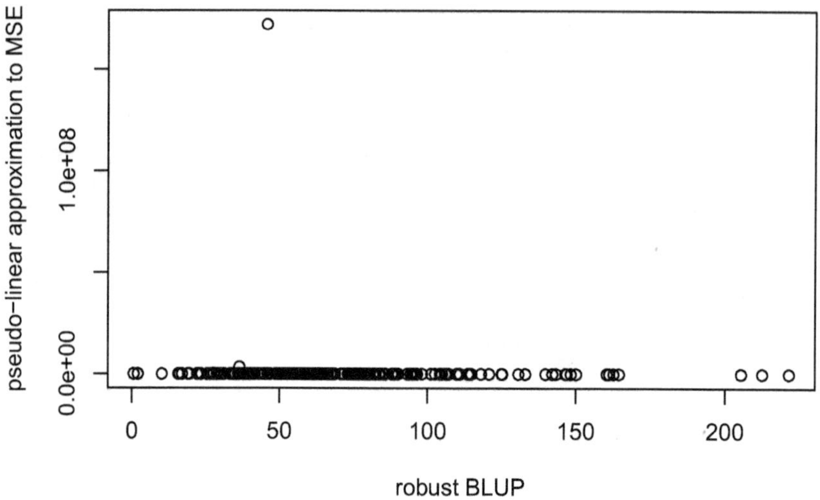

FIGURE 2.3
Pseudo-linear approximations to the MSE for the REBLUPBCs for the synthetic data grape production example. (Taken from [27].)

presents the scatterplot for coefficient of variations (CV) of the direct estimate and REBLUPBC for each domain sorted by CV of the direct estimators. Figure 2.3 shows the pseudo-linear approximations to the MSPE for the REBLUPBCs plotted against the REBLUPBCs.

Some re-sampling methods for MSPE estimation has required weaker assumptions. For example, the jackknife method [118]; see §1.8.1 is known to produce a second-order unbiased MSPE estimator under the posterior linearity assumption [72]; see §2.2. The latter assumption is weaker than the normality assumption. It should be noted that the jackknife method proposed by [118] is not restricted to work under the posterior linearity assumption. However, the method is more convenient to use because, under posterior linearity, the MSPE of the BP has an analytic expression.

A question then arises on what to do if the MSPE of the BP does not have an analytic form. The McJack method, proposed by [117], provides an answer to the question for situations where the number of small areas, m, is at most moderately large. Instead of estimating the MSPE directly, [117] proposes to estimate the logarithm of the MSPE to avoid negative MSPE estimates. Also note that, in [118], the authors decomposed the MSPE into two terms [see (1.40)], and proposed to jackknife the two terms separately [see (1.41), (1.42)]. No such a decomposition is needed in [117], who proposed to Evaluate the logarithm of MSPE of an EBP using a Monte-Carlo method. Specifically, the Monte-Carlo is done via a parametric bootstrap procedure, that is, one uses the estimated model parameters, treated as the true parameters, to generate the data repeatedly. For each generated data set, the EBP is computed. The bootstrapped MSPE is computed as the mean squared difference between the bootstrapped EBP and the true mixed effect with respect to the bootstrapped samples. The logarithm of the bootstrapped MSPE is then used in a jackknife formula similar to (1.41), where the b function is replaced by the logarithm of the bootstrapped MSPE, under the full data or data without those from the ith small area. On the other hand, the McJack method can be computationally intensive, if m is large (this is why it was assumed above that the number of small areas is 'at most moderately large').

Ref. [88] proposed a bootstrap method for estimating the MSPE of EBLUP under a unit-level model, which is an extension of the NER model [8]. The authors proposed to do it in a non-parametric way in the sense that the between-area random effects and within-area errors are not assumed to be normally distributed. Instead, the random effects are assumed to be i.i.d., with mean 0 and finite fourth moment; similar assumptions are made for the errors; no specific distributional assumption is made regarding the form of the distributions. Note that, with i.i.d. data, one may bootstrap nonparametrically as in [51]; however, there does not seem to be an easy way to do this under an NER model, or LMM in general. Of course, parametric or model-based bootstrap methods are available; see, for example, [19], [89], but here the goal is to bootstrap nonparametrically, that is, without specifying a particular form of the distribution, such as the normal distribution.

To achieve the latter goal, [88] notes that, without the normality assumption, the MSPE of the EBLUP may depend on higher moments of the random effects and errors, in addition to the variance components (i.e., the variances). Specifically, the authors obtained an approximation to the MSPE such that the approximation error is $o(m^{-1})$. The approximation depends on the higher moments through the kurtoses of the random effects and errors. This affects a $O(m^{-1})$ term in the approximation, and therefore cannot be ignored if one is to obtain a second-order unbiased MSPE estimator.

A good news is that the MSPE, up to a difference of $o(m^{-1})$ term, depends only on the kurtoses; in other words, no other feature of the unknown distribution does one need to know, if the goal is to estimate the MSPE second-order correct, or second-order unbiasedly. In view of this, [88] proposes to generate the random effects and errors from some known distributions that match the estimated second and fourth moments of the random effects and errors. This is called a moment-matching bootstrap. The idea leads to a second-order unbiased MSPE estimator, if double bootstrap is used. On the other hand, double bootstrapping is known to be computationally intensive. As noted in [133], the double-bootstrap methods [88], [89] are computationally more intensive than the McJack method [117]; see the next section for a real-data example illustrating the McJack method.

2.7 Example: Income data

Ref. [47] considered a data set regarding the median income of four-person families for the fifty states of U.S. and the District of Columbia (D.C.) using cross-sectional and time series modeling. The primary source of data is the annual supplement to the March Sample of the Current Population Survey (CPS), which provides individual annual income data categorized into intervals of $2500. The direct survey estimates were obtained from the CPS using linear interpolation. Two secondary sources of data were also available. The first is the U.S. decennial censuses (Census) which produce median incomes for the 50 states and D.C. based on the "long form" filled out by approximately one-sixth of the U.S. population. These census median income estimates are believed to be free of sampling errors. The second is per capita income estimates produced by the Bureau of Economic Analysis (BEA) division of the U.S. Department of Commerce. Since the per-capita income estimates are not based on any sampling techniques, they do not have any sampling errors associated with them. From the Census and BEA data, an adjusted census median income (adjusted Census) is obtained by multiplying the preceding census median income by the ratio of BEA per-capita income for the current year to that of the preceding census year.

Following [47], we consider the four-person families data for the years 1979, and 1981–1989, that is, a total of 10 years. The direct survey estimates are denoted by y_{it}, where $i = 1, \ldots, 51$ corresponding to the 50 states and D.C., and $t = 1, \ldots, 10$, corresponding to the 10 years. The Census variable is denoted by $x_{1,i}$ (note that this variable is at the state level only, i.e., does not change with the year), and the adjusted Census variable is denoted by $x_{2,it}$ [note that this variable is at both state and time (year) levels]. The goal is to estimate the median income of four-person families for all 50 states of U.S. and the D.C. for the year 1989. Ref. [47] used x_2 as the only covariate in their modeling for the mean. As for the variance-covariance structure, the authors proposed a random walk model for the state-level (vector-valued) random effects, in addition to the sampling errors whose variances are known. We refer this model as DLM model. In a similar context, [201] proposed a cross-sectional/time series model which is the same as the DLM model except that the variance-covariance structure of the random effects follows that of a stationary AR(1) model. This model is referred to as R-Y model. We intend to compare the DLM model and R-Y model in this particular application. In addition, we are interested in entertaining an additional covariate variable, namely, x_1. Thus, we set up a model-selection framework as follows. See Chapter 6 for a more systematic account of model selection and related topics.

Let $y_i = (y_{it})_{1 \le t \le 10}$ denote the vector of direct survey estimators for state i (including D.C.). Let $X_i = (x'_{it})_{1 \le t \le 10}$ denote the matrix of covariates for state i, where the first component of x_{it} is 1, corresponding to the intercept, and the rest of the components are subject to selection from $x_{1,it} = x_{1,i}$, $x_{2,it}$, or both. A candidate model can be expressed as

$$y_i = X_i\beta + u_i + e_i, \quad i = 1, \dots, 51, \tag{2.14}$$

where the vectors u_i, e_i are independent with $u_i \sim N(0, \Sigma)$ and $e_i \sim N(0, \Psi_i)$. Here Ψ_i is a diagonal matrix whose diagonal elements are known and corresponding to the sampling variances (strictly speaking, those are estimated). The covariance matrix Σ is either $\Sigma_{(1)}$, corresponding to the DLM model, or $\Sigma_{(2)}$, corresponding to the R-Y model. More specifically, $\Sigma_{(1)} = \sigma_v^2 J_{10} + \sigma_\epsilon^2 \Gamma\Gamma'$, with J_{10} being the 10×10 matrix of 1's, and Γ being the 10×10 lower triangular matrix with diagonal and non-zero off-diagonal elements equal to 1, and $\sigma_v^2, \sigma_\epsilon^2$ are unknown variance components. $\Sigma_{(2)} = \sigma_v^2 J_{10} + \sigma_\epsilon^2 \Sigma(\rho)$, with

$$\Sigma(\rho) = (1 - \rho^2)^{-1}[\rho^{|t-s|}]_{1 \le s,t \le 10},$$

and $\sigma_v^2, \sigma_\epsilon^2, \rho$ are unknown variance components with $|\rho| < 1$.

In all, (2.14) includes six candidate models: x_1, x_2, or both, for the covariates, and $\Sigma_{(1)}$ or $\Sigma_{(2)}$ for Σ. A model selection procedure is carried out to select an "optimal" model before applying the SAE methods. Note that, here, we have a post-model selection SAE, or SAE after model selection, problem.

A main challenge is to obtain the measures of uncertainty, which need to taken into consideration of the additional uncertainty in model selection. McJack is applicable to such a problem. To apply McJack, we need to have a full model (for bootstrapping) that covers all of the candidate models as special cases, and there is no such a model among the candidate models. Note that, although x_1, x_2 is a full model for the covariates, the DLM and R-Y covariance models do not cover each other as special cases. Therefore, we consider the following full model which is not a candidate model–it is only used for the McJack computation: A convex combination of $\Sigma_{(1)}$ and $\Sigma_{(2)}$:

$$\Sigma_{\rm f} \quad = \quad \sigma_v^2 J_{10} + \sigma_\epsilon^2\{\lambda\Gamma\Gamma' + (1-\lambda)\Sigma(\rho)\},$$

where $\lambda \in [0, 1]$ is an additional variance component. It is clear that, $\Sigma_{\rm f}$ includes $\Sigma_{(1)}$ and $\Sigma_{(2)}$ as special cases. The full model is the one with x_1, x_2 as the covariates and $\Sigma_{\rm f}$ as Σ.

The BIC procedure [e.g., [112], sec. 6.1] is used to select the optimal model among the candidates. The selected model has both x_1, x_2 as the covariates, and $\Sigma_{(1)}$ as Σ. This model is denoted by M^*. Based on the selected model, maximum likelihood estimation (MLE) is used to obtain estimates of the model parameters, and the EBLUPs for the 1989 median income of 4-person families are computed and presented in Table 2.3. Also presented are the corresponding square roots of the MSPE estimates using McJack, taking into account of the model selection, denoted by $\widehat{\rm MSPE}_1^{1/2}$. As a comparison, we also computed the McJack MSPE estimates based on M^* (assumed known), denoted by $\widehat{\rm MSPE}_2^{1/2}$. All numbers are rounded to the nearest integers. The Monte-Carlo sample size for computing the McJack estimates is $K = 2000$.

TABLE 2.3
Estimation of 1989 median income of 4-person families.

State	EBLUP	$\widehat{\text{MSPE}}_1^{1/2}$	$\widehat{\text{MSPE}}_2^{1/2}$	State	EBLUP	$\widehat{\text{MSPE}}_1^{1/2}$	$\widehat{\text{MSPE}}_2^{1/2}$
ME	38360	2554	2729	NC	38482	682	663
NH	48938	360	370	SC	36039	1092	1130
VT	39056	1892	1841	GA	40113	1487	1326
MA	52018	851	885	FL	37294	3148	3102
RI	44918	337	360	KY	33156	579	586
CT	55118	687	678	TN	34231	1577	1637
NY	44199	240	238	AL	33394	3116	3058
NJ	54902	438	430	MS	30196	2350	2412
PA	39871	1224	1316	AR	29874	1529	1423
OH	40629	418	435	LA	34659	1606	1606
IN	38085	1616	1548	OK	35330	695	669
IL	43540	7236	7194	TX	36216	722	771
MI	43070	975	957	MT	32835	5667	5892
WI	41108	756	757	ID	32335	1900	1970
MN	42940	273	265	WY	36368	1084	1093
IA	36765	457	471	CO	40329	124	126
MO	38425	952	946	NM	29449	3581	3543
ND	34019	1400	1167	AZ	37962	1306	1330
SD	32323	1685	1723	UT	36856	1039	1007
NE	35819	207	218	NV	40622	876	926
KS	38774	1273	1306	WA	42001	1489	1526
DE	42791	463	470	OR	39566	1748	1699
MD	52395	468	489	CA	42586	717	737
DC	40232	7093	7304	AK	47359	990	1004
VA	46561	2156	2115	HI	45780	2360	2273
WV	29813	829	874				

It is seen that the two MSPE estimates are very close; for some states one MSPE estimate is slightly larger while for the other states it is slightly smaller. At first, this might seem a little surprising, as one would expect that $\widehat{\text{MSPE}}_1$, which takes into account the potentially additional variation in model selection, to be larger than $\widehat{\text{MSPE}}_2$. However, MSPE is not just the variation. This can be seen from the equation

$$\text{MSPE} = (\text{prediction bias})^2 + \text{prediction variance}. \tag{2.15}$$

Although model selection may increase the second term on the right side of (2.15), it may reduce the first term for selecting the optimal model, which fits the data better. Note that, although M^* is the optimal model for this particular data set, it may not always be the optimal model, if the data are repeatedly generated, even under the same (estimated) parameters. For example, out of the 2,000 Monte-Carlo samples generated under the M-estimate based on the full data set, $\hat{\psi}$, 85.3% selected M^*; another 14.1% selected the model with only x_2 as the covariate, and the same $\Sigma = \Sigma_{(1)}$. Note that this is the same model as the DLM, denoted by M_{dlm}.

Take a look at another example by considering the M-estimate based on the data with the jth state deleted, $\hat{\psi}_{-j}$, where $j = 20$ (the number 20 is randomly chosen from $1, \dots, 51$ by a computer). This time, out of the 2,000 Monte-Carlo samples, 81.8% selected M^*, and another 17.8% selected $M_{\rm dlm}$. In the end, it all depends on the relative contributions of the two terms on the right side of (2.15), when it comes to the MSPE measure.

It appears that, for this application, the overall MSPE is about the same for the two McJack estimates. This may also be explained by the fact that there is not much variation in model selection after all. Once again, using the above examples, for the majority of the Monte-Carlo samples (85.3% and 81.8%) one has M^* as the selected model, and almost all of the rest (14.1% and 17.8%) $M_{\rm dlm}$ as the selected model. Note that the two models, M^* and $M_{\rm dlm}$, are actually very close, especially in terms of the prediction performance, and the two models combined counted for 99.4% and 99.6% of the selected models, respectively, for those two examples.

For a similar reason, the MSPE estimates are not necessarily smaller than the corresponding $\Psi_{i,10}$, which is the (estimated) sampling variance of the direct estimator. Namely, MSPE is not just a measure of variation, as noted above. Thus, for example, for the state of FL, bias appears to be a more significant factor than variation.

3

Outlier Robustness

In the previous chapter, we discussed several SAE methods that were developed under weaker assumptions. In a way, such a robustness was against violation of a model assumption. Examples of such assumptions include normality, mean and variance functions, etc. There are other concerns of robustness that are not in terms of violation of model assumptions. These are related to outliers. The appearance of an outlier does not necessarily violate any model assumption. For example, even under the normality assumption, which is often considered the most "ideal" model assumption, an extremely large value (in absolute value) can occur, at least theoretically. In real life, such extreme values can occur more than theoretically and, when they do occur, there are undesirable consequences. In this chapter, we first introduce some preliminaries regarding outlier robustness. We then discuss approaches to robustize SAE methods. For the most part, the approach is to modify an existing method so that the outcome is robust to the impact of the outliers.

3.1 Preliminaries

Most of the concepts and methods reviewed in this section can be found from Huber's classic text, *Robust Statistics* (see [100]).

The classic example of a statistic not robust to outliers is the sample mean. Suppose a random sample of $n = 50$ annual incomes are drawn from the U. S. population and the sample mean is computed. If the sample mean is used to estimate the "typical" annual income of the people in the U.S., the estimate would be seriously off if one of the samples happened to be Elon Musk. The reason is simple, the total annual income of 49 "ordinary" people, plus Musk, is almost no difference from the annual income of Musk himself. When this total is divided by 50, and used as an estimate of the typical annual income, one gets impression that everyone in the U.S. is making a huge amount of money, which is not true.

A statistic that is known to be robust against the outliers is the sample median. Take a look again at the above hypothetical example. Obviously, the sample median is not affected by Musk, because he makes no difference when computing the median. This gives a much more realistic estimate of

DOI: 10.1201/9781003395171-3

the typical annual income. It is known that the distribution of the incomes is skewed to the right. For such a distribution, the population median is smaller than the population mean. The sample median would be a better estimate for the population median than it is for the population mean. Nevertheless, the sample median does have the robustness feature that it is not affected by an extreme value or outlier.

As another example, consider a problem of robust regression [99]. In classical linear regression, one fits the model

$$y_i = x_i'\beta + \epsilon_i, \quad i = 1, \dots, n \tag{3.1}$$

via the least squares (LS). This amounts to minimize

$$\sum_{i=1}^{n} \psi(\tilde{\epsilon}_i), \tag{3.2}$$

with respect to β, where $\tilde{\epsilon}_i = y_i - x_i'\beta$ is the "residual" and $\psi(u) = u^2$. This function ψ is not robust to outliers in that a large residual (in absolute value) can significantly impact the sum in (3.2). This is why the LS method is not robust to outliers. One way to robustize the LS method is to use a robust ψ function in (3.2) so that it is "resistant" to outliers. Some well-known functions that are outlier-resistant include Andrew's sine:

$$\psi(u) = \begin{cases} c\{1 - \cos(u/c)\}, & \text{if } |u| < \pi c; \\ 2c, & \text{if } |u| \ge \pi c \end{cases} \tag{3.3}$$

for some constant c; the Huber function:

$$\psi(u) = \begin{cases} u^2, & \text{if } |u| < c; \\ 2c|u| - c^2 & \text{if } |u| \ge c, \end{cases} \tag{3.4}$$

for some constant c; and Tukey's biweight:

$$\psi(u) = \begin{cases} (c^2/6)[1 - \{1 - (u/c)^2\}^3], & \text{if } |u| < c; \\ c^2/6 & \text{if } |u| \ge c, \end{cases} \tag{3.5}$$

for some constant c. It can be seen that Huber's function is a modified version of the LS ψ in (3.2). The functions (3.3)–(3.5) are all robust in the sense that they are either constant or changing more slowly, when $|u|$ exceeds certain bounds. Also, all of these functions are also even, that is, $\psi(-u) = \psi(u)$, so their derivatives are good choices for "errors". Note that Huber's function has derivative everywhere; the other two functions have derivatives at all except two points. Thus, if the distribution of ϵ_i is symmetric about 0 in the sense that $\epsilon_i \sim -\epsilon_i$ ($\sim$ means having the same distribution), then, we have $\mathrm{E}\{\psi'(\epsilon_i)\} = \mathrm{E}\{\psi'(-\epsilon_i)\} = -\mathrm{E}\{\psi'(\epsilon_i)\}$, implying

$$\mathrm{E}\{\psi(\epsilon_i)\} = 0. \tag{3.6}$$

In robust regression, the regression errors, ϵ_i, are assumed to satisfy (3.6). It is seen that, in the special case of $\psi(u) = u^2$ [hence $\psi'(u) = 2u$], (3.6) is equivalent to the classical assumption of $\mathrm{E}(\epsilon_i) = 0$ for linear regression.

Assumption (3.6) ensures that the estimating equation

$$\sum_{i=1}^{n} \psi'(\tilde{\epsilon}_i)x_{ij} = 0, \tag{3.7}$$

obtained by differentiating (3.2) with respect to β_j, the jth component of $\beta = (\beta_j)_{1\leq j\leq p}$, is unbiased for every $1 \leq j \leq p$. From there, consistency and asymptotic normality of the estimator of β, which is the solution to (3.7), can be established under regularity conditions.

We mentioned the sample median earlier. In fact, this estimator may be viewed as a special case of robust regression. We illustrate this as an example.

Example 3.1 (sample median). If $x_i'\beta = \mu$, an unknown constant, in (3.1), the regression has no real predictors. This is a special case, in which we are estimating the population mean, μ, from independent observations, $y_i, i = 1, \ldots, n$ with mean μ and variance σ^2, which is also unknown. In this case, the minimizer of (3.2) is $\hat{\mu} = \bar{y}$, the sample mean of the y's; thus, the LS estimator is the same as the sample mean. Now let us consider a different ψ function, $\psi(u) = |u|$. It can be shown that the minimizer of (3.2) with the new ψ is the sample median of the y's, $q_y(0.5)$, where $q_y(p)$ denotes the pth quantile of the y's. Here, the new ψ is not one of the (3.3)–(3.5), but it gets the robustness job done anyway. As noted, the sample median is not affected by a few extreme values, provided that the sample size is not too small.

Now let us think a little more about this. The sample median is robust because, perhaps, it intends to estimate a population parameter that itself is robust. This population parameter is the population median, that is, the point such that 50% of the population distribution is to the left. Clearly, the population median does not change if we change the values in the upper or lower tails of the distribution. In this sense, the population median is a robust quantity, or parameter.

More generally, a population pth quantile is a number so that the area under the density curve of the population distribution and to the left of the number is equal to p. Mathematically, let Y be a random variable. The pth quantile of Y, denoted by $q_Y(p)$, is defined as the number q such that

$$\mathrm{P}(Y \leq q) = p.$$

It follows that the population median of Y is $q_Y(0.5)$. The point is if we target a robust quantity to estimate, a good estimator is also likely to have robustness features. Quantile regression is a subject area in which the interest is on estimating the regression quantiles, rather than the regression means. Note that the $x_i'\beta$ in (3.1) is the mean of y_i; so, here, the mean of y_i is modeled as $x_i'\beta$. In quantile regression, the pth quantile is modeled as $x_i'\beta$; for different

p there is a different quantile regression model. The parameters, β, in the quantile regression can be estimated by minimizing (3.2) with

$$\psi(u) = \psi_p(u) = u\{p - 1_{(u<0)}\},$$

the so-called check function. See [139]. We consider a special case below.

Example 3.1 (continued). Consider ψ_p with $p = 0.5$. It follows that

$$p_{0.5}(u) = \begin{cases} -u/2, & u < 0, \\ 0, & u = 0, \\ u/2, & u > 0. \end{cases}$$

This is the same as $|u|/2$. Thus, minimizing (3.2) with $\psi(u) = \psi_{0.5}(u)$ is equivalebt to minimizing (3.2) with $\psi(u) = |u|$. In particular, when $x_i'\beta = \mu$, the estimator of μ is the sample median.

There are different population parameters that provide information about the "location". This is similar to the "typical annual income" in the income example discussed earlier. The population mean is one of them, and the population median is another; the different population quantiles also provide information about the location. If, instead of the population mean, we consider the population median, or the quantiles, as the location measures, their estimators are expected to have robustness features.

Although the approaches described above are under the regression setting, the ideas can be generalized. In fact, most existing approaches in SAE to outlier robustness are extensions, or modifications, of the above.

3.2 Robustizing the EBLUP

The EBLUP is a standard method in SAE (e.g., [199]; also see §1.7). However, the value of EBLUP can be heavily influenced by outliers in the data. Following a similar approach as described above, we can modify the EBLUP procedure to make it more robust.

Recall the EBLUP is obtained by replacing the (unknown) variance components involved in the BLUP with their estimators. There are different ways of obtaining the BLUP (e.g., [123], Sec. 2.3.2). One method, proposed by C. R. Henderson [94], is called Henderson's mixed model equation (MME). Let us consider a more general setting, an LMM expressed as

$$y = X\beta + Zv + e, \tag{3.8}$$

where $v \sim N(0, G)$, $e \sim N(0, R)$, and v and e are independent. Suppose that both G and R are non-singular. Here, β is the vector of fixed effects and α the vector of random effects. It can be shown that the logarithm of the joint pdf of v and y can be expressed as

$$c - \frac{1}{2}\left\{(y - X\beta - Zv)'R^{-1}(y - X\beta - Zv) + v'G^{-1}v\right\}, \tag{3.9}$$

where c is a constant. Ref. [94] proposed to find the "maximum likelihood estimates" of β and v, treating the latter as (fixed) parameters, by differentiating (3.9) with respect to β and v and setting the partial derivatives equal to zero. This leads to Henderson's MME:

$$\begin{pmatrix} X'R^{-1}X & X'R^{-1}Z \\ Z'R^{-1}X & G^{-1} + Z'R^{-1}Z \end{pmatrix}\begin{pmatrix} \tilde{\beta} \\ \tilde{v} \end{pmatrix} = \begin{pmatrix} X'R^{-1} \\ Z'R^{-1} \end{pmatrix} y, \tag{3.10}$$

the solution to which leads to the BLUE and BLUP (see §1.7).

Except for the case with spatial correlations (see §1.6.2), an LMM for SAE typically has the following specifications: $X = (X_i)_{1\le i\le m}$, $Z = \mathrm{diag}(Z_i, 1 \le i \le m)$, $G = \mathrm{diag}(G_i, 1 \le i \le m)$ and $R = \mathrm{diag}(R_i, 1 \le i \le m)$, where the index i corresponds to the ith small area, and data from different small areas are assumed to be independent. The LMM (3.10) can now be written as

$$\sum_{i=1}^{m} X_i'R_i^{-1/2}\tilde{e}_i = 0, \tag{3.11}$$

$$Z_i'R_i^{-1/2}\tilde{e}_i - G_i^{-1/2}\tilde{v}_i = 0, \quad 1 \le i \le m, \tag{3.12}$$

where $\tilde{e}_i = R_i^{-1/2}(y_i - X_i'\beta - Z_iv_i)$ and $\tilde{v}_i = G_i^{-1/2}v_i$. The terms $\tilde{e}_i$ and $\tilde{v}_i$ may be viewed as "standardized residuals". Thus, following the previously described approach, one may replace these terms in (3.11) and (3.12) with $\psi'(\tilde{e}_i)$ and $\psi'(\tilde{v}_i)$, respectively, where $\psi(\cdot)$ is a function with robust features, such as the Huber function. This leads to the robustized MME:

$$\sum_{i=1}^{m} X_i'R_i^{-1/2}\psi'(\tilde{e}_i) = 0, \tag{3.13}$$

$$Z_i'R_i^{-1/2}\psi'(\tilde{e}_i) - G_i^{-1/2}\psi'(\tilde{v}_i) = 0, \quad 1 \le i \le m. \tag{3.14}$$

Ref. [60] proposed robust estimation of variance components by utilizing (3.13) and (3.14). The ψ function suggested by the author is

$$\psi(u) = \psi_b(u) = u \min\left(1, \frac{b}{|u|}\right), \tag{3.15}$$

where $b > 0$ is a tuning constant. When $b = \infty$, one has $\psi_b(u) = u$, in which case (3.7) and (3.14) are the same as Henderson's MME. As noted by [215], other ψ functions can also be used in this procedure.

The main objective of [60] was to estimate the variance components. Note that Henderson proposed the MME for estimating the fixed and random effects. If the variance components were of primary interest, Henderson's MME might not be the best choice. This was confirmed by a study of [215], who proposed to robustize the maximum likelihood (ML) equations instead of the MME. The ML equations can be written as (e.g., [123], sec. 1.3)

$$\sum_{i=1}^{m} X_i'V_i^{-1}(y_i - X_i\beta) = 0, \tag{3.16}$$

$$\sum_{i=1}^{m}\left\{(y_i - X_i\beta)'V_i^{-1}\frac{\partial V_i}{\partial \theta_s}V_i^{-1}(y_i - X_i\beta) - \text{tr}\left(V_i^{-1}\frac{\partial V_i}{\partial \theta_s}\right)\right\} = 0,\ \ 1 \le s \le q, \tag{3.17}$$

where $V_i = \text{Var}(y_i)$, and $\theta = (\theta_s)_{1\le s\le q}$ is the vector of variance components. Next, we modify the ML equations to come up with a robust version of them. Ref. [215] proposed to do it as follows. First focus on the term $y_i - X_i\beta$, which looks like a residual, and write it as $U_i^{1/2}U_i^{-1/2}(y_i - X_i\beta) = U_i^{1/2}r_i$, where U_i is the diagonal matrix with the same diagonal elements as V_i, and r_i is defined in an obvious way. Then, following a similar approach to [60], we replace r_i with $\Psi(r_i)$, where the $j$the component of $\Psi(r_i)$ is $\psi(r_{ij})$ with r_{ij} being the jth component of r_i. After making these changes, the ML equations become

$$\sum_{i=1}^{m} X_i'V_i^{-1}U_i^{1/2}\Psi(r_i) = 0, \tag{3.18}$$

$$\sum_{i=1}^{m}\left\{\Psi(r_i)'U_i^{1/2}V_i^{-1}\frac{\partial V_i}{\partial \theta_s}V_i^{-1}U_i^{1/2}\Psi(r_i) - \text{tr}\left(V_i^{-1}\frac{\partial V_i}{\partial \theta_s}\right)\right\} = 0,\ \ 1 \le s \le q. \tag{3.19}$$

We are not yet done with the modification because the new equations, (3.19) are not unbiased. An estimation equation is unbiased if its expected value, when the parameters involved are replaced by the true parameters, is zero. The unbiasedness of the estimating equation is a desirable property that is often needed to ensure the consistency of the estimator. The original ML equations, (3.16) and (3.17), are unbiased. The new equation (3.18) is unbiased if $\psi(\cdot)$ is symmetric about 0 and the distribution of $y_i - X_i\beta$ is symmetric about 0. However, the new equations (3.19) are generally not unbiased. To make the latter equations unbiased, we multiply the trace on the left side of (3.19) by $c = \text{E}\{\psi^2(\xi)\}$ with $\xi \sim N(0,1)$. This leads to the final modified equations:

$$\sum_{i=1}^{m} X_i'V_i^{-1}U_i^{1/2}\Psi(r_i) = 0, \tag{3.20}$$

$$\sum_{i=1}^{m}\left\{\Psi(r_i)'U_i^{1/2}V_i^{-1}\frac{\partial V_i}{\partial \theta_s}V_i^{-1}U_i^{1/2}\Psi(r_i) - c\,\text{tr}\left(V_i^{-1}\frac{\partial V_i}{\partial \theta_s}\right)\right\} = 0,\ \ 1 \le s \le q. \tag{3.21}$$

Note that (3.20) and (3.21) are a system of nonlinear equations for solving for β and θ. [215] suggests that the equations can be solved iteratively. Starting with initial values β^0, θ^0, one uses a Newton-Raphson iteration to update β from (3.20), and another Newton-Raphson to update θ based on the updated β; repeat until convergence. The estimates of β and θ at convergence are then used to estimate the random effects v_i using, again, Newton-Raphson in solving equation (3.14). The robust estimates of β and v_i are then used to replace the estimates of β and v_i in the EBLUP of the small area means. This leads to the robust EBLUP (REBLUP) of the small area means.

As for measures of uncertainty, [215] proposed to the use of a parametric bootstrap method to estimate the MSPE of the REBLUP. The procedure is as follows. Based on the robust estimates of β and θ, one can generate data under the assumed LMM, (3.8). This involves generating the random effects, v, in the first place. The generated random effects lead to the "true" small area means in the bootstrapped world. Those small area means are recorded. One then uses the bootstrapped data, following the same procedure as described above, to obtain the predicted small area means. Now we have the predicted small area means, and the recorded true small area means (in the bootstrapped world), which leads to the squared prediction error. We can do this repeatedly (because it is bootstrapping), and compute the average squared prediction error, which is the bootstrap MSPE estimate.

In a series of simulation studies, [215] investigated the estimation and prediction performance of the proposed estimators of the model parameters and predictors of small area means, as well as the proposed MSPE estimators. The simulation was carried out under several settings. Here, we mainly focus on their first simulation setting called contaminated distributions. The simulated data were generated under an NER model with $m = 40$, $n_i = 4$, and a single auxiliary variable generated from $N(1, 1)$ (see §1.5.2). The random effects and errors are generated from a normal mixture distribution expressed as $(1-\gamma)N(0, \sigma_1^2) + \gamma N(0, \sigma_2^2)$, where $\sigma_1^2 = 1$ and $\sigma_2^2 = 25$. Here, $N(0, \sigma_1^2)$ corresponds to the case when "business is as usual", that is, there is no outlier, and $N(0, \sigma_2^2)$ to the distribution of the outlier. Furthermore, γ is either 0, meaning there is no outlier, or 0.1, meaning a draw from the outlying distribution with a 10% probability (note that this does not mean the outlier occurs with a 10% chance, because the outlying distribution itself has a chance error). This naturally brings four scenarios for v and e under the NER model: $(0, 0)$—no outlier in v or e; $(1, 0)$—outlier in v but not in e; $(0, 1)$—outlier in e but not in v; and $(1, 1)$—outliers in both v and e.

In terms of estimation performance, [215] reported results of the simulated MSE in estimating the fixed effects and variance components. Three estimators were compared: ML, RML—the robust estimator proposed by [215], and Fellner—the robust estimator proposed by [60]. In the $(0, 0)$ case, ML outperformed RML and Fellner, which is not surprising. In the other three cases, RML performed the best in terms of variance components estimation while Fellner performed the best in estimating the fixed effects.

TABLE 3.1
Average simulated MSPE of small area predictors.

Contamination	(0, 0)	(1, 0)	(0, 1)	(1, 1)
EBLUP	0.2066	0.2290	0.5055	0.6578
REBLUP	0.2140	0.2235	0.2993	0.3226
Fellner	0.2122	0.2229	0.2950	0.3425

In view of the special interest in SAE, it is more important to compare the performance of different methods in estimating the small area means. Here, the small area means are $\theta_i = \beta_0 + \beta_1 \bar{X}_i + v_i$, $i = 1, \dots, m$, and $\bar{X}_i$ is assumed to be 1 for simplicity [recall the x_{ij}'s are generated from $N(1, 1)$]. Table 3.1, taken from Table 2 of [215], reports the simulated MSPE of EBLUP, REBLUP and Fellner's robust modification of the EBLUP (Fellner; [60]), averaged over the small areas. It is seen that, in the case of no contamination [(0, 0)], EBLUP performs slightly better than the robust modifications, which is not surprising. In the case of contamination in v only [(1, 0)], the two robust versions of EBLUP perform only slightly better than the EBLUP, with Fellner slightly outperforming REBLUP. In the case of contamination in e only [(0, 1)], both robust versions of the EBLUP significantly outperform the EBLUP, with Fellner slightly outperforming REBLUP. Finally, when contaminations are in both v and e, both robust modifications of the EBLUP significantly outperform the EBLUP, with REBLUP slightly outperforming Fellner. Overall, the results seem to suggest that contamination in e has much greater impact on the performance of the EBLUP than contamination in v.

Finally, [215] presented simulation results on the evaluation of the proposed bootstrap MSPE estimator. The authors considered three different bootstrap sample sizes, $B = 100, 200, 500$. Table 3.2, again copied from [215], reported the simulated and actual MSPEs averaged over the small areas. It is seen that the accuracy of the MSPE estimator improves as the bootstrap sample size increases only in the case of (0, 0), that is, when there is no contamination in either v or e; in all other three cases, larger bootstrap samples do not seem to improve the accuracy. Note that, however, the sample size is fixed at $m = 40$ (and $n_i = 4$). Typically, bootstrap MSPE estimators without bias-correction is first-order unbiased [i.e.,the bias is $O(m^{-1})$] rather than second-order unbiased [i.e., the bias is $o(m^{-1})$; e.g., [199]].

3.3 M-quantile regression

This section considers a quantile-based method for SAE. Quantile regression is a well-known classical statistical method (e.g., [140]). In the standard linear

TABLE 3.2
Simulated and estimated MSPEs, averaged over small areas.

Bootstrap	(0,0)		(1,0)		(0,1)		(1,1)	
Sample Size	Sim.	Est.	Sim.	Est.	Sim.	Est.	Sim.	Est.
100	.2134	.2156	.2259	.2321	.2925	.2824	.3333	.3315
200	.2140	.2130	.2321	.2235	.2824	.2993	.3315	.3226
500	.2147	.2146	.2245	.2292	.2946	.2832	.3345	.3294

regression, one models the mean of y_i as a linear function of x_i, that it, $\mathrm{E}(y_i) = x_i'\beta$, for $i = 1, \dots, n$ (n is the sample size), where β is a vector of unknown regression coefficients. In quantile regression, one models a given quantile, say, $Q_q(y_i)$, where q is a given number between 0 and 1, as a linear function of x_i, that is, $Q_q(y_i) = x_i'\beta$. For example, when $q = 1/2$, one is modeling the median as $x_i'\beta$. In linear regression, β is typically estimated via the LS, that is, by minimizing

$$\sum_{i=1}^{n}(y_i - x_i'\beta)^2 \tag{3.22}$$

over β. This is equivalent to maximum likelihood estimation of β under normal distribution of the regression errors. In quantile regression, the function (3.22) is modified to

$$\sum_{i=1}^{n}\rho_q(y_i - x_i'\beta), \tag{3.23}$$

where $\rho_q(\cdot)$ is some "distance" function measuring the errors of model fitting. In particular, [140] proposed the following function:

$$\rho_q(u) = |u|\{(1-q)1_{(u\le 0)} + q1_{(u>0)}\}. \tag{3.24}$$

Thus, if $q = 1/2$, the function (3.24) reduces to $|u|$, the absolute value. This is equivalent to the maximum likelihood estimation of β under the Laplace distribution of the regression errors. Here, the Laplace distribution for the regression errors, denoted by $L(0, \tau)$, has the pdf

$$f(x) = \frac{1}{2\tau}e^{-|x|/\tau}, \quad -\infty < x < \infty, \tag{3.25}$$

where $\tau > 0$ is a scale parameter.

An M-quantile is an extension of the quantile. For a given $q \in (0, 1)$, the qth M-quantile depends on an influence function, $\psi_q(\cdot)$, which depends on q, and is defined as the solution Q to the equation

$$\int \psi_q(y - Q)f(y)dy = 0, \tag{3.26}$$

where $f(\cdot)$ is the probability density function (pdf) of Y. For example, if one takes $\psi_q(u) = 1_{(u \le 0)} - q$, then the left side of (3.26) is equal to

$$\int \{1_{(y-Q\le 0)} - q\} f(y) dy = F(Q) - q,$$

where $F(\cdot)$ is the cumulative distribution function (cdf) of Y. It is seen that, in this case, the qth M-quantile is the same as the qth quantile. M-quantile regression is an extension of the quantile regression, in which case the $f(\cdot)$ in (3.26) is replaced by $f(\cdot|x)$, the conditional pdf of Y given x, and the M-quantile is modeled as

$$Q = Q_q(x_i) = x_i'\beta. \tag{3.27}$$

Ref. [29] considered M-quantile models for SAE. Unlike the majority of the SAE models (see §1), random effects are not present in their M-quantile models. The authors argued that there are other ways to characterize the inter-area differences. Here, the authors focus on a particular influence function:

$$\begin{aligned} \psi_q(u) &= 2\psi\left(\frac{u}{\tau}\right)\{q1_{(u>0)} + (1-q)1_{(u\le 0)}\} \\ \text{with } \psi(u) &= u1_{(-c\le u\le c)} + c\text{sign}(u), \end{aligned} \tag{3.28}$$

where τ is a scale parameter and $c > 0$ is a constant. The function $\psi(\cdot)$ in (3.28) is called Huber's proposal 2 influence function.

To apply M-quantile regression to SAE, [29] associates each unit, (x_i, y_i), a quantile coefficient, q_i, via the equation

$$Q_{q_i}(x_i) = y_i. \tag{3.29}$$

For the jth small area, let θ_j denote the corresponding small area mean, and $q_{[j]}$ be the average of the quantile coefficients over the small area; also let N_j be the population size, and s_j be the set of sampled indexes. It is assumed that the following approximation holds:

$$\begin{aligned} \theta_j &= \frac{1}{N_j}\left(\sum_{i\in s_j} y_i + \sum_{i\notin s_j} y_i\right) \\ &= \frac{1}{N_j}\left\{\sum_{i\in s_j} y_i + \sum_{i\notin s_j} x_i'\beta(q_i)\right\} \\ &\approx \frac{1}{N_j}\left\{\sum_{i\in s_j} y_i + \sum_{i\notin s_j} x_i'\beta(q_{[j]})\right\}. \end{aligned} \tag{3.30}$$

Here, the second equation in (3.30) follows from (3.29) and (3.27), where the β depends on q_i, hence is denoted by $\beta(q_i)$. Furthermore, the $q_{[j]}$ in the third line of (3.30) is the average of the q_i's over the small area. Because $q_{[j]}$ is unknown, [29] suggests that an estimator of $q_{[j]}$ is the average of $q_i, i \in s_j$; alternatively, it can be estimated by the median of $q_i, i \in s_j$. Let $\hat{q}_{[j]}$ denote the estimator of $q_{[j]}$. However, the vector of regression coefficients, $\beta(q_{[j]})$, is estimated differently as described below.

For a given q, [29] suggests that $\beta(q)$ can be estimated by solving

$$\sum_{i \in s_j} \psi_q(y_i - x_i'\beta)x_i = 0 \tag{3.31}$$

for β, where the function $\psi_q(\cdot)$ is defined in (3.28), with τ replaced by a robust estimator of scale. The solution to (3.31) can be expressed as

$$\hat{\beta}(q) = \{X_s'W_s(q)X_s\}^{-1}X_s'W_s y_s, \tag{3.32}$$

where $X_s = (x_i')_{i \in s}$, $y_s = (y_i)_{i \in s}$, and W_s is the weighting matrix used to solve (3.31) via iterative weighted least squares.

The estimate of the small area mean, θ_j, is given by the right side of (3.30) with $\beta(q_{[j]})$ replaced by $\hat{\beta}(q_{[j]})$, that is,

$$\hat{\theta}_j = \frac{1}{N_j}\left\{\sum_{i \in s_j} y_i + \sum_{i \notin s_j} x_i'\hat{\beta}(q_{[j]})\right\}. \tag{3.33}$$

A key question regarding the M-quantile regression SAE is how to interpret the quantity $q_{[j]}$. Ref. [29] suggests that $q_{[j]}$ is, in a way, similar to an area-specific random effect. For example, if all area M-quantile coefficients are equal to 0.5, one concludes "there is no between-area variation beyond that explained by the model covariates". However, the interpretation seems to be less straightforward than that for the random effects. Nevertheless, the authors claimed that a main advantage of the M-quantile regression over a mixed-effects model approach is that the former allows for outlier-robust inference using "widely available M-estimation software".

In terms of a measure of uncertainty, the authors suggested using the mean squared error (MSE). This seems a bit unnatural as the point was thought to avoid using the mean-based approach, which leads to the best predictor under the MSE. Also, unlike the mean-based approach, such as the EBLUP, there was no optimality consideration under the M-quantile framework, whose theoretical foundation appears to be lacking.

In empirical studies, [29] carried out both model-based and design-based simulation studies. In the model-based simulation, the data were generated under a NER model with bivariate area-specific random effects. Once the data were generated, they were fixed to form the finite populations, from which the samples were drawn. In the design-based simulation, the data were drawn from a real data set involving 1,652 broadacre farms of Australia. The authors compared the performance of the mixed model prediction (MMP) based method (i.e., EBLUP) and the M-quantile regression based method in terms of bias and MSE for estimating the small area means and medians. The model-based simulation study showed little difference between the two methods. In the design-based simulation study, the two methods performed similarly, overall, in terms of estimating the small area means; however, in terms of estimating the small area medians, the M-quantile regression method was shown to outperform the MMP-based method.

3.4 Density power divergence and related topics

The use of density divergence measures to develop estimators with good robust properties may be traced back to [13], who introduced the minimum Hellinger distance estimator for continuous distribution models. Ref. [1] introduced the minimum density power divergence (MDPD) estimator in the case of i.i.d. observations. Here, the density power divergence (DPD) measures the distance, or discrepancy, between two density functions, say, f and g, defined as

$$d_\gamma(f,g) = \int \left\{ g^{1+\gamma}(y) - \left(1+\frac{1}{\gamma}\right) f(y) g^\gamma(y) + \frac{1}{\gamma} f^{1+\gamma}(y) \right\} dy, \qquad (3.34)$$

where $\gamma > 0$ is a divergence parameter. For $\gamma = 0$, the DPD is defined as the limit of $d_\alpha(f,g)$ as $\gamma \to 0$, which is the Kullback-Leibler divergence:

$$d_0(f,g) = \int f(y) \log\left\{\frac{f(y)}{g(y)}\right\} dy. \qquad (3.35)$$

The divergence parameter γ offers the flexibility of choices. Note that (3.34) can be written alternatively as

$$\begin{aligned} d_\gamma(F,G) &= \int g^{1+\gamma}(y)dy - \left(1+\frac{1}{\gamma}\right) \int g^\gamma(y) dF(y) \\ &\quad + \frac{1}{\gamma} \int f^\gamma(y) dF(y), \end{aligned} \qquad (3.36)$$

where $F(\cdot), G(\cdot)$ are the cdfs corrresponding to $f(\cdot), g(\cdot)$, respectively.

To apply the DPD for parameter estimation in the i.i.d. case, suppose that $Y_1, \dots, Y_n$ are i.i.d. observations from F_θ, a cdf depending on a parameter vector, θ, of interest. If, in (3.36), one replaces the G by F_θ, and the F by the empirical cdf of the observations, one has an empirical DPD in the form of

$$d_\gamma(\hat{F}, F_\theta) = \int f_\theta^{1+\gamma}(y)dy - \left(1+\frac{1}{\gamma}\right) \frac{1}{n} \sum_{i=1}^n f_\theta^\gamma(Y_i) + s(Y,\gamma), \qquad (3.37)$$

where f_θ is the pdf of F_θ and $s(Y,\gamma)$ does not depend on θ. The minimum DPD (MDPD) estimator, $\hat{\theta}$, is defined as the θ that minimizes (3.37). Clearly, this is equivalent to minimizing the following over θ:

$$\int f_\theta^{1+\gamma}(y)dy - \left(1+\frac{1}{\gamma}\right)\frac{1}{n}\sum_{i=1}^n f_\theta^\gamma(Y_i). \tag{3.38}$$

An extension of (3.38) to independent not-identically distributed (i.n-i.d.) observations $Y_i, i = 1,\dots,n$ is relatively straightforward. Note that minimizing (3.38) is equivalent to maximizing

$$\frac{1}{\gamma}\sum_{i=1}^n f_\theta^\gamma(Y_i) - \frac{n}{1+\gamma}\int f_\theta^{1+\gamma}(y)dy. \tag{3.39}$$

In the i.n-i.d. case, (3.39) is replaced by

$$l_\gamma(\theta) = \frac{1}{\gamma}\sum_{i=1}^n f_{i,\theta}^\gamma(Y_i) - \frac{1}{1+\gamma}\sum_{i=1}^n \int f_{i,\theta}^{1+\gamma}(y)dy, \tag{3.40}$$

where $f_{i,\theta}(\cdot)$ is the pdf of Y_i under the parameter vector θ.

Ref. [217] adopted (3.40) to SAE under the Fay-Herriot model, (1.16). Here, we make a notational change: Let ψ denote the vector of parameters instead of θ. Let $f_{i,\psi}(\cdot)$ be the pdf of the y_i in (1.16), that is, the pdf of $N(x_i'\beta, A + D_i)$, and $n = m$. The parameter vector $\psi = (\beta', A)'$ is then estimated via MDPD.

Next, the BP (or Bayes estimator) of θ_i, (1.25), is modified under a robust consideration. According to Tweedie's formula (see [52]), the BP (1.25) can be expressed as

$$\tilde{\theta}_i = y_i + D_i\frac{\partial l}{\partial y_i}, \tag{3.41}$$

where $l = l(\psi)$ is the log-likelihood function under the Fay-Herriot model,

$$l(\psi) = -\frac{1}{2}\left\{m\log 2\pi + \sum_{i=1}^m \log(A+D_i) + \sum_{i=1}^m \frac{(y_i - x_i'\beta)^2}{A+D_i}\right\}.$$

A robust modification of the BP is obtained by replacing the l in (3.41) by the l_γ function in (3.40). This leads to

$$\tilde{\theta}_i^{\rm R} = y_i - \frac{D_i}{A+D_i}(y_i - x_i'\beta)s_i(y_i,\psi), \tag{3.42}$$

where $s_i(y_i,\psi) = a_i^{-\gamma/2}e^{-\gamma(y_i-x_i'\beta)^2/2(A+D_i)}$ with $a_i = 2\pi(A+D_i)$. The robust BP is defined as (3.42) with ψ replaced by its MDPD estimator.

An issue with the DPD method is how to choose γ, which may be viewed as a tuning parameter. Ref. [217] followed an approach of [73], which shows that

$${\rm MSPE}(\tilde{\theta}_i^{\rm R}) = {\rm E}(\tilde{\theta}_i^{\rm R} - \theta_i)^2 = g_{1i}(A) + g_{2i}(A), \tag{3.43}$$

where $g_{1i}(A) = AD_i/(A + D_i)$ and

$$g_{2i}(A) = \frac{D_i^2}{A + D_i}\left\{1 + \frac{1}{(2\gamma+1)^{3/2}a_i^{\gamma}} - \frac{2}{(\gamma+1)^{3/2}a_i^{\gamma/2}}\right\}.$$

Note that $g_{1i}(A)$ does not depend on γ. On the other hand, it can be shown (see [217]) that $g_{2i}(A)$ is non-decreasing in γ. This seems to suggest that the optimal value of γ is $\gamma = 0$, which corresponds to the non-robust MLE. However, expression (3.43) is the MSPE of $\tilde{\theta}_i^{\mathrm{R}}$, which is not computable due to the unknown β and A. In practice, one replaces β and A by their MDPD estimators, which also depend on γ. Thus, $\gamma = 0$ is not necessarily the best choice for minimizing the MSPE of

$$\hat{\theta}_i^{\mathrm{R}} = \tilde{\theta}_i^{\mathrm{R}}|_{\beta=\hat{\beta}_\gamma, A=\hat{A}_\gamma}, \tag{3.44}$$

where $\hat{\beta}_\gamma$, $\hat{A}_\gamma$ are the MDPD estimators of β, A, respectively with the given γ. (3.44) is called robust empirical Bayes (EB) estimator [217]. The latter suggests that this shows the choice of γ is a trade-off between robustness and efficiency. The author suggested choosing γ the following way. First define

$$\mathrm{Exc}(\gamma) = 100 \times \frac{\sum_{i=1}^m g_{2i}(\hat{A}_\gamma)}{\sum_{i=1}^m g_{1i}(\hat{A}_\gamma)} \tag{3.45}$$

as the percentage relative excess mean squared error. It was proposed to choose γ so that $\mathrm{Exc}(\gamma)$ is bounded by $c\%$, where c is a pre-specified number, say, $c = 5$. In other words, one solves the equation $\mathrm{Exc}(\gamma) = c$ to determine γ.

As for measures of uncertainty, [217] derived a second-order approximation to the MSPE of $\hat{\theta}_i^{\mathrm{R}}$. However, the author indicated that the derivation of a Prasad-Rao type second-order unbiased MSPE estimator [196]] is tedious. As an alternative, the author proposed to use the parametric bootstrap method of [19] to estimate the MSPE.

In a simulation study, [217] compared empirical performance of several methods for estimating the small area means, including the traditional EBLUP method, the robust EBLUP method of [215] (see §3.2), and the proposed robust EB method. The simulation setting is (1.16) with $x_i'\beta = \beta_0 + \beta_1 x_i$, where $\beta_0 = 0$ and $\beta_1 = 2$, and x_i are generated from the Uniform$(0,1)$ distribution. Furthermore, $v_i = \sqrt{A}u_i$, where $A = 0.5$ and u_i are generated from the normal-mixture distribution $(1-\rho)N(0,1) + \rho N((0,100)$, where three values of ρ are considered: (I) $\rho = 0$; (II) $\rho = 0.15$; and (III) $\rho = 0.30$. As for the e_i, it is assumed to be independent $N(0, D_i)$, where D_i has five different values, 0.2, 0.4, 0.6, 0.8, 1.0, distributed evenly among the $m = 30$ small areas. As for the value of c [see below (3.45)], the author considered $c = 1$ and $c = 5$. The simulation results suggest that, in case (II), EBLUP performed the best, which is not surprising, but the proposed robust EB method performed competitively in this case. In cases (II) and (III), the proposed robust EB method performed better than the EBLUP and other comparing methods. The simulation results also showed that the performance of the robust EB method depends on the choice of c. Specifically, the robust EB method with $c = 1$ performed better than that with $c = 5$ in case (I), while the robust EB method with $c = 5$ performed considerably better than that with $c = 1$ in cases (II) and (III). This suggests that the choice of c is important in practice, and there is not a satisfactory answer in this regard at this point.

In the case of the unit-level model, [187] adopted the DPD approach directly for parameter estimation under an extended NER model. The latter is the same as the NER model in §1.5.2 but with $e_{ij} = k_{ij}\epsilon_{ij}$, where k_{ij} are known constants and ϵ_{ij}'s are independent $N(0, \sigma_e^2)$. The $f_\theta(Y_i)$ in (3.38) is replaced by $f_\theta(y_i|X_i)$, which is the pdf of $N(X_i\beta, V_i)$. Here $y_i = (y_{ij})_{1\le j\le n_i}$, $X_i = (x'_{ij})_{1\le j\le n_i}$, and $V_i = \sigma_v^2 1_{n_i}1'_{n_i} + \sigma_e^2\text{diag}(k_{ij}^2, 1 \le j \le n_i)$. The parameter vector θ is the combination of β, σ_v^2 and σ_e^2. It can be shown that the objective function, (3.38), now has the expression

$$\hat{d}(\theta) = \sum_{i=1}^{m} \frac{1}{(2\pi)^{n_i\gamma/2}|V_i|^{\gamma/2}} \left\{\frac{1}{(1+\gamma)^{n_i/2}} - \left(1 + \frac{1}{\gamma}\right) e^{-(\gamma/2\sigma_e^2)B_i}\right\}, \quad (3.46)$$

where $|V_i|$ is the determinant of V_i and

$$B_i = \sum_{j=1}^{n_i} a_{ij}(y_{ij} - x'_{ij}\beta)^2 - \frac{\rho_i}{a_{i\cdot}}\left\{\sum_{j=1}^{n_i} a_{ij}(y_{ij} - x'_{ij}\beta)\right\}^2$$

with $a_{ij} = k_{ij}^{-2}$, $a_{i\cdot} = \sum_{j=1}^{n_i} a_{ij}$, and $\rho_i = \sigma_v^2/(\sigma_v^2 + \sigma_e^2/a_{i\cdot})$. It is seen that, if there is an outlier from $y_{ij}, 1 \le j \le n_i$, the sampled outcomes from the ih small area, in the sense that the difference $y_{ij} - x'_{ij}\beta$ is large in absolute value, hence the value of B_i will be large. Then, in (3.46), the corresponding term $e^{-(\gamma/2\sigma_e^2)B_i}$ will be small. This means that the outlier has small contribution to the objective function. Thus, intuitively, the MDPD estimator should be robust to the outliers.

As for the area-level model, here we also have the issue of choosing the parameter γ, which may be viewed as a tuning parameter. Ref. [187] noted that the choice of γ is a trade-off between robustness and efficiency. When $\gamma = 0$, the choice is equivalent to maximum likelihood (ML), which is efficient but not robust to outliers. As γ increases from 0, one gains in robustness but loses in efficiency. The authors then proposed a method of choosing γ to minimize the MSE of parameter estimation. As noted by [126], the primary interest of SAE is the estimation of small area characteristics such as small area means and proportions, which is a problem of mixed model prediction under a mixed-effects model. While parameter estimation is involved in the prediction, it is not the predictor itself. It would be more relevant to choose γ in view of minimizing the MSPE of the prediction, in some way. In this regard, the approach of [217] seems more natural.

In simulation studies, [187] compared the performance of different methods of parameter estimation, including ML, the robust estimator of θ proposed by [215], and the MDPD estimator with different values of γ, including the optimal choice based on MSE of parameter estimation. The simulation setting was the same as the contaminated distribution scenario considered in [215] (see §3.2). It was found that the performance of the MDPD estimator heavily depends on the choice of γ, and MDPD with the proposed optimal choice of γ has better performance, in terms of the MSE, than the few fixed values of γ under consideration, namely, $\gamma = 0.1, 0.2, 0.3$, in nearly all of the cases considered.

As noted, it is more important to compare the predictive performance of different methods in estimating the small area mean, which are the end points. In their simulation studies, [187] also compared the performance of different estimators of the small area means, including the traditional EBLUPs, the REBLUP of [215], the M-quantile estimator of [29] (see §3.3), and EBLUPs based on MDPD estimators of parameters with $\gamma = 0.1, 0.2, 0.3$, and the optimal choice based on the MSE of parameter estimation. The simulation results showed that the MDPD-based EBLUPs perform better than the traditional EBLUP, REBLUP, and M-quantile methods in terms of the MSPE in the case of contaminated distributions, and the MDPD-based EBLUP with the optimal choice of γ performed the best. In the case of non-contaminated distributions, the traditional EBLUP performs the best, but the other methods performed competitively.

Similar to the DPD, [147] considered so-called γ-divergence under the area-level model for both point and interval estimation of the small area means, $\theta_i = x_i'\beta + v_i$. The authors noted that, under the Fay-Herriot model, one has

$$\mathrm{E}(\theta_i|y_i) = y_i - \frac{D_i}{A+D_i}(y_i - x_i'\beta), \quad \mathrm{var}(\theta_i|y_i) = \frac{AD_i}{A+D_i}. \tag{3.47}$$

The conditional variance, in addition to the conditional mean, is involved because it plays an important role in the interval estimation. Once again, by Tweedie's formula (e.g., [52]), one can write (3.47) as

$$\mathrm{E}(\theta_i|y_i) = y_i + D_i\frac{\partial}{\partial y_i}\log f_\psi(y_i), \ \mathrm{var}(\theta_i|y_i) = D_i + D_i^2\frac{\partial^2}{\partial y_i^2}\log f_\psi(y_i), \tag{3.48}$$

where $f_\psi(y_i) = \{2\pi(A+D_i)\}^{-1/2}e^{-(y_i - x_i'\beta)^2/2(A+D_i)}$ is the pdf of y_i. A similar technique to (3.42) is then used, to replace the log-pdf in (3.48) with a robust modification. Here, the robust modification is the γ-divergence, defined as

$$d_\gamma(y_i) = \frac{c_\gamma(A)}{\gamma}f_\psi^\gamma(y_i), \tag{3.49}$$

where $c_\gamma(A) = \{2\pi(A+D_i)\}^{\gamma^2/2(1+\gamma)}$. It is easy to see that the modified conditional mean is

$$\tilde{\theta}_i^\gamma = y_i - \frac{D_i}{A+D_i}(y_i - x_i'\beta)g_i(y_i, \psi), \tag{3.50}$$

where $g_i(y_i, \psi) = a_i^{-\gamma/2(1+\gamma)} e^{-\gamma(y_i - x_i'\beta)^2/2(A+D_i)}$ and a_i is defined below (3.42). Comparing (3.50) and (3.42), it is clear that the two are only slightly different (especially when the value of γ is small) in the power of a_i, namely, one is $-\gamma/2$ and the other is $-\gamma/2(1+\gamma)$.

A bigger difference, however, is made by carefully choosing the tuning parameter γ. We have seen in the DPD case that the choice of c [see below (3.45)] can affect the performance of the robust EP method. Here, the interest is not just to provide point estimates of the small area means; interval estimates are also of interest. The confidence (or prediction) interval for θ_i based on the direct estimate is given by

$$(y_i - z_{\alpha/2}\sqrt{D_i},\ y_i + z_{\alpha/2}\sqrt{D_i}), \tag{3.51}$$

where $z_{\alpha/2}$ is the $\alpha/2$ critical value of the standard normal distribution and $(1-\alpha)\%$ is the confidence level. On the other hand, the EB confidence interval for θ_i is given by

$$(\hat{\theta}_i - z_{\alpha/2}\hat{s}_i,\ \hat{\theta}_i - z_{\alpha/2}\hat{s}_i), \tag{3.52}$$

where $\hat{\theta}_i$ and $\hat{s}_i$ are the conditional mean and variance in (3.47) with β, A replaced by their MLEs. One can similarly construct the γ-divergence EB confidence interval by replacing the MLEs in the EB confidence interval by the corresponding γ-divergence estimator, which maximizes

$$\sum_{i=1}^{m} d_\gamma(y_i). \tag{3.53}$$

Let $\hat{s}_{i,\gamma}$ denote $\hat{s}_i$ with the MLE of A, $\hat{A}$, replace by the γ-divergence estimator, $\hat{A}_\gamma$. The optimal γ is chosen as the minimizer of

$$\sum_{i=1}^{m} w_i \hat{s}_{i,\gamma}^2, \tag{3.54}$$

over $\gamma \in \{\gamma_1, \dots, \gamma_M\}$, a grid of candidate γ values. Here, $w_i, i = 1, \dots, m$ are a set of given weights. This is called adaptive tuning parameter (ATP).

ATP is a date-drive procedure for choosing the tuning parameter, γ. Although it seems to raise another tuning-parameter type issue, that is, the selection of the weights, w_i. Ref. [147] suggests $w_i = 1$ works well in simulation studies. Specifically, the authors consider an area-level model, which can be expressed as the Fay-Herriot model, (1.16), with $x_i'\beta = \beta_0 + \beta_1 x_{i1} + \beta_2 x_{i2}$, where x_{i1} were generated from a standard normal distribution and x_{i2} generated from the Bernoulli(0.5) distribution, $\beta_0 = 0$, $\beta_1 = -1$ and $\beta_2 = 1$, and $m = 100$. Also $v_i = \sqrt{A}u_i$, where A is either 1 or 0.5; u_i were generated under one of the following five distributions:
(i) $u_i \sim N(0, 1)$;
(ii) $u_i \sim LN(0, 1)$, where $LN(a, b)$ denotes the log-normal distribution with log-mean a and log-variance b;

(iii) $u_i \sim$ Cauchy$(0, 1)$, where the Cauchy(a, b) distribution has pdf $f_\lambda(x) = (b/\pi)/\{(x - a)^2 + b^2\}$, $-\infty < x < \infty$;
(iv) $u_i \sim 0.95N(0, 1) + 0.05N(10, 1)$;
(v) $u_i \sim 0.9N(0, 1) + 0.1N(10, 1)$.
The last two scenarios correspond to some contaminated normal mixture distributions. However, unlike in [217] (see above), in which the contamination is in the variance, here the contamination is in the mean. The D_i's are evenly distributed in five groups with values 0.2, 0.6, 1.0, 1.4 and 2.0, respectively. In terms of point estimation, the authors compared their proposed γ-divergence (GD) method with ATP with the EB method, the DPD method of [217] discussed above with $c = 5$, the robust EBLUP method of [215], discussed in §3.2, and another robust EB method proposed by [73]. The results show that, in case (i), in which the EB method is known to be the best, the GD method with ATP was identical to the EB; in the other four cases, where outliers occur in some ways, the GD method performed the best, overall.

In terms of interval estimation, [147] compared GD, EB confidence interval, (3.52), and the direct confidence interval (DR), (3.51), in terms of the performance of the 95% confidence intervals. The results showed that all three methods provided coverage probabilities close to the nominal level in all scenarios considered. However, in terms of the average length of the confidence intervals, GD provided more efficient intervals in terms of shorter average length than EB and DR. It was also found that, with the exception of case (i), the average length of EB is close to that of DR.

3.5 Characterizing the outliers

In an earlier paper, Chambers [30] introduced the concept of "representative outlier", which means a sample outlier that is potentially drawn from a group of population outliers existing in the population. In other words, a representative outlier is representative of the non-sampled part of the population, and therefore "legitimate" [136].

The point is that a representative outlier is a reflection of "life", and therefore should not be treated the same way as, for example, a recording error. Ref. [31] notes that the traditional approaches in SAE to outlier robustness, such as those discussed in the previous sections, are to down-weight the importance of the outliers, in some way, in order to reduce the impact of the outliers. The authors believed, however, that a representative outlier should not be treated the same way. Instead, the authors suggested bias-correct the bias in the small-area predictors induced by the representative outliers; see below. Furthermore, the authors distinguished what they call the robust projective approach and robust predictive approach. In their distinction, the traditional approaches based on down-weighting the importance of the outliers, such as

[60], [215], [29], and [217], are robust projective because they "project sample non-outlier (i.e. working model) behavior onto the non-sampled part of the survey population". Alternatively, a robust predictive approach would "attempt to predict the contribution of the population outliers to the population quantity of interest".

Recall the unit-level NER model introduced in §1.5.2. Here, the objective is to estimate the population mean, $\bar{Y}_i = N_i^{-1} Y_{i\cdot}$, where $Y_{i\cdot}$ is the total of the ith small-area population. Note that part of the population is already observed in the samples, $y_{ij}, j \in s_i$, from the ith small area, where s_i denotes the subset of sampled indexes. Thus, we can express the population mean as

$$\bar{Y}_i = \frac{1}{N_i}\left(\sum_{j\in s_i} y_{ij} + \sum_{j\notin s_i} y_{ij}\right). \tag{3.55}$$

One knows the first sum inside the $(\cdot)$ in (3.55) but not the second, which corresponds to the non-sampled part of the population. A model-based SAE method would then approximate the second sum by its BP under the assumed model. Ref. [235] considered a similar approach to estimating the empirical distribution of the unit values in the ith area, in which a robust model-based predictor is used for the non-sampled part. If the non-sampled part were observed, the empirical distribution would be given by

$$\tilde{F}(t) = \frac{1}{N_i}\left(\sum_{j\in s_i} 1_{(y_{ij}\le t)} + \sum_{j\notin s_i} 1_{(y_{ij}\le t)}\right). \tag{3.56}$$

Because the non-sampled part is not available, one replaces the corresponding $1_{(y_{ij}\le t)}$ by a model-based predictor. In [235], the underlying model is

$$y_{ij} = x_{ij}\beta + \sigma v^{1/2}(x_{ij}) e_{ij}, \tag{3.57}$$

where x_{ij} is an auxiliary variable whose population values are known, β is an unknown coefficient, σ is an unknown standard deviation, $v(\cdot)$ is a known variance function, and e_{ij} are i.i.d. errors with mean 0 and variance 1. Under (3.57), a natural estimator of $1_{(y_{ij}\le t)}$ for the non-sampled y_{ij} would be $1_{(\hat{\beta}x_{ij}\le t)}$, where $\hat{\beta}$ is an estimator of β. Ref. [235] suggests to use the bi-weight estimator proposed by [9]. Following the idea of [30], a robust bias-correction (RBC) term is needed to address the representative outliers. In the case of estimating the population total, the RBC estimate of the population total is in the form of

$$\hat{Y}_{\text{rbc}} = \sum_{j\in s_i} y_{ij} + \hat{\beta}\sum_{j\notin s_i} x_{ij} + \hat{B}_c, \tag{3.58}$$

where $\hat{B}_c$ is the RBC term, with the expression

$$\hat{B}_c = \frac{c\hat{\sigma}\sum_{j\notin s_i} x_{ij}}{\sum_{j\notin s_i} x_{ij}^2/v(x_{ij})} \sum_{j\in s_i} \frac{x_{ij}}{\sqrt{v(x_{ij})}}\psi\left(\frac{y_{ij}-\hat{\beta}x_{ij}}{c\hat{\sigma}\sqrt{v(x_{ij})}}\right), \tag{3.59}$$

$\hat{\sigma}$ is a robust estimator of σ, $\psi(\cdot)$ is an influence function, and $c \geq 0$ is a scale-tuning parameter. Ref. [235] suggests to use Huber's ψ-function, $\psi(u) = \max(-1, \min(u, 1))$. The scale-tuning parameter strikes a trade-off between linearity ($c = \infty$) and boundedness. In particular, [30] suggests $c = 1.34$. Combining (3.58) and (3.59), an RBC estimator of β can be derived, which has the expression

$$\hat{\beta}_{\mathrm{rbc}} = \hat{\beta} + \frac{c\hat{\sigma}}{\sum_{j\in s_i} x_{ij}^2/v(x_{ij})} \sum_{j\in s_i} \frac{x_{ij}}{\sqrt{v(x_{ij})}} \psi\left(\frac{y_{ij} - \hat{\beta}x_{ij}}{c\hat{\sigma}\sqrt{v(x_{ij})}}\right). \tag{3.60}$$

As noted by [31], an estimator of the population mean is obtained by taking the mean of the empirical distribution, that is, $\int t d\hat{F}(t)$. Furthermore, the method of [235] can be extended to the NER model, with $\hat{\beta}x_{ij}$ replaced by an robust predictor of the small-area mean. Here, the scale-tuning parameter, c, depends on i and j as well as ψ, that is, $c = c_{ij}^{\psi}$. The bias-corrected EBLUP of the population mean has a similar expression to (3.60), namely,

$$\hat{Y}_i^{\mathrm{reblup\_bc}} = \hat{Y}_i^{\mathrm{reblup}} + \frac{1 - r_i}{n_i} \sum_{j\in s_i} c_{ij}^{\psi} \phi\left(\frac{y_{ij} - x_{ij}'\hat{\beta}^{\mathrm{r}} - \hat{v}_i^{\mathrm{r}}}{c_{ij}^{\psi}}\right), \tag{3.61}$$

where $\hat{Y}_i^{\mathrm{reblup}}$ is the robust EBLUP (REBLUP) of Y_i, $r_i = n_i/N_i$, $n_i = |s_i|$ (cardinality), $\phi(\cdot)$ is another influence function satisfying $|\phi| \geq |\psi|$, and $\hat{\beta}^{\mathrm{r}}$ and $\hat{v}_i^{\mathrm{r}}$ are the robust estimator of β and REBLUP of v_i, respectively (see Section 3.2). The second term on the right side of (3.61) may be viewed as an RBC to the REBLUP. Ref. [31] called (3.61) a robust predictive estimator.

Ref. [136] considered an alternative approach, called conditional bias, to characterize a representative outlier. The conditional bias was introduced by [10], which, in the case of the NER model is defined as

$$B_{iaj}(y_{aj}, v_a, \beta, \gamma) = \mathrm{E}\{\tilde{\theta}_i(\gamma) - \theta_i | s, y_{aj}, v_a\}, \tag{3.62}$$

where $\theta_i = \bar{Y}_i$ is the small area mean (we use θ_i instead of $\bar{Y}_i$ for the ease of adding the hat to the top), γ denotes the vector of variance components, $\tilde{\theta}_i(\gamma)$ is the BLUP of θ_i given γ, $s = \cup_{i=1}^m s_i$ is the set of sampled indexes from all m small areas, and E denotes the model-based expectation. The conditional bias characterizes the contribution, or impact, of a unit from a potentially different small area, if $a \neq i$, on the prediction error.

With the conditional bias, the prediction error can be nicely expressed as

$$\begin{aligned}
\tilde{\theta}_i(\gamma) - \theta_i &= \sum_{a=1}^m \sum_{j=1}^{N_i} B_{iaj}(y_{aj}, v_a, \beta, \gamma) - \frac{1}{N_i}\sum_{a=1}^m (N_a - 1)W_{ia}(\gamma)v_a \\
&= \sum_{a=1}^m \sum_{j\in s_a} B_{iaj}(y_{aj}, v_a, \beta, \gamma) + \sum_{a=1}^m \sum_{j\notin s_a} B_{iaj}(y_{aj}, v_a, \beta, \gamma) \\
&\quad - \frac{1}{N_i}\sum_{a=1}^m (N_a - 1)W_{ia}(\gamma)v_a,
\end{aligned} \tag{3.63}$$

where $W_{ia}(\gamma)$ are some weights whose definition depends on γ (see [136]). A robust modification of the prediction error is obtained by modifying the first and third term on the right side of (3.63). Specifically, the $B_{iaj}(y_{aj}, v_a, \beta, \gamma)$ inside the first summation is replaced by $B^{\mathrm{r}}_{iaj,d_1,d_2}(y_{aj}, v_a, \beta, \gamma)$, and the $W_{ia}(\gamma)v_a$ in the third summation is replaced by $\psi_{d_2}(W_{ia}(\gamma)v_a)$, where $B^{\mathrm{r}}_{iaj,d_1,d_2}(y_{aj}, v_a, \beta, \gamma)$ is a robustified version of $B_{iaj}(y_{aj}, v_a, \beta, \gamma)$ depending on two scale-tuning parameters, d_1 and d_2, through the bounded influence functions $\psi_{d_1}(\cdot)$ and $\psi_{d_2}(\cdot)$, where $\psi_d(u) = \min(|d|, \max(-|d|, u))$. See Section 3.2 of [136] for details. Thus, the robustified version of the prediction error is

$$\begin{aligned}\tilde{\theta}^{\mathrm{r}}_i(\gamma) - \theta_i &= \sum_{a=1}^{m}\sum_{j\in s_a} B^{\mathrm{r}}_{iaj,d_1,d_2}(y_{aj}, v_a, \beta, \gamma) + \sum_{a=1}^{m}\sum_{j\notin s_a} B_{iaj}(y_{aj}, v_a, \beta, \gamma) \\ &\quad - \frac{1}{N_i}\sum_{a=1}^{m}(N_a - 1)\psi_{d_2}(W_{ia}(\gamma)v_a). \qquad (3.64)\end{aligned}$$

By taking the difference of (3.63) and (3.64), hence canceling the unknown θ_i, an expression of the difference between the BLUP and its robustified version is obtained. Ref. [136] noted that considerable simplification occurs when $d_2 = \infty$. In this case, denoting the corresponding B_{iaj}, $B_{iaj,d_1,\infty}$ terms, with β, γ replaced by $\hat{\beta}, \hat{\gamma}$, respectively, by $\hat{B}_{iaj}$, $\hat{B}^{\mathrm{r}}_{iaj,d_1,\infty}$, respectively, one obtains

$$\hat{\theta}^{\mathrm{r}}_{i,\mathrm{eblup}} = \hat{\theta}_{i,\mathrm{eblup}} - \sum_{a=1}^{m}\sum_{j\in s_a} \hat{B}_{iaj} + \sum_{a=1}^{m}\sum_{j\in s_a} \hat{B}^{\mathrm{r}}_{iaj,d_1,\infty}. \qquad (3.65)$$

In a way, expression (3.65) is similar to (3.61) in that the last two terms together on the right side of (3.65) may be viewed as a bias correction to the standard EBLUP. The bias correction takes into account the contributions from different small areas (sum over $a = 1, \ldots, m$) and samples within the areas (sum over $j \in s_a$.

Results of simulation studies demonstrated gains of the conditional bias-correction method over several existing approaches, including the EBLUP, the robust EBLUP of [215], and some alternative approaches following [30], under certain circumstances.

In addition, [136] proposed a bootstrap approach to measure of uncertainty of their proposed bias-corrected predictors, which modified the bootstrap method of [215]. The main idea is to take into consideration the representative outliers in the non-sampled part of the population. The performance of the 95% prediction intervals using different bootstrap strategies was evaluated via a Monte-Carlo simulation study.

4

Observed Best Prediction

This and the next two chapters are concerned about robustness to model misspecification. The current chapter focuses on the OBP method and its extensions. The next two chapters discuss topics regarding semi-parametric and non-parametric method, and model selection.

Model-based SAE methods are typically developed under certain model assumptions. As quoted by George Box, "All models are wrong". It is therefore natural to ask whether an SAE method is robust against model misspecification; if not, whether there is an alternative way of model fitting that is more robust against model misspecification. The latter type of methods is discussed here in this chapter.

There are, of course, other ways of dealing with model misspecification by carefully choosing the assumed model. These alternative approaches are discussed in the next two chapters. Furthermore, the assumed model, such as a linear model, may be adopted for practical reasons. For example, a linear model is attractive due to its simplicity and ease to interpret, and it utilizes auxiliary variables in a simple way. Note that the auxiliary data may be collected using taxpayer's money; therefore, it may be "politically incorrect" not to use such information. Statistically, one could run a model selection procedure, which may end up with a more sophisticated model with some of the auxiliary variables eliminated. But, in the end, no one is going to use such a model, practically, due to its complexity and not using information collected via taxpayer's money. Therefore, here we focus on a situation that the assumed model is not subject to change; all we could do is to find a better way of fitting the assumed model so that the resulting SAE method is more robust against potential model misspecification.

4.1 Observed best prediction

Ref. [126] proposed an SAE method called observed best prediction (OBP) that is remarkably different from the traditional EBLUP method. The starting point of OBP is the same as that of EBLUP. Consider a general LMM, (1.20). When the fixed effects and variance components are known, the BP of a linear mixed effect, $\eta = b'\beta + a'v$, where a and b are known vectors, is its conditional

DOI: 10.1201/9781003395171-4

expectation given the data, y, that is,

$$\tilde{\eta} = \mathrm{E}(\xi|y) = b'\beta + a'\mathrm{E}(v|y). \tag{4.1}$$

Here, the BP is in the sense of minimum MSPE. Under the normality assumption, one has the well-known expression for the conditional expectation:

$$\mathrm{E}(v|y) = GZ'V^{-1}(y - X\beta), \tag{4.2}$$

where $V = \mathrm{Var}(y) = ZGZ' + R$. Thus, if β, G, R are known, the BP of η is given by (4.1) and (4.2).

The β, G, R involved in the BP are typically unknown in practice. The traditional method is to replace those with their MLE or restricted maximum likelihood (REML) estimators. This leads to the EBLUP.

The difference between OBP and EBLUP is all about how to estimate the unknown parameters involved in the BP. Ref. [126] proposed a different way of estimating those parameters. A motivation is that the BLUP, which leads to the EBLUP, may be viewed as a hybrid of optimal prediction, that is, BP, and optimal estimation, that is, MLE. Recall the BLUP is obtained by replacing the β in the BP with its MLE, assuming that G, R are known. However, if the prediction of the mixed effect is of primary interest, it may be wondered why one has to hybridize. Perhaps, the parameter estimation should also be derived from a predictive perspective?

4.1.1 The case of known G and R

To make it simpler, let us assume, for now, that G, R are known. A key to the predictive approach is to entertain two models. One is the assumed LMM, and the other is called a broader model, or true model. For simplicity, let us consider, for now, the case that the potential model misspecification is only in terms of the mean function; in other words, the true model is

$$y = \mu + Zv + e, \tag{4.3}$$

where $\mu = \mathrm{E}(y)$. Here, E denotes expectation under the true distribution of y, which may be unknown but is not model-dependent. Our interest is prediction of a vector of mixed effects, expressed as

$$\zeta = F'\mu + C'v, \tag{4.4}$$

where F, C are known matrices. We illustrate with an example.

Example 4.1 (Fay–Herriot model). Recall the Fay-Herriot model of Section 1.5.1. Here, the interest is to estimate the small area mean, $\zeta = (\zeta_i)_{1 \le i \le m} = \mu + v = \mathrm{E}(y|v)$, which can be expressed as (4.4) with $F = C = I_m$.

Similar to (4.1), (4.2), the BP of ζ is given by

$$\begin{aligned} \mathrm{E}_M(\zeta|y) &= F'X\beta + C'\mathrm{E}_M(v|y) \\ &= F'X\beta + C'GZ'V^{-1}(y - X\beta) \\ &= F'y - \Gamma(y - X\beta), \end{aligned} \tag{4.5}$$

where $\Gamma = F' - B$ with $B = C'GZ'V^{-1}$. Here, unlike E, E_M denotes conditional expectation under the assumed LMM, (1.20), denoted by M. Thus, the BP is model-based in the sense that its expression depends on the assumed LMM, including the normality assumption.

Let $\check{\zeta}$ denote the right side of (4.5) with a fixed, arbitrary β. Then, by (4.4), (4.5), we have $\check{\zeta} - \zeta = H'v + F'e - \Gamma(y - X\beta)$, where $H = Z'F - C$. Thus, we have

$$\begin{aligned}\mathrm{MSPE}(\zeta) &= \mathrm{E}(|\check{\zeta} - \zeta|^2)\\ &= \mathrm{E}(|H'v + F'e|^2) - 2\mathrm{E}\{(v'H + e'F)\Gamma(y - X\beta)\}\\ &\quad +\mathrm{E}\{(y - X\beta)'\Gamma'\Gamma(y - X\beta)\}\\ &= I_1 - 2I_2 + I_3.\end{aligned}$$

It is easy to see that I_1 does not depend on β. In fact, I_2 does not depend on β either, because we have

$$\begin{aligned}I_2 &= \mathrm{E}\{(v'H + e'F)\Gamma(y - \mu)\} + \{\mathrm{E}(v'H + e'F)\}\Gamma(\mu - X\beta)\\ &= \mathrm{E}\{(v'H + e'F)\Gamma(Zv + e)\},\end{aligned}$$

by (4.3). Thus, we can express the MSPE as

$$\begin{aligned}\mathrm{MSPE}(\check{\zeta}) &= \mathrm{E}(|\check{\zeta} - \zeta|^2)\\ &= \mathrm{E}\{(y - X\beta)'\Gamma'\Gamma(y - X\beta) + \cdots\},\end{aligned} \tag{4.6}$$

where $\cdots$ does not depend on β. The idea is to estimate β by minimizing the observed MSPE, which is the expression inside the expectation in (4.6). Then, because $\cdots$ does not depend on β, this is equivalent to minimizing

$$(y - X\beta)'\Gamma'\Gamma(y - X\beta).$$

The solution is what we call best predictive estimator, or BPE, given by

$$\hat{\beta} = (X'\Gamma'\Gamma X)^{-1}X'\Gamma'\Gamma y, \tag{4.7}$$

assuming non-singularity of $\Gamma'\Gamma$ and that X is full rank. The resulting predictor of ζ is the OBP, given by the right side of (4.5) with β replaced by $\hat{\beta}$. The term OBP is due to the fact that the BPE is obtained by minimizing the observed MSPE, that is, the expression inside the expectation on the right side of (4.6); if the observed MSPE were the true MSPE, the same procedure would lead to the BP.

A key technique used to derive the OBP is coined as *thurstoning*. Howard Thurston (1869–1936) was an American magician famous for disappearing. Imagine the mixed effect ζ is the magician; he appeared inside the first expectation on the right side of (4.6), but not inside the second expectation [and no, he is not hiding inside the $\cdots$; see, e.g., (4.17) for a detailed expression of the $\cdots$]. This way, the magician has disappeared.

Let us use an example to illustrate the difference between BPE and MLE, and OBP and BLUP.

Example 4.1 (continued). It is easy to verify that, in this case,

$$\Gamma = \text{diag}(1 - B_i, 1 \le i \le m) \tag{4.8}$$

with $B_i = A/(A + D_i)$, and the BPE of β is given by

$$\hat{\beta} = \left\{\sum_{i=1}^{m}\left(\frac{D_i}{A + D_i}\right)^2 x_i x_i'\right\}^{-1} \sum_{i=1}^{m}\left(\frac{D_i}{A + D_i}\right)^2 x_i y_i. \tag{4.9}$$

The BP of ζ_i, the ith component of ζ, is given by (1.25), that is,

$$\tilde{\zeta}_i = \frac{A}{A + D_i} y_i + \frac{D_i}{A + D_i} x_i'\beta, \tag{4.10}$$

where β, A are the true parameters. Thus, assuming that A is known, the OBP of ζ_i is given by (4.10) with β replaced by $\hat{\beta}$ of (4.9). On the other hand, it can be shown that the MLE of β, assuming that A is known, is

$$\tilde{\beta} = \left(\sum_{i=1}^{m} \frac{x_i x_i'}{A + D_i}\right)^{-1} \sum_{i=1}^{m} \frac{x_i y_i}{A + D_i}. \tag{4.11}$$

The BLUP of ζ_i is given by (4.10) with β replaced by $\tilde{\beta}$ of (4.11).

It is interesting to note that the MLE (4.11) and BPE (4.9) are both in the form of a weighted average, the only difference being the weights. The BPE gives more weight to areas with larger sampling variances, D_i, while the MLE does just the opposite, assigning more weight to areas with smaller sampling variances. The question is: Who is right?

Ref. [126] interpreted this difference as follows. First note that, by the expression of the BP, (4.10), it is evident that the assumed model is involved only through $x_i'\beta$, whose corresponding weight, $D_i/(A + D_i)$, is increasing with D_i. In other words, the model-based BP is more relevant to those areas with larger D_i. Imagine that there is a meeting of representatives from the different small areas to discuss what estimate of β is to be used in the BP. The areas with larger D_i think that their "voice" should be heard more (i.e., they should receive more weight), because the BP is more relevant to their "business". Their request is reasonable (although, politically, this may not work out under a democratic voting system).

Not only the OBP has an intuitive explanation, it also has an attractive property, that is, it is more robust to model misspecification than the BLUP, in terms of the MSPE. This is not surprising, because OBP is derived based on the true expectation, E, not the expectation E_M that is model dependent. This robust feature can also be justified theoretically.

Consider an empirical best predictor (EBP), $\check{\zeta}$, which is the BP [i.e., right side of (1.25)] with β replaced by a weighted least squares (WLS) estimator (recall that G, R are assumed known),

$$\check{\beta} = (X'WX)^{-1}X'Wy, \tag{4.12}$$

where W is a positive definite weighting matrix. Clearly, the BPE and MLE are all special cases of the WLS estimator, with $W = \Gamma'\Gamma$ for the BPE and $W = V^{-1}$ for the MLE. Thus, the OBP and BLUP are special cases of the EBP. Also recall $\mu = \mathrm{E}(y)$. The theorem below is due to [126].

Theorem 4.1. The MSPE of the EBP can be expressed as

$$\mathrm{MSPE}(\check{\zeta}) = a_0 + \mu' A_1(W)\mu + \mathrm{tr}\{A_2(W)\}, \tag{4.13}$$

where a_0 is a non-negative constant that does not depend on W; $A_j(W), j = 1, 2$ are non-negative definite matrices that depend on W. Furthermore, $\mu' A_1(W)\mu$ is minimized by the OBP, with $W = \Gamma'\Gamma$, and $\mathrm{tr}\{A_2(W)\}$ is minimized by the BLUP, with $W = V^{-1}$. The expressions of $a_0, A_j(W), j = 1, 2$ are given below. Define $L = F' - \Gamma\{I - X(X'WX)^{-1}X'W\}$, where I denotes the identity matrix, and recall $B = C'GZ'V^{-1}$. Then, we have

$$\begin{aligned}
a_0 &= \mathrm{tr}\{C'(G - GZ'V^{-1}ZG)C\}, \\
A_1(W) &= (F' - L)'(F' - L), \\
A_2(W) &= (L - B)V(L - B)'.
\end{aligned}$$

An important implication of Theorem 4.1 is the following. First note that the second term on the right side of (4.13) disappears when the mean of y is correctly specified, that is, $\mu = X\beta$ for some β. So, when the underlying model is correct, BLUP usually wins the battle of MSPE, because it minimizes the only (remaining) term that depends on W. On the other hand, when the mean is misspecified, that is, $\mu \neq X\beta$ for any β, the second term is often of higher order than the third term. Thus, in this situation, OBP is likely to win, because it minimizes a higher order term. We consider an example.

Example 4.2. Ref. [126] used the following simple example to illustrate the potential advantage of OBP over BLUP. Consider a special case of Example 4.1, in which $x_i'\beta = \beta$, an unknown mean. To make it even simpler, suppose that A is known so that one can actually compute the BLUP. Furthermore, suppose that $m = 2n$, $D_i = a, 1 \leq i \leq n$, and $D_i = b, n+1 \leq i \leq m$, where a, b are positive known constants. Now suppose that the true underlying model is

$$y_i = \begin{cases} c + v_i + e_i, & 1 \leq i \leq n, \\ d + v_i + e_i, & n+1 \leq i \leq m, \end{cases}$$

where $c \neq d$; thus, the mean function is misspecification by assuming $c = d$. The vector of true small area means is $\zeta = (\zeta_i)_{1 \leq i \leq m}$, where

$$\zeta_i = \begin{cases} c + v_i, & 1 \leq i \leq n, \\ d + v_i, & n+1 \leq i \leq m. \end{cases}$$

For this special case, one can actually derive exact expressions of the MSPEs. It can be shown that

$$\begin{aligned}
\mathrm{MSPE}(\hat{\zeta}_{\mathrm{OBP}}) &= \left\{\left(\frac{a}{A+a}+\frac{b}{A+b}\right)A+\frac{a^2b^2(c-d)^2}{a^2(A+b)^2+b^2(A+a)^2}\right\}n \\
&\quad+\frac{a^4(A+b)^3+b^4(A+a)^3}{(A+a)(A+b)\{a^2(A+b)^2+b^2(A+a)^2\}} \\
&= g_1 n + h_1, \qquad (4.14) \\
\mathrm{MSPE}(\hat{\zeta}_{\mathrm{BLUP}}) &= \left\{\left(\frac{a}{A+a}+\frac{b}{A+b}\right)A+\frac{(a^2+b^2)(c-d)^2}{(2A+a+b)^2}\right\}n \\
&\quad+\frac{a^2(A+b)^2+b^2(A+a)^2}{(A+a)(A+b)(2A+a+b)} \\
&= g_2 n + h_2. \qquad (4.15)
\end{aligned}$$

If $c \neq d$, then $g_1 \leq g_2$ and $h_1 \geq h_2$ with equality holding in both cases if and only if the following identity holds:

$$a^2(A+b) = b^2(A+a). \qquad (4.16)$$

Now suppose that (4.16) does not hold, then we have $g_1 < g_2$, and $h_1 > h_2$, but the latter is not important when n is large. In fact, we have

$$\lim_{n\to\infty} \frac{\mathrm{MSPE}(\hat{\zeta}_{\mathrm{OBP}})}{\mathrm{MSPE}(\hat{\zeta}_{\mathrm{BLUP}})} = \frac{g_1}{g_2} < 1.$$

For example, suppose that $A/(c-d)^2 \approx 0$, $A/b \approx 0$ and $b/a \approx 0$. Then it is easy to show that $g_1/g_2 \approx 0.5$. Thus, in this case, the MSPE of the OBP is asymptotically about half of that of the BLUP.

On the other hand, if $c = d$, that is, if the underlying model is correctly specified, then, we have $g_1 = g_2$ while, still, $h_1 \geq h_2$. Therefore, in this case, we have $\mathrm{MSPE}(\hat{\zeta}_{\mathrm{OBP}}) \geq \mathrm{MSPE}(\hat{\zeta}_{\mathrm{BLUP}})$; however, we also have

$$\lim_{n\to\infty} \frac{\mathrm{MSPE}(\hat{\zeta}_{\mathrm{OBP}})}{\mathrm{MSPE}(\hat{\zeta}_{\mathrm{BLUP}})} = 1;$$

thus, the MSPEs of the OBP and BLUP are asymptotically the same.

Another theoretical result, also established in [126], is regarding asymptotic behavior of the BPE. Note that, under the misspecification, the BPE cannot be consistent (because the true parameter vector does not exist). However, it was shown that, as $m \to \infty$, the BPE still converges in probability to "something". In fact, the something is the minimizer of some MSPE, which is a function of the parameter vector. See [126] (Theorem 2) and also §4.1.5 (above Theorem 4.2). In a way, this result is similar to a well-known theory regarding asymptotic behavior of the MLE under a misspecified model [236].

So far we have assumed that G, R are known. We now consider several extensions of the OBP beyond such an assumption, and beyond LMM.

4.1.2 OBP under Fay-Herriot model: A unknown

The OBP method can be extended to the Fay-Herriot model without assuming A known. In fact, the thurstoning technique applies, in this case, with (4.6) now written as

$$\mathrm{MSPE}(\tilde{\zeta}) = \mathrm{E}\{(y - X\beta)'\Gamma^2(y - X\beta) + 2A\mathrm{tr}(\Gamma) - \mathrm{tr}(D)\}, \tag{4.17}$$

where Γ is given by (4.8). The BPE of $\psi = (\beta', A)'$ is obtained by minimizing the expression inside the expectation on the right side of (4.17), which is equivalent to minimizing

$$Q(\psi) = (y - X\beta)'\Gamma^2(y - X\beta) + 2A\mathrm{tr}(\Gamma). \tag{4.18}$$

Note that, in (4.18), ψ is treated as an unknown parameter vector, rather than the true parameter vector. Let $\tilde{Q}(A)$ be $Q(\psi)$ with $\beta = \tilde{\beta}$ given by the right side of (4.9), considered as a function of A. Then, we have the following expression:

$$\tilde{Q}(A) = y'\Gamma P_{(\Gamma X)^\perp}\Gamma y + 2A\mathrm{tr}(\Gamma), \tag{4.19}$$

where for any matrix M, $P_{M^\perp} = I - P_M$ with $P_M = M(M'M)^{-1}M'$, assuming non-singularity of $M'M$. Thus, we have

$$P_{(\Gamma X)^\perp} = I_m - \Gamma X(X'\Gamma^2 X)^{-1}X'\Gamma,$$

where I_m is the m-dimensional identity matrix. The BPE of A, denoted by $\hat{A}$, is the minimizer of $\tilde{Q}(A)$ over $A \geq 0$. Once $\hat{A}$ is obtained, the BPE of β, $\hat{\beta}$, is given by (4.9) with A replaced by $\hat{A}$.

Given the BPE of ψ, $\hat{\psi} = (\hat{\beta}', \hat{A})'$, the OBP of ζ is given by the BP (4.10), with ψ replaced by $\hat{\psi}$.

In [126], a simulated example was carried out to compare the performance of the OBP with several versions of the EBLUP. The example was the same as Example 4.2 except that A is unknown. Four different estimators of A were considered, which have been traditionally used. These are the MLE, REML estimator, the Fay-Herriot estimator [F-H [59]] and the Prasad-Rao estimator [P-R; [196]]. These lead to four different EBLUPs, denoted by EBLUP-1, EBLUP-2, EBLUP-3, and EBLUP-4, respectively.

The simulation setting had $a = 4$, $b = 1$, $c = 0$, $d = 1$ or 5; $m = 50$, 100, or 200; and $A = 0.2$.

The total empirical MSPE, that is, the sum of the empirical MSPEs over the m small areas, for the four EBLUPs and the OBP, based on $K = 500$ simulation runs, are reported in Table 4.1, taken from [126]. The % in the parentheses is the percentage increase in MSPE by the corresponding EBLUP over the OBP. It is seen that all of the EBLUPs perform very similarly, while the OBP is a distance away from the EBLUPs. The MSPEs of the EBLUPs are somewhere between 11% to 44% higher than that of the OBP. The percentage increase in MSPE by the EBLUPs is more substantial for $d = 5$ than for $d = 1$.

TABLE 4.1
Empirical MSPE (% increase over OBP).

m	d	EBLUP-1	EBLUP-2	EBLUP-3	EBLUP-4	OBP
50	1	28.76 (28%)	27.94 (25%)	25.00 (11%)	25.87 (15%)	22.43
100	1	51.05 (26%)	50.20 (24%)	47.74 (18%)	49.02 (21%)	40.42
200	1	94.22 (26%)	93.95 (26%)	92.52 (24%)	93.87 (25%)	74.86
50	5	95.83 (42%)	95.05 (41%)	93.55 (39%)	93.12 (38%)	67.41
100	5	189.93 (44%)	189.22 (43%)	186.51 (41%)	185.61 (41%)	132.01
200	5	372.59 (44%)	371.92 (44%)	366.96 (42%)	365.21 (41%)	258.60

This makes sense because $d = 5$ features a more serious model misspecification than $d = 1$ does, and the OBP shines, relative to the EBLUP, when the underlying model is misspecified.

A more explicit explanation of this pattern may be seen from the comparison of the formulae for the exact MSPEs, that is, (4.14) and (4.15) (with $c = 0$), although the latter are derived for the case that A is known. The formulae show that the difference in MSPE between the OBP and any of the EBLUPs gets larger, even proportionally, as d increases.

Of primary interest to SAE is the area-specific MSPEs. This is reported below using graphical tools, again taken from [126]. Specifically, we summarize the results using boxplots and histograms, as shown in Figures 4.1 and 4.2. The figures reveal some untold stories by the overall MSPEs. First, the boxplots show a significant difference in the distributions of the empirical MSPEs between the OBP and the EBLUPs. Not only does the OBP have smaller median empirical MSPE in each case, what is more apparent is the range, or variation, of the empirical MSPEs overwhelmingly in favor of the OBP.

Second, the histograms exhibit quite different shapes between the OBP and EBLUPs. A closer look at the numbers shows that the empirical MSPEs of the EBLUPs are somewhere between slightly to moderately smaller than those of the OBP for half of the small areas. As for the other half, the empirical MSPEs of the EBLUPs are much larger than those of the OBP. This pattern may be interpreted. Recall the assumed model has a common mean for all of the small areas, while the true model has one mean for half of the areas and another mean for the other half. Apparently, what the EBLUP does is to "side with" one mean while "abandoning" the other. The OBP, on the other hand, uses a rather different strategy, by "staying in the middle" or "balancing" between the two means. This explains the bimodal histograms for the EBLUPs, compared to the fairly normal-look-like histograms for the OBP (with much narrower spreads). Overall, the simulation results show a much more robust performance of the OBP in terms of the area-specific MSPE as compared to the EBLUPs.

It should be noted that the case considered above is rather extreme in that, in real-life, the assumed model may be partially correct, in some sense. Ref. [126] also studied a case, in which the underlying model was only slightly

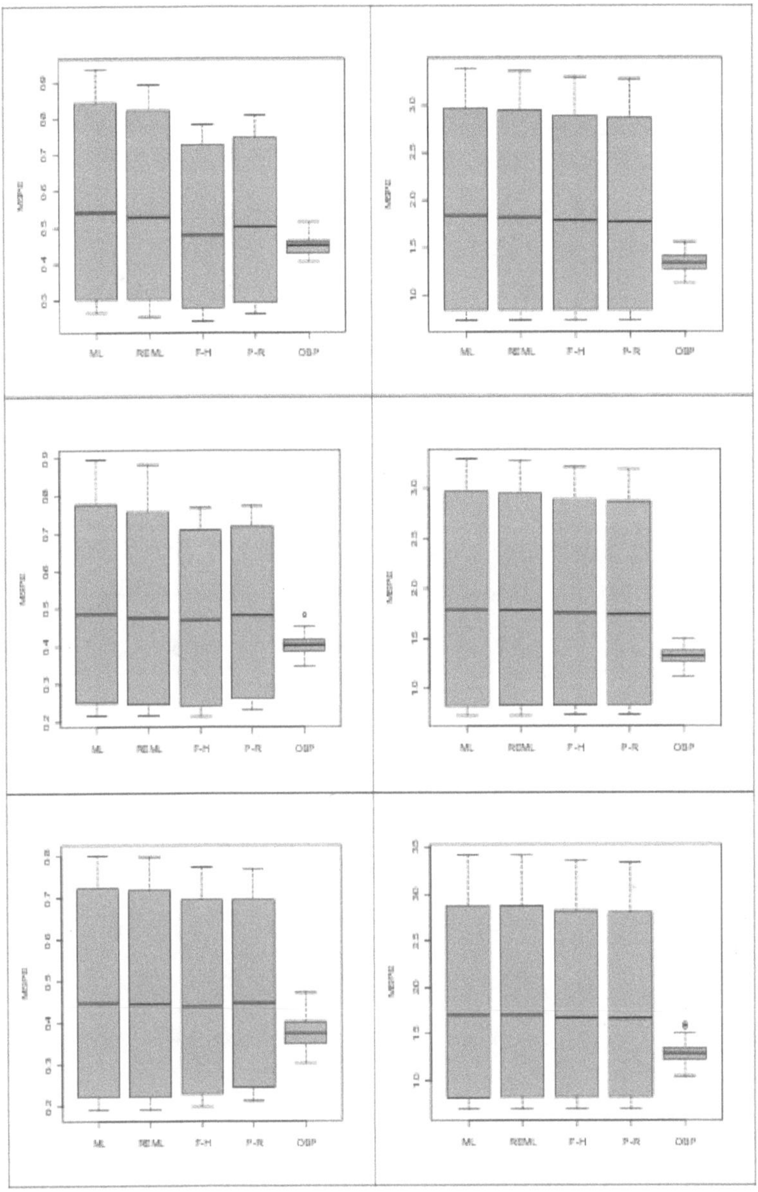

FIGURE 4.1
Boxplots of $\text{MSPE}_i^*, i = 1, \ldots, m$. Upper Left: $m = 50, d = 1$; Upper Right: $m = 50, d = 5$; Middle Left: $m = 100, d = 1$; Middle Right: $m = 100, d = 5$; Lower Left: $m = 200, d = 1$; Lower Right: $m = 200, d = 5$. Within each plot from left to right: EBLUP-1, EBLUP-2, EBLUP-3, EBLUP-4, OBP.

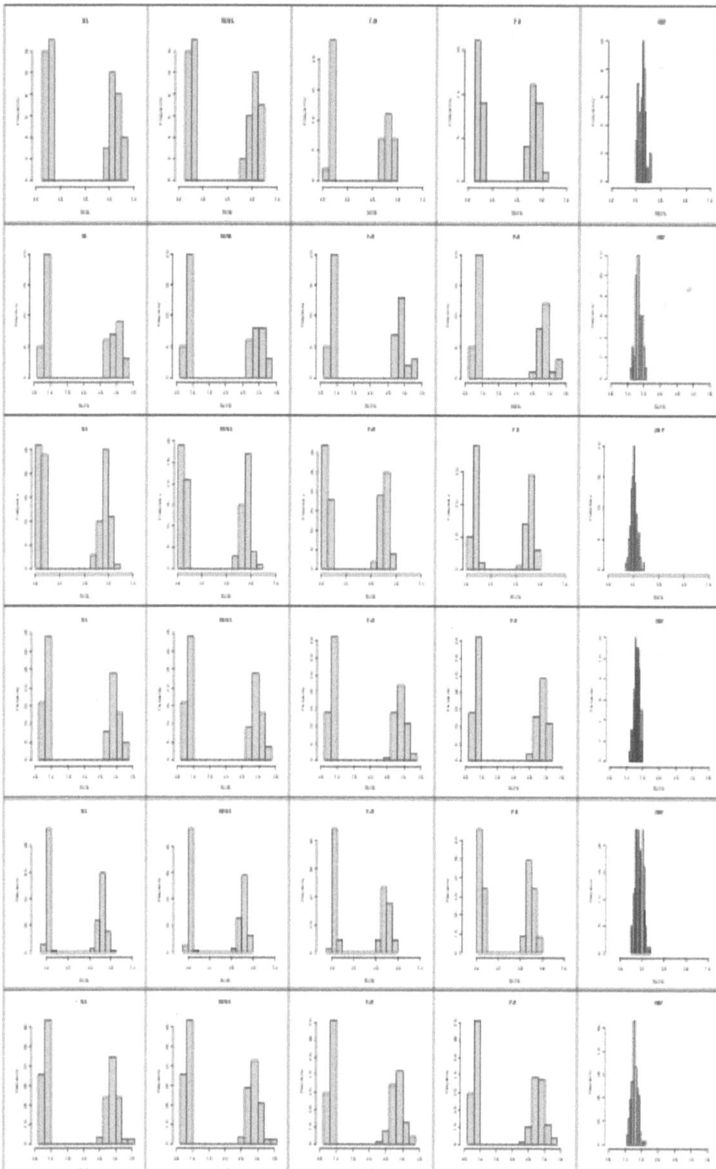

FIGURE 4.2
Histograms of $\text{MSPE}_i^*, i = 1, \ldots, m$. The rows, from top to bottom, correspond to $m = 50, d = 1$; $m = 50, d = 5$; $m = 100, d = 1$; $m = 100, d = 5$; $m = 200, d = 1$; and $m = 200, d = 5$, respectively. Within each row from left to right: EBLUP-1, EBLUP-2, EBLUP-3, EBLUP-4, OBP. Within each row the x-axes have the same range.

misspecified. It showed that, in this case, the gain of OBP over EBLUP is very marginal, or even negative (i.e., the EBLUP has smaller empirical MSPE). The reported gain/loss were all within 1-3%.

4.1.3 OBP for small area counts

The Fay-Herriot model is ideally for outcomes that are Gaussian, or approximately Gaussian. In some cases, however, the outcome variable is a count. Such an outcome variable is typically modeled as having a Poisson distribution, instead of normal distribution. For example, [183] used both the binomial and Poisson mixed models for estimating population counts in the 2011 German census; similarly, [61] proposed a Poisson log-normal model, which is similar to the Poisson mixed model except formulated as a hierarchical model, for estimation of the number of recruits by firms; [83] proposed a two-level hierarchical Poisson model for estimation of infant mortality rates using a Bayesian approach.

The OBP method can be extended to such type of outcomes via a two-stage model. Again, we consider the case where the outcome variable is at the area-level, denoted by y_i. In addition, a vector of covariates, x_i, is also available at the area-level. In the first-stage model, it assumes that, given the area-specific means, $\mu_i, 1 \leq i \leq m$, where m is the number of small areas, $y_i, 1 \leq i \leq m$ are conditionally independent such that $y_i|\mu_i \sim \text{Poisson}(\mu_i)$.

In the second-stage model, a distribution is assumed for the μ_i's. Here, we focus on two models. The first is a mixed-effect log-normal (LN) model, expressed through a logarithm link function as

$$\log(\mu_i) = x_i'\beta + v_i, \tag{4.20}$$

where β is a vector of fixed effects and v_i is an area-specific random effect that is assumed to be independent (for different i's) and distributed as $N(0, \sigma^2)$ for some unknown variance σ^2. The second is a Gamma (GM) model, which assumes that the μ_i's are independent with

$$\mu_i \sim \text{Gamma}(x_i'\beta/\phi, \phi), \tag{4.21}$$

where $\text{Gamma}(\alpha, \phi)$ is the Gamma distribution with shape parameter α, scale parameter ϕ, and probability density function (pdf) given by

$$f(u|\alpha, \phi) = \frac{1}{\Gamma(\alpha)\phi^\alpha} u^{\alpha-1} e^{-u/\phi}, \quad u > 0.$$

Under either model, the BP of μ_i can be expressed as

$$\text{E}_{M,\psi}(\mu_i|y) = g_i(\psi, y_i), \tag{4.22}$$

where $\text{E}_{M,\psi}$ denotes conditional expectation under the assumed model, M, and parameter vector, ψ, under M. Here, $g_i(\cdot, \cdot)$ is a function that also depends on the assumed model. Note that, from a Bayesian point of view, the BP is simply the posterior mean of μ_i under the model $y_i|\mu_i \stackrel{\text{ind}}{\sim} \text{Poisson}(\mu_i)$ and treating the distribution of μ_i as a prior.

The first-stage model is assumed to be correctly specified. The second-stage model is potentially misspecified, which is of main concern.

The LN model is a special case of GLMM [e.g.,[123], ch. 3]. In particular, the random effect v_i often has an interpretation, which is sometimes attractive to the practitioners. It may also be expressed without explicitly involving the random effects, that is,

$$\mu_i \quad \sim \quad \text{LN}(x_i'\beta, \sigma^2), \tag{4.23}$$

where LN stands for the log-normal distribution [$\xi \sim \text{LN}(\mu, \sigma^2)$ iff $\log(\xi) \sim N(\mu, \sigma^2)$]. Expression (4.23) is, in a way, similar to the GM model (4.21). The BP under the LN model does not have an analytic expression. In fact, using the expression $v_i = \sigma\xi_i$, where $\xi_i \sim N(0, 1)$, and letting $\psi = (\beta', \sigma)'$, it can be derived that, under the LN model, we have

$$g_i(\psi, y_i) = e^{x_i'\beta} \frac{\int \exp\{(y_i + 1)\sigma u - \mu_i(u) - u^2/2\} du}{\int \exp\{y_i \sigma u - \mu_i(u) - u^2/2\} du}, \tag{4.24}$$

where $\mu_i(u) = \exp(x_i'\beta + \sigma u)$. The one-dimensional integrals in (4.24) can be approximated by numerical integration. Alternatively, if y_i is large, the integrals in (4.24) can be approximated via Laplace approximation [e.g., [123], sec. 3.5.1], leading to the following approximation: $g_i(\psi, y_i) = a_i$ if $\sigma = 0$ (this result is exact rather than approximate); and

$$\begin{aligned}
& g_i(\psi, y_i) \\
\approx\ & a_i \sqrt{\frac{\sigma^2 y_i + 1}{\sigma^2(y_i + 1) + 1}} \exp\left[(y_i + 1)\log\left(\frac{y_i + 1}{a_i}\right)\right. \\
& -y_i \log\left(\frac{y_i}{a_i}\right) - 1 + \frac{y_i}{2(\sigma^2 y_i + 1)}\left\{\log\left(\frac{y_i}{a_i}\right)\right\}^2 \\
& \left. -\frac{y_i + 1}{2\{\sigma^2(y_i + 1) + 1\}}\left\{\log\left(\frac{y_i + 1}{a_i}\right)\right\}^2\right]
\end{aligned} \tag{4.25}$$

if $\sigma > 0$, where $a_i = e^{x_i'\beta}$. The Laplace approximation is computationally more efficient than numerical integration.

As for the GM model, it has a major advantage in that, in this case, the BP has a simple analytic expression. This follows as a result of Gamma being the conjugate prior, which makes the GM model attractive from a Bayesian perspective. If we let $\psi = (\beta', \phi)'$, then for the GM model we have

$$g_i(\psi, y_i) = \frac{\phi y_i + x_i'\beta}{\phi + 1}. \tag{4.26}$$

We now proceed to derive the OBP. Again, we entertain two models, the assumed model and a broader model. Here, the broader model is simply the first-stage model, without the distributional assumption about the μ_i in the second-stage model. In other words, under the broader model, the μ_i's are completely unspecified. Write $\mu = (\mu_i)_{1\leq i\leq m}$, and $\tilde{\mu} = (\tilde{\mu}_i)_{1\leq i\leq m}$, where $\tilde{\mu}_i$ is the right side of (4.22), where ψ is treated as an unknown parameter vector. As noted, the key is to find a thurstoning expression of the MSPE. We have

$$\begin{aligned}
\text{MSPE} &= \text{E}(|\tilde{\mu} - \mu|^2) \\
&= \sum_{i=1}^{m} \text{E}\{g_i(\psi, y_i) - \mu_i\}^2 \\
&= \text{E}\left\{\sum_{i=1}^{m} g_i^2(\psi, y_i)\right\} - 2\sum_{i=1}^{m} \text{E}\{g_i(\psi, y_i)\mu_i\} + \sum_{i=1}^{m} \text{E}(\mu_i^2) \\
&= I_1 - 2I_2 + I_3, \qquad (4.27)
\end{aligned}$$

where E denotes expectation under the broader model. Note that I_3 does not involve ψ, even though it may be completely unknown. Also note that

$$\begin{aligned}
\text{E}\{g_i(\psi, y_i)\mu_i\} &= \text{E}[\mu_i \text{E}\{g_i(\psi, y_i)|\mu\}] \\
&= \sum_{k=0}^{\infty} g_i(\psi, k)\text{E}\left(\frac{e^{-\mu_i}\mu_i^{k+1}}{k!}\right) \\
&= \sum_{k=0}^{\infty} g_i(\psi, k)(k+1)\text{E}\left\{\frac{e^{-\mu_i}\mu_i^{k+1}}{(k+1)!}\right\} \\
&= \sum_{k=0}^{\infty} g_i(\psi, k)(k+1)\text{E}\{1_{(y_i=k+1)}\},
\end{aligned}$$

where $1_{\mathcal{A}}$ is the indicator of event $\mathcal{A}$ (= 1 if $\mathcal{A}$ occurs, and 0 otherwise). If we define $g_i(\psi, -1) = 0$, we can write

$$\begin{aligned}
\text{E}\{g_i(\psi, y_i)\mu_i\} &= \text{E}\left\{\sum_{k=0}^{\infty} g_i(\psi, k)(k+1)1_{(y_i=k+1)}\right\} \\
&= \text{E}\{g_i(\psi, y_i - 1)y_i\}. \qquad (4.28)
\end{aligned}$$

Combining (4.28) with (4.27), we have the following expression:

$$\text{MSPE} = \text{E}\left\{\sum_{i=1}^{m} g_i^2(\psi, y_i) - 2\sum_{i=1}^{m} g_i(\psi, y_i - 1)y_i + \cdots\right\}, \qquad (4.29)$$

where $\cdots$ does not depend on ψ. Clearly, thurstoning has succeeded, as it is clear that μ, the "magician", does not appear inside the expectation on the right side of (4.29). This leads to the BPE of ψ by minimizing the expression inside the expectation without $\cdots$, that is,

$$Q(\psi, y) = \sum_{i=1}^{m} g_i^2(\psi, y_i) - 2\sum_{i=1}^{m} g_i(\psi, y_i - 1)y_i. \tag{4.30}$$

Once the BPE of ψ is obtained, denoted by $\hat{\psi}$, the OBP of the mean small area count, μ_i, is obtained by (4.22) with ψ replaced by $\hat{\psi}$, that is,

$$\hat{\mu}_i = g_i(\hat{\psi}, y_i), \quad 1 \le i \le m. \tag{4.31}$$

[34] discussed several extensions of the above OBP development as well as some computational strategies. It was noted that the GM model has a significant computational advantage over the LN model both in terms of computing the BPE and in terms of obtaining the OBP.

In terms of the theoretical result, the general result of [126] regarding the asymptotic behavior of the BPE also applies to the current situation.

We conclude this subsection by reporting some empirical results from simulation studies. In the first simulation study, the performance of OBP is compared with that EBP under a GM model. The EBP is also obtained via (4.31) except that $\hat{\psi}$ is the MLE. Specifically, we consider (4.21) with $x_i'\beta = \beta_1 x_i, i = 1, \dots, 20$, where the x_i's are generated from the Uniform$[0, 1]$ distribution, and fixed throughout the simulation. Here, β_1 and ϕ are the unknown parameters to be estimated. The true underlying model is

$$\mu_i \sim \text{Gamma}(\beta_0 + \beta_1 x_i/\phi, \phi).$$

Thus, when $\beta_0 = 0$, the assumed model holds; otherwise, the model is misspecified. We consider the (overall) simulated MSPE (same as in Table 4.1) as a measure of predictive performance. The results under different parameter settings, based on 1,000 simulation runs, are presented in Table 4.2 [34].

It is seen that the EBP performs better than OBP when the assumed model holds, that is, $\beta_0 = 0$, which is not surprising. The trend is reversed, though, when the assumed model is misspecified. A similar pattern is observed when considering a LN model with a similar structure (results omitted).

The OBP method can be extended to more complex area-level models. Here we consider one of such cases under an LN model. The outcomes are

TABLE 4.2
OBP vs EBP under a GM model. Simulated MSPEs (based on $K = 1000$ simulation runs) under different parameter settings.

β_0	β_1	ϕ	OBP	EBP
0.0	2.0	2.0	15.81	15.18
2.0	2.0	2.0	79.86	80.73
8.0	2.0	2.0	302.63	309.83
8.0	1.0	2.0	296.57	300.71
8.0	2.0	1.0	146.10	150.62

TABLE 4.3
OBP vs EBP under a LN model. Simulated MSPEs (based on $K = 500$ simulation runs) under different true models and sample sizes.

	Assumed Model		Model (i)		Model (ii)		Model (iii)	
m	OBP	EBP	OBP	EBP	OBP	EBP	OBP	EBP
8	64	56	241	258	527	649	5500	5739
32	218	204	811	841	2481	3277	35640	36445
128	856	839	3421	4090	11049	14979	178534	183780

y_{ij}, $i = 1, \dots, m, j = 1, \dots, s$, where the index i corresponds to the area (e.g., county), and the index j to a classification of demographic groups (e.g., age groups) within the area. The first-stage model assumes that, given μ_{ij}, $1 \leq i \leq m$, $1 \leq j \leq s$, the y_{ij}'s are conditionally independent such that $y_{ij}|\mu_{ij} \sim \text{Poisson}(\mu_{ij})$. The second-stage models stays that $\mu_{ij} = \exp(\eta_{ij} + v_i)$, where the area-specific random effects v_i, $1 \leq i \leq m$ are independent $N(0, 1)$. Here, the random effect, v_i, is at the area-level, but the interest is the area/demographic mean counts, μ_{ij}.

Another simulation study was carried out in [34], in which the assumed model has $\eta_{ij} = \log(\mu_{ij}) = x_{ij}\beta$, where the x_{ij}'s are generated independently from the Uniform(0, 2) distribution, then fixed throughout the simulation study. Three true underlying distributions were considered: (i) $\eta_{ij} = x_{ij} + 1$, (ii) $\eta_{ij} = 2x_{ij}^2/3 + 1$; and (iii) $\eta_{ij} = x_{ij} + z_{ij}$, where the z_{ij}'s are generated independently from the Uniform(0, 2) distribution, then fixed throughout the simulation study. It follows that, in each case, the assumed model is misspecified. An additional case, in which the assumed model holds, is also considered for comparison. Three different samples sizes are considered: $m = 8$, $m = 32$, and $m = 128$, with $s = 6$ in all of the cases. Once again, the overall simulated MSPE (summed over the small areas and demographic groups) is considered. The results, based on 500 simulation runs and rounded to the nearest integers, are presented in Table 4.3 [34].

It is seen that, when the underlying model is correctly specified, the EBP performs better than the OBP, although the proportionate difference is diminishing as m increases. On the other hand, when the underlying model is misspecified, the OBP performs better than the EBP, and the proportionate difference is not going away as m increases. The patterns (also see Table 4.4, again from [34]) are consistent with those found in [108] and [128].

TABLE 4.4
Ratio of simulated MSPEs (OBP/EBP) from Table 4.3.

m	Assumed Model	Model (i)	Model (ii)	Model (iii)
8	1.14	0.93	0.81	0.96
32	1.07	0.96	0.76	0.98
128	1.02	0.84	0.74	0.97

TABLE 4.5
LN model vs GM model. Simulated MSPEs (based on $K = 400$ simulation runs) under different cases; Case I is when the LN model holds; Case II is when the GM model holds; and Case III is when neither model hold.

Case	OBP/LN	OBP/GM
I	12.35	16.84
II	23.07	18.20
III	13.47	30.61

It may be wondered what happens if one of the two models under consideration, the LN model and the GM model, is misspecified as the other, or both are misspecified. Below we consider a case where a LN model is misspecified as a GM model, or vice versa, or both model are misspecified.

There are three cases. In the first case (case I), the true underlying model has (4.20) with $x_i'\beta = \beta_1 x_i$, where $\beta_1 = 0.5$, and $x_i = 1 + \xi_i$ with ξ_i being independent Bernoulli(0.5); also $\sigma = 0.2$. In the second case (case II), the true underlying model is (4.21) with $x_i'\beta$ being the same as in case I, and $\phi = 1.0$. In the third case (case III), the true second-stage model is $\mu_i = 2, 1 \leq i \leq m$. The sample size m is 40 in all of the cases. The settings made sure that the range of y_i is roughly the same in all of the cases.

In each case, the OBPs under the LN model and the GM model were obtained, and their performances are compared in terms of the overall simulated MSPE based on 400 simulation runs. The results are presented in Table 4.5, where OBP/LN and OBP/GM stand for OBP under the LN model and OBP under the GM model, respectively.

It is not surprising that each model wins the case when the model holds. As for the case that neither model holds, OBP/LN seemed to perform (much) better than OBP/GM. However, one should be cautious in making a conclusion because this could be a case where the LN model is closer to the truth, in some way, than the GM model.

The development above can be extended to allow an off-set, which is often used for count data. Specifically, the first-stage model can be extended to $y_i \sim \text{Poisson}(\lambda_i \mu_i)$, where $\log(\lambda_i)$ is a known off-set. The rest of the development is similar to the special case of $\lambda_i = 1$ (i.e., no off-set) considered above. For example, under the GM model, the OBP of μ_i is given by

$$\hat{\mu}_i = \frac{\hat{\phi} y_i + x_i'\hat{\beta}}{1 + \hat{\phi}\lambda_i}, \tag{4.32}$$

where $\hat{\beta}$, $\hat{\phi}$ are BPE of β, ϕ, respectively. See [34] (also [184]) for details.

4.1.4 OBP under a unit-level model

So far the application of OBP has been limited to area-level models. To extend OBP to unit-level models, let us consider sampling from finite sub-populations, which are considered samples from some super-populations. Suppose that the sub-populations of outcomes, $\{Y_{ik}, k = 1, \dots, N_i\}$, and auxiliary data, $\{X_{ikl}, k = 1, \dots, N_i\}, l = 1, \dots, p$, are realizations from corresponding super-populations that are assumed to satisfy an NER model,

$$Y_{ik} = X_{ik}'\beta + v_i + e_{ik},\ i = 1, \dots, m, k = 1, \dots, N_i, \tag{4.33}$$

where β is a vector of unknown regression coefficients, v_i is a sub-population-specific random effect, and e_{ik} is an error. It is assumed that the v_i's and e_{ik}'s are independent with $v_i \sim N(0, \sigma_v^2)$ and $e_{ik} \sim N(0, \sigma_e^2)$.

Under the finite-population setting, the true small area mean is

$$\theta_i = \bar{Y}_i = \frac{1}{N_i} \sum_{k=1}^{N_i} Y_{ik}$$

for $1 \leq i \leq m$. Furthermore, write $r_i = n_i/N_i$. Suppose that simple random samples [SRS, e.g., [163]] $(y_{ij}, x_{ij}), 1 \leq j \leq n_i$ are drawn from the finite population, $\{(Y_{ik}, X_{ik}), 1 \leq k \leq N_i\}$, $1 \leq i \leq m$. Then, a finite-population version of the BP has the following expression:

$$\tilde{\theta}_i = \mathrm{E}_M(\theta_i|y_i) = \bar{X}_i'\beta + \left\{ r_i + (1 - r_i)\frac{n_i\gamma}{1 + n_i\gamma} \right\} (\bar{y}_{i\cdot} - \bar{x}_{i\cdot}'\beta), \tag{4.34}$$

where $y_i = (y_{ij})_{1 \leq j \leq n_i}$, $\bar{y}_{i\cdot} = n_i^{-1}\sum_{j=1}^{n_i} y_{ij}$, $\bar{x}_{i\cdot} = n_i^{-1}\sum_{j=1}^{n_i} x_{ij}$. Here, E_M denotes the (conditional) expectation under the assumed super-population NER model, and β and $\gamma = \sigma_v^2/\sigma_e^2$ are the true parameters under the NER model. Again, note that the BP is model-dependent.

In practice, an assumed model is subject to misspecification. So far we have only considered misspecification of the mean function, while assuming that the variance-covariance structure of the data is correctly specified. However, the variance-covariance structure may be misspecified in practical. In this subsection, we consider potential model misspecification to both the mean function and the variance-covariance structure.

One possible misspecification of the variance-covariance structure is heteroscedasticity in the sense that $\mathrm{var}(e_{ij}) = \sigma_i^2$ for area i, $1 \leq i \leq m$, where the σ_i^2's are unknown and possibly different. For example, [121] used the well-known Iowa crops data [8] to illustrate that the homoscedastic variance assumption of the NER model may not hold in practice. The latter authors proposed a HNER model that extends the NER model; see Section 2.5.

Again, we need to find a better way to estimate the parameters, $\psi = (\beta', \gamma)'$, under the assumed model, so that the resulting BP (4.34) is more robust against the model misspecifications. We do so by considering a design-based MSPE that is not model-dependent. Let $\theta = (\theta_i)_{1 \leq i \leq m}$ denote the vector of small area means, and $\tilde{\theta} = [\tilde{\theta}_i]_{1 \leq i \leq m}$ the vector of BPs, considered as functions of ψ, that is, $\tilde{\theta}_i = \tilde{\theta}_i(\psi)$. The design-based (e.g., [66]) MSPE can be expressed as

$$\mathrm{MSPE}(\tilde{\theta}) = \mathrm{E}(|\tilde{\theta} - \theta|^2) = \sum_{i=1}^{m} \mathrm{E}\{\tilde{\theta}_i(\psi) - \theta_i\}^2 = \mathrm{E}\{Q(\psi) + \cdots\}, \tag{4.35}$$

where $\cdots$ does not depend on ψ. Note that the E in (4.35) is different from the E_M in (4.34) in that E is completely model-free; namely, the expectation in (4.35) is with respect to the SRS from the finite populations, which has nothing to do with the assumed model. The third equation in (4.35) shows there is a thurstoning expression; in fact, we have

$$Q(\psi) = \sum_{i=1}^{m} \left\{ \tilde{\theta}_i^2(\psi) - 2\frac{1 - r_i}{1 + n_i\gamma}\bar{y}_{i\cdot}\bar{X}_i'\beta + b_i(\gamma)\hat{\mu}_i^2 \right\} = \sum_{i=1}^{m} Q_i, \tag{4.36}$$

where $b_i(\gamma) = 1 - 2a_i(\gamma)$ with $a_i(\gamma) = r_i + (1 - r_i)n_i\gamma(1 + n_i\gamma)^{-1}$, and Q_i is defined in an obvious way. Furthermore, $\hat{\mu}_i^2$ is a design-unbiased estimator of $\bar{Y}_i^2$ under the SRS, which has the following expression:

$$\hat{\mu}_i^2 = \frac{1}{n_i}\sum_{j=1}^{n_i} y_{ij}^2 - \frac{N_i - 1}{N_i(n_i - 1)}\sum_{j=1}^{n_i}(y_{ij} - \bar{y}_{i\cdot})^2, \tag{4.37}$$

assuming that $n_i > 1$. The BPE of ψ, $\hat{\psi}$, is obtained as the minimizer of $Q(\psi)$ with respect to ψ.

The OBP of θ_i is given by (4.34) with ψ replaced by its BPE.

Ref. [128] carried out a series of simulation studies to evaluate the performance of the OBP and its comparison with the EBLUP. They first considered a simple case, in which a single covariate, x_{ij}, is thought to be linearly associated with the response y_{ij} through the following NER model:

$$y_{ij} = \beta x_{ij} + v_i + e_{ij}, \quad i = 1, \ldots, m, \ j = 1, \ldots, 5,$$

where β is an unknown coefficient, and v_i, e_{ij} are the same as in the NER model. Thus, in particular, there is a belief that the mean response should be zero when the value of the covariate is zero.

Three different sample sizes, $m = 50$, 100 or 400, were considered in conjunction with two different true values of b: $b = 0.5$ or 1.0, where b is defined below. Thus, there are six cases, each being a combination of the sample size and b value. In each case, an x sub-population is generated from the normal distribution with mean equal to 1 and standard deviation equal to $\sqrt{0.1} \approx 0.32$. The y sub-population is then generated from the following super-population heteroscedastic NER model:

$$Y_{ik} = b + v_i + e_{ik}, \quad i = 1, \ldots, m, \ k = 1, \ldots, 1000$$

(so the sub-population size is $N_i = 1000$, $1 \le i \le m$), where v_i is generated from the normal distribution with mean 0 and standard deviation $\sqrt{0.1} \approx 0.32$;

TABLE 4.6
Overall empirical MSPE (% increase is EBLUP over OBP).

m	b	OBP	EBLUP	% Increase
50	0.5	0.130	0.161	24
50	1.0	0.503	0.598	19
100	0.5	0.076	0.277	264
100	1.0	0.396	1.077	172
400	0.5	0.096	0.965	905
400	1.0	0.393	4.046	930

e_{ij} is generated from the normal distribution with mean 0 and standard deviation σ_i, where σ_i^2 are generated independently from the Uniform$[0.05, 0.15]$ distribution (so that range for σ_i is approximately from 0.22 to 0.39); and the v_i's and e_{ik}'s are generated independently. It is seen that the assumed NER model is misspecified in terms of both the mean and the variance. The generated x and y sub-populations were fixed throughout the simulations.

In each simulation, an SRS of size 5 was drawn from $\{1, \dots, 1000\}$ that determines the samples x_{ij} and y_{ij}, $j = 1, \dots, 5$, for each i. This is repeated for $K = 1000$ simulation runs. Same-data comparisons of the OBP and EBLUP were made, with the ML estimator of γ for the latter, in terms of both the overall and area-specific MSPEs, with the overall MSPE defined similarly as in the previous subsections. Table 4.6 reports the overall empirical MSPE. It is seen that the percentage increase in the overall MSPE of the EBLUP over the OBP ranges between around 20% to almost 1000%, depending on the sample size and value of b. The patterns shown here are consistent with those in Table 4.1, where model-based predictive performances were evaluated. However, the gain by the OBP is much more significant, for $m = 100$ and $m = 400$, than those reported in Table 4.1.

As for the area-specific MSPEs, we again use boxplots to exhibit the distributions of the area-specific MSPEs associated with both methods. See Figure 4.3. The plots reveal details not shown by the overall MSPEs. For example, it might be wondered whether the percentage increase by the EBLUP in the overall MSPE is simply due to the increased number of areas adding together. A simple calculation suggests that this may not be true, for example, $(400/50) \times 19\%$ is only 152% (not 930%).

A more explicit explanation is seen in Figure 4.3. For example, comparing the case of $m = 50, b = 1$ with that of $m = 400, b = 1$, it is seen that while there is a considerable overlap between the boxplots of OBP and EBLUP in the former case, the boxplots are completely separated in the latter case; in other words, the largest area-specific MSPE of the OBP is smaller than the smallest area-specific MSPE of the EBLUP. This pattern cannot be simply credited to adding or duplicating the areas. In fact, in the latter case, the OBP is doing much better than the EBLUP not just overall, but also for every one of the 400 small areas.

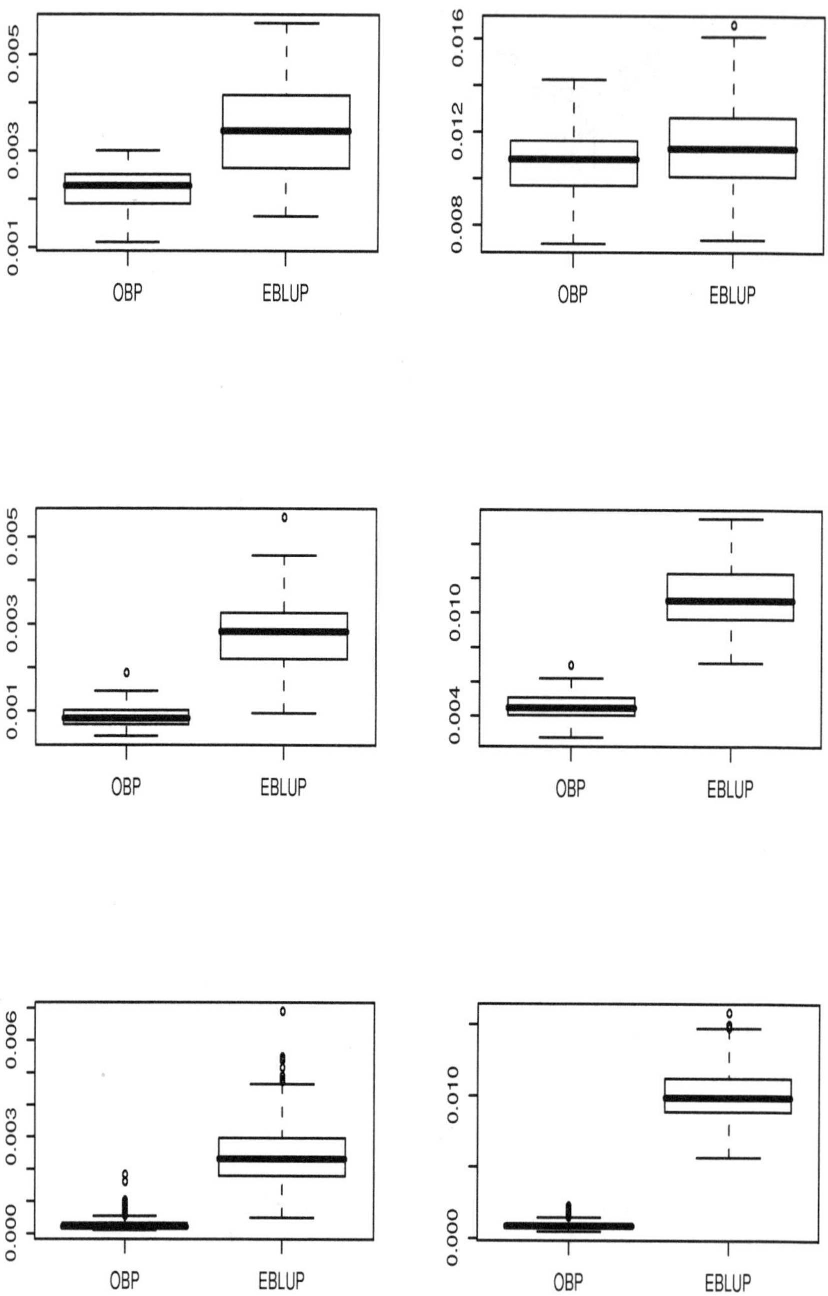

FIGURE 4.3
Area-specific Empirical MSPEs (Boxplots). Upper Left: $m = 50, b = 0.5$; Upper Right: $m = 50, b = 1.0$; Middle Left: $m = 100, b = 0.5$; Middle Right: $m = 100, b = 1.0$; Lower Left: $m = 400, b = 0.5$; Lower Right: $m = 400, b = 1.0$.

Ref. [128] also considered a case where the assumed model is partially correct. In this case, the true model has a nonzero slope and also a nonzero intercept. The value of the intercept is rather small though; in this sense, the assumed model is wrong but not terribly wrong. More specifically, the true model is

$$Y_{ij} = b_0 + b_1 X_{ik} + v_i + e_{ik}, \quad i = 1, \ldots, m, \ k = 1, \ldots, 1000, \tag{4.38}$$

where the X sub-population is generated from the normal distribution with mean 1 and standard deviation $\sqrt{0.1} \approx 0.32$; $b_0 = 0.2$, $b_1 = 0.1$; the v_i are generated independently from $N(0, 0.01)$; the e_{ik} are generated from the heteroscedastic normal distribution so that $e_{ik} \sim N(0, \sigma_i^2)$, where σ_i^2 are generated independently from the Uniform$[0.05, 0.15]$ distribution; and the v_i's and e_{ik}'s are generated independently.

In addition to the overall MSPE, we also report contribution to the MSPE due to "bias" and "variance" (recall the MSPE is equal to the sum of the squared bias and variance). Let $d_i = \hat{\theta}_i - \theta_i$, and $d_i^{(k)}$ be d_i based on the kth simulated data set, $1 \le k \le K$. We define the empirical bias and variance for the ith small area as

$$\bar{d}_i = \frac{1}{K} \sum_{k=1}^{K} d_i^{(k)} \quad \text{and} \quad v_i^2 = \frac{1}{K-1} \sum_{k=1}^{K} \{d_i^{(k)} - \bar{d}_i\}^2,$$

respectively. Let MSPE$_i$ denote the empirical MSPE for the ith small area. It is easy to show that the overall empirical MSPE is

$$\sum_{i=1}^{m} \text{MSPE}_i = \frac{K-1}{K} \sum_{i=1}^{m} v_i^2 + \sum_{i=1}^{m} (\bar{d}_i)^2.$$

Thus, the bias and variance contribution to the overall MSPE are defined as $\sum_{i=1}^{m} (\bar{d}_i)^2$ and $\sum_{i=1}^{m} v_i^2$, respectively.

Results based on 1,000 simulation runs are presented in Table 4.7. As we can see, for the smaller m, $m = 50$, OBP performs (slightly) worse than the EBLUP, but for the larger m, $m = 100$ and $m = 400$, OBP performs (slightly) better, and its advantage increases with m. As for the bias, variance contribution, OBP seems to have smaller bias, and smaller variance for larger m ($m = 100, 400$).

Ref. [128] also considered a case where the assumed model is actually correct. Namely, the true underlying model is (4.38) with $b_0 = 0$; and the errors e_{ik} are homoscedastic with variance equal to 0.1, and everything else is the same as the case considered above. Results based on 1,000 simulation runs are presented in Table 4.8. This time, we see that the EBLUP performs slightly better than OBP under different m, but the difference is diminishing as the sample size increases. As for the bias, variance contribution, EBLUP seems to have smaller variance, and smaller bias for larger m ($m = 100, 400$), but its advantages in both bias and variance shrink as m increases.

TABLE 4.7
Overall empirical MSPE (bias, variance contribution): assumed model is partially correct; % Increase is MSPE of EBLUP over MSPE of OBP (negative number means decrease).

m	OBP	EBLUP	% Increase
50	0.421 (0.224, 0.197)	0.405 (0.238, 0.167)	-4.0
100	0.733 (0.448, 0.285)	0.748 (0.457, 0.291)	2.1
400	2.745 (1.847, 0.899)	2.848 (1.878, 0.971)	3.8

Recently, [35] observed in their simulation studies that, in some cases, the BPE under an NER model can be numerically unstable; as a result, the OBP under an NER model may perform much worse than the EBLUP. An explanation is that although the the thurstoning equation (4.35) holds for a fixed ψ, due to the numerical instability, it may not hold for a random ψ, that is, when ψ is replaced by an estimator such as the BPE. However, the identity holds approximately if the estimator of ψ has small variation, which seemed to be the case for the MLE but not the case for the BPE.

One way to avoid such an instability is to minimize (4.36) over a restricted parameter space around the MLE. This leads to a restricted BPE, and the corresponding restricted OBP, whose performance may be much more stable; furthermore, a model-based OBP approach was proposed; see [243].

4.1.5 Estimation of MSPE of OBP

Measure of uncertainty is important in SAE, as is in other fields of statistics. Obtaining a measure of uncertainty for OBP is particularly challenging. This is because the OBP is derived by taking into account of the potential model misspecification. Thus, to derive a measure of uncertainty, the potential model misspecification needs to be taken into consideration. More importantly, it is desirable to evaluate uncertainty due to the potential model misspecification. So far, this element of uncertainty has not been considered.

A standard measure of uncertainty is the MSPE. Let us first consider this under a Fay-Herriot model (see §4.1.3). In proposing the OBP, [126] also

TABLE 4.8
Overall empirical MSPE (bias, variance contribution): assumed model is correct; % increase is MSPE of EBLUP over MSPE of OBP (negative number indicates decrease).

m	OBP	EBLUP	% Increase
50	0.335 (0.204, 0.131)	0.330 (0.205, 0.125)	-1.4
100	0.749 (0.457, 0.292)	0.746 (0.456, 0.290)	-0.4
400	2.796 (1.800, 0.997)	2.794 (1.799, 0.996)	-0.1

proposed an MSPE estimator under potential model misspecification, which we call the JNR estimator in the sequel. The authors showed that the JNR estimator is second-order unbiased, that is, its bias is $o(m^{-1})$. However, the estimator is known to have large variation. To see why, note that the leading term of the JNR estimator has the expression

$$(\hat{\theta}_i - y_i)^2 + D_i(2\hat{B}_i - 1), \tag{4.39}$$

where $\hat{\theta}_i$ is the OBP of θ_i, $B_i = A/(A + D_i)$, and $\hat{B}_i$ is B_i with A replaced by $\hat{A}$, the BPE of A. It is seen that the direct estimator, y_i, is involved in (4.39), which has large variation compared to, for example, $\hat{A}$. The latter is an estimator based on all of the data, $y_i, x_i, 1 \leq i \leq m$, which has relatively small variation. As a result, the JNR estimator has a significantly nonzero chance of taking negative values.

In addition to the JNR estimator, [126] also proposed a bootstrap MSPE estimator. Although the bootstrap estimator is guaranteed non-negative, its bias was shown to be significantly larger than the JNR estimator. The method also seemed lack of theoretical justification, in which the bootstrap samples were drawn independently under the model $y_i^* \sim \hat{\theta}_i + e_i^*$, where $\hat{\theta}_i$ is the OBP and $e_i^* \sim N(0, D_i)$, $1 \leq i \leq m$.

Ref. [161] proposed a OBOR estimator for the MSPE of OBP. Here, OBOR is an abbreviation of one-bring-one-route. It is called OBOR because the estimator consists of averages of terms, where each term involves y_i, plus one other y_j for $j \neq i$. The average is over $m - 1$ such y_j's for $j \neq i$. The idea can be generalized to one-bring-two, one-bring-three, etc., but the computational burden mounts as it moves on. In this regard, the JNR estimator may also be viewed as a special case of one-bring-none. Although the OBOR estimator reduces the variation of the JNR estimator, the result was not all satisfactory, compared to a much better estimator found later (see below).

A well-known method for obtaining a second-order unbiased MSPE estimator is the Prasad-Rao (PR) linearization method [196]. The method is developed under the assumption that the underlying model is correct. In fact, the assumed model is substantially used in the derivation of the PR MSPE estimator. Given that, it would be surprising to learn that, in spite of the model misspecification, the PR MSPE estimator for OBP is, still, mostly correct. In fact, [162] found that the PR MSPE estimator remains first-order unbiased in the sense that the bias of the estimator is $O(m^{-1})$, even if the underlying model is misspecified in its mean function. Furthermore, the same authors showed the PR MSPE estimator can be modified to achieve the second-order unbiasedness, again under the potential model misspecification in the mean function. These are the main stories that are about to become unfold.

In the SAE literature, different types of MSPEs, including conditional and unconditional ones, have been considered (e.g., [43]). Ref. [162] argued that, in the current situation, there is good rationale for the unconditional MSPE. Note that the conditional MSPE is the MSPE conditioning on the auxiliary data, X. When the underlying model is correctly specified, the relationship between the outcome variable and the auxiliary variables is well understood; therefore, given the auxiliary variables, there is no more uncertainty due to the latter. However, when the underlying model is misspecified, knowing the auxiliary variables does not necessarily take away the uncertainties associated with the auxiliary variables, because one is not sure in what way the outcome variable is associated with the auxiliary variables. Furthermore, the auxiliary data are typically sampled along with the outcome data; hence, they should be treated as random variables as well.

The basic assumptions defining the unconditional MSPE are the following. Suppose the true underlying model can be expressed as

$$y_i = x_i'\beta + \Delta_i + v_i + e_i, \quad i = 1, \dots, m. \tag{4.40}$$

Here, Δ_i is an unobserved (random) variable satisfying the following conditions: Let $z_i = (x_i', \Delta_i)'$ and $\delta_i = \Delta_i - x_i'\{\mathrm{E}(x_1 x_1')\}^{-1}\mathrm{E}(x_1\Delta_1)$.

A1. $z_i, v_i, e_i, i = 1, \dots, m$ are independent.

A2. $\mathrm{E}(z_i)$, $\mathrm{E}(z_i z_i')$ and $\mathrm{E}(\delta_i^4)$ are finite (component-wise for a vector or a matrix) and do not depend on i, $v_i \sim N(0, A)$, and $e_i \sim N(0, D_i)$, where $A > 0$ is unknown, and $D_i > 0, 1 \le i \le m$ are known.

The unconditional, area-specific, MSPE is then defined as

$$\mathrm{MSPE}(\hat{\theta}_i) = \mathrm{E}(\hat{\theta}_i - \theta_i)^2, \tag{4.41}$$

where the expectation is taken with respect to the joint distribution under the above model.

According to [126], the BPE of $\psi = (\beta', A)'$, denoted by $\hat{\psi} = (\hat{\beta}, \hat{A})'$, converges in probability to the minimizer of

$$q(\psi) = \mathrm{E}\{|\tilde{\theta}(\psi) - \theta|^2\} = \sum_{i=1}^{m} \mathrm{E}\{\tilde{\theta}_i(\psi) - \theta_i\}^2, \tag{4.42}$$

where $\tilde{\theta}_i(\psi)$ is the BP, $\tilde{\theta}_i$, of θ_i considered as a function of ψ. Let $\psi_* = (\beta_*', A_*)'$ denote the minimizer. An amazing property of ψ_* is the following, which we state as a theorem.

Theorem 4.2. The parameter vector ψ_* minimizes $\mathrm{E}\{\tilde{\theta}_i(\psi) - \theta_i\}^2$ over all ψ for every $1 \le i \le m$.

The reason this is amazing is because, by definition, ψ_* only needs to minimize $\sum_{i=1}^m \mathrm{E}\{\tilde{\theta}_i(\psi) - \theta_i\}^2$, not necessarily each term in the sum. Of course, if it minimizes each term in the sum, it also minimizes the sum, but not the other way around. Some may wonder if there are other things in the world that also possess this amazing property. The answer is yes: If the model is correctly specified, then the true ψ, ψ_0, also has this property, although no one would be surprised in that case.

We now show, with a heuristic argument, that the PR estimator remain first-order unbiased despite the model misspecification. A rigorous proof can be found in [162]. It can be shown that the leading term of the MSPE, that is, a first-order approximation so that the difference of the approximation is $O(m^{-1})$, defined by (4.41), is $c_i(A_*)$, where

$$c_i(A) = \frac{AD_i}{A + D_i}. \tag{4.43}$$

On the other hand, it can be shown that, under the same distribution that defines the unconditional MSPE, the leading term of the PR MSPE estimator is also $c_i(A_*)$, with $c_i(A)$ given by (4.43). It then follows that the PR MSPE estimator is first-order unbiased, that is, its bias is $O(m^{-1})$.

Furthermore, the PR estimator can be modified to achieve second-order unbiasedness. To do so, one starts with the following decomposition:

$$\begin{aligned} & \text{MSPE}(\hat{\theta}_i) \\ = \;& c_i(A_*) + 2\text{E}[\{g_i(\psi_*, y_i) - \theta_i\}\{g_i(\hat{\psi}, y_i) - g_i(\psi_*, y_i)\}] \\ & + \text{E}\{g_i(\hat{\psi}, y_i) - g_i(\psi_*, y_i)\}^2. \end{aligned} \tag{4.44}$$

Note that, under correct model specification, the second (cross) term on the right side of (4.44) vanishes. However, under potential model misspecification, this term is $O(m^{-1})$, which cannot be ignored. One then derives a plug-in estimator of the cross term. In a similar way, one derives a plug-in estimator of the last term on the right side of (4.44). Finally, one needs to bias-correct a plug-in estimator of the first term on the right side of (4.44).

Regarding the bias-correction, here is something to keep in mind. If a term is already $O(m^{-1})$, one does not need a bias-correction—a plug-in estimator will "do the trick". This is because a $O(m^{-1})$ term typically can be expressed as $m^{-1}a(\psi_*)$, where $a(\cdot)$ is a continuous function. The plug-in estimator is $m^{-1}a(\hat{\psi})$. The difference between the two is $m^{-1}\{a(\hat{\psi}) - a(\psi_*)\} = o(m^{-1})$, because $\hat{\psi}$ converges to ψ_* and $a(\cdot)$ is continuous.

Any term of $o(m^{-1})$ can be ignored without affecting the second-order unbiasedness. Thus, when everything is put together, the combined estimator is second-order unbiased, that is, its bias is $o(m^{-1})$. This is, of course, not a rigorous proof. The latter can be found in [162].

A simulation study was carried out that compared four MSPE estimators, the PR estimator (PR), modified PR estimator (MPR), the JNR estimator, and a naive MSPE estimator given by (4.39) with A replaced by $\hat{A}$, the BPE of A. The assumed model is a Fay-Herriot model, (1.16), with $x_i'\beta = \beta_0 + x_i'\beta$. Three true underlying models were considered:

(I). $y_i = \beta_0 + x_i'\beta + v_i + e_i, \quad i = 1, 2, \cdots, m$;

(II). $y_i = \beta_0 + \beta_1 x_{1i} + \arctan(z_i) + v_i + e_i, \quad i = 1, 2, \cdots, m$;

(III). $y_i = \arctan(4x_i'\beta) + v_i + e_i, \quad i = 1, 2, \cdots, m$.

Here, $\beta_0 = 0.2$, $\beta = (\beta_1, \beta_2)'$ with $\beta_1 = \beta_2 = 0.5$, $x_i = (x_{1i}, x_{2i})'$ were generated such that $(x_i', z_i)'$ has a tri-variate normal distribution with means 0, variances 2, $\text{cov}(x_{1i}, x_{2i}) = \text{cov}(x_{2i}, z_i) = 0.4$, and $\text{cov}(x_{1i}, z_i) = 0$; $v_i \sim N(0, A)$ with $A = 1$, and $e_i \sim N(0, D_i)$ with $D_i = 0.5 + (i-1)/(m-1)$,

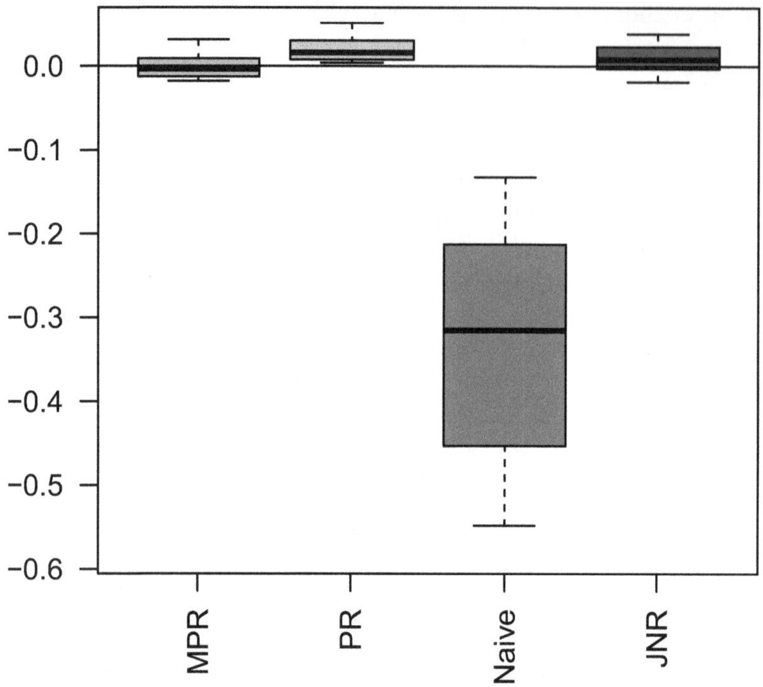

FIGURE 4.4
Box plot of simulated %RB: Case (II), $m = 20$.

$i = 1, 2, \cdots, m$. The results, based on 10,000 simulation runs are presented in figures (for the case $m = 20$); see Figures 4.3 and 4.4, taken from [162].

Figure 4.4 presents the simulated percentage relative bias, defined as

$$\%\text{RB} = 100 \times \left\{ \frac{\text{E}(\widehat{\text{MSPE}}) - \text{MSPE}}{\text{MSPE}} \right\}, \tag{4.45}$$

where MSPE is the simulated true MSPE, $\text{E}(\widehat{\text{MSPE}})$ is the simulated mean of $\widehat{\text{MSPE}}$, an MSPE estimator, which can be one of the four estimators under comparison. The %RB is a measure of performance in terms of the bias. It is seen that, in terms of %RB, PR, MPR, and JNR are quite close, with MPR seemingly doing the best among the three. On the other hand, the naive estimator is doing much worse compared to the other three.

However, in terms of the variation, measured by the standard deviation (s.d.) of the MSPE estimator, we observe a quite different pattern; see Figure 4.5. Again, PR and MPR are quite close, while the naive estimator and JNR are doing much worse in this regard. In fact, it appears that JNR is doing even worse than the naive estimator in terms of the s.d.

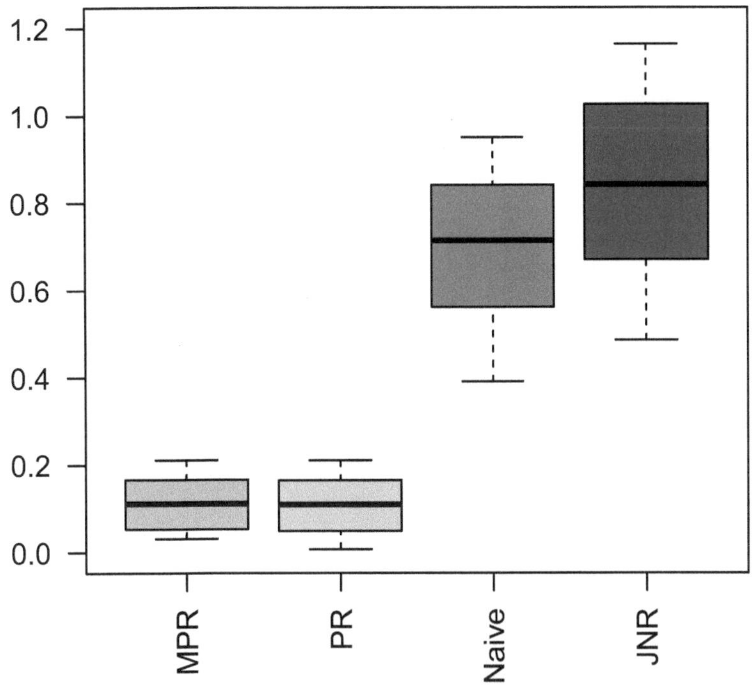

FIGURE 4.5
Box plot of simulated s.d.: case (II), $m = 20$.

A real-data example associated with OBP and its MSPE estimator under the Fay-Herriot model is discussed in the next subsection.

Next, we consider MSPE estimation for OBP for small area counts (see §4.1.3). Ref. [34] proposed a bootstrap MSPE estimator; hereafter CJN. One concern about CJN is that the bootstrap sample, y_i^*, is drawn under a Poisson distribution whose mean is y_i, a single outcome from area i. Such a single observation may have large variation, which may result in a large sampling variation for the bootstrap MSPE estimator. In a way, this is similar to the JNR estimator discussed above. CJN also lacks theoretical justification.

As noted, estimation of the MSPE of OBP is known to be difficult due to the potential model misspecification. Because the OBP is derived with the understanding that the underlying model is potentially misspecified, it is reasonable to develop a method of MSPE estimation under the same framework. This adds challenges to obtaining a second-order unbiased MSPE estimator, which can be a daunting task even under correct model specification.

A well-known method of deriving a second-order unbiased MSPE estimator is the Prasad-Rao method [196]. This was executed above for OBP under the Fay-Herriot model. In fact, the development of the MPR estimator was quite tedious, although the tediousness was not shown in the above heuristic description. The process of deriving a MPR type estimator is even more challenging when it comes to the count data.

More importantly, mathematical errors likely occur in the process of derivation. Specifically, such derivations involve procedures to break out some expressions, collect similar terms and put them together, and simplify the final expression. Analytical errors can occur in each of these steps, so it is very difficult to achieve the goal error-free, even for well-trained mathematical statisticians. Furthermore, even if one can come up with a analytical expression error-free, the expression is likely to be lengthy; as a result, coding errors often occur based on such length expressions, and it may not be easy to debug the code once such errors occur.

Often times, a practitioners may not have the mathematical skills, time, patience, or carefulness, to carry out such derivations error-free. To them, an expression, or expressions, that are easy to implement from a computational standpoint is good enough, or even better than a lengthy, all-in-one formula. The latter consideration was closely followed in [184], which proposes a computationally oriented approach to second-order unbiased MSPE estimation for OBP for small area counts.

Let us focus on the GM model for its simplicity, including an analytic expression of the BP. We consider the extended model, which allows an off-set (see the end of §4.1.3), with the OBP for μ_i given by (4.32). As for the MSPE, [184] considered an unconditional MSPE under the following framework. It is assumed that the true second-stage model is

$$\mu_i \sim \text{Gamma}\left(\frac{x_i'\beta + \Delta_i}{\phi}, \phi\right), \tag{4.46}$$

where Δ_i is an unknown quantity representing the amount of misspecification. Furthermore, it is assumed that $(x_i, \Delta_i), i = 1, \dots, m$ are i.i.d. with finite second moments such that $\text{E}(x_1 x_1')$ is non-singular. Accordingly, the first-stage model is extended to that, give $\mu = (\mu_i)_{1\le i\le m}, X = (x_i')_{1\le i\le m}$, $\Delta = (\Delta_i)_{1\le i\le m}$, $y_i, 1 \le i \le m$ are conditionally independent with $y_i \sim \text{Poisson}(\lambda_i\mu_i)$; the second-stage model is extended to that, given X and Δ, $\mu_i, 1 \le i \le m$ are conditionally independent with the conditional distribution given by the right side of (4.46). The unconditional MSPE is defined as

$$\text{MSPE}(\hat{\mu}_i) = \text{E}(\hat{\mu}_i - \mu_i)^2, \tag{4.47}$$

where the E denotes the true expectation with respect to the joint distribution of y, μ, X, Δ defined above.

Under this framework, [184] showed that a first-order unbiased estimator of the unconditional MSPE of $\hat{\mu}_i$ is given by

$$\widehat{\text{MSPE}}_1(\hat{\mu}_i) = \frac{\hat{\phi}\overline{y/\lambda}}{1 + \hat{\phi}\lambda_i}, \tag{4.48}$$

TABLE 4.9
Performance of MSPE estimators.

Model Misspecification	True MSPE	$\widehat{\text{MSPE}}_1$ %RB	$\widehat{\text{MSPE}}_1$ CV	$\widehat{\text{MSPE}}$ %RB	$\widehat{\text{MSPE}}$ CV	$\widetilde{\text{MSPE}}$ %RB	$\widetilde{\text{MSPE}}$ CV
None	0.537	-16.02	0.44	1.29	0.39	30.09	1.81
$\delta = 0.0$	0.556	-18.66	0.44	-1.83	0.38	31.01	1.84
$\delta = 0.1$	0.599	-16.99	0.42	-0.29	0.37	33.01	1.72
$\delta = 0.2$	0.688	-20.24	0.41	-4.73	0.36	28.74	1.72
$\delta = 0.5$	0.824	-16.48	0.38	-1.10	0.33	26.79	1.59

where $\overline{y/\lambda} = m^{-1}\sum_{i=1}^m y_i/\lambda_i$ and $\hat{\phi}$ is the BPE of ϕ. Furthermore, a second-order unbiased MSPE estimator is given with the following expression:

$$\widehat{\text{MSPE}}(\hat{\mu}_i) = h_i(\hat{\phi}, \hat{\tau}) + t_i(\hat{\phi}, \hat{\tau}, \hat{V}, \hat{\mathcal{E}}), \tag{4.49}$$

where $\tau = m^{-1}\sum_{u=1}^m y_u/\lambda_u$, $\hat{V}$ is an estimated covariance matrix, and $\hat{\mathcal{E}}$ is a collection of some estimated quantities. The functions h_i and t_i are not designed to be "pretty"; however, they are computationally easy to implement, that is, expressed according to the flow of a computing algorithm.

Ref. [184] carried out a serious of simulation studies to evaluation the empirical performance of the MSPE estimators introduced above, including the CJN, $\widehat{\text{MSPE}}_1$, and $\widehat{\text{MSPE}}$. We present a simple-case scenario here and refer more results to the paper. The simple case has $x_i'\beta = \beta$ and $\lambda_i = 1$, $1 \le i \le m$.

We consider a case with no model misspecification, that is, $\Delta_i = 0$, and cases with model misspecification. In the latter cases, the Δ_i's are generated independently from a normal distribution with mean δ and standard deviation 0.1. Four different values of δ are considered: $\delta = 0.0, 0.1, 0.2, 0.5$.

Two performance measures are considered. The first is the %RB defined above. The second is coefficient of variation (CV), defined as the standard deviation (s.d.) of the MSPE estimator divided by the absolute value of the mean of the MSPE estimator, both evaluated based on the simulation runs. The results, based on 10,000 simulation runs, are presented in the Table 4.9, where $\widetilde{\text{MSPE}}$ denotes the CJN estimator.

It is seen that $\widehat{\text{MSPE}}$ performs overwhelmingly the best in terms of %RB; it also performs the best in terms of CV, although $\widehat{\text{MSPE}}_1$ comes quite close under this measure. The CJN estimator is seen to be inferior to both $\widehat{\text{MSPE}}$ and $\widehat{\text{MSPE}}_1$ not only in %RB but also, more remarkably, in CV, which confirms our earlier speculation that the CJN estimator has large variation.

A real-data example is discussed in the next subsection.

Finally, we consider estimation of the design-based MSPE for OBP under the NER model (see §4.1.4). Ref. [128] proposed a non-parametric bootstrap MSPE estimator following Efron's original idea [51]. The method does not rely on the NER model, hence is not affected by the model misspecification.

Furthermore, the proposed MSPE estimator is guaranteed non-negative and positive with probability one.

Suppose that the small area sub-populations, or the N_i's are large enough, so that the sampling from the sub-populations can be treated approximately as with replacement. Let $z_{ij} = (x'_{ij}, y_{ij})', j = 1, \dots, n_i$ denote the (original) samples from the ith small area, $1 \le i \le m$. We then draw samples,

$$z_{ij}^{(b)} = [\{x_{ij}^{(b)}\}', y_{ij}^{(b)}]', j = 1, \dots, n_i,$$

with replacement, from $\{z_{ij}, j = 1, \dots, n_i\}$, independently for $1 \le i \le m$. Suppose that B bootstrap samples are drawn, yielding samples

$$z^{(b)} = \{z_{ij}^{(b)}, 1 \le j \le n_i, 1 \le i \le m\}, \quad 1 \le b \le B.$$

The bootstrapped version of the BP (4.34) is

$$\tilde{\theta}_i^{(b)} = \bar{X}'_{i\cdot}\beta + \left\{r_i + (1 - r_i)\frac{n_i\gamma}{1 + n_i\gamma}\right\}[\bar{y}_{i\cdot}^{(b)} - \{\bar{x}_{i\cdot}^{(b)}\}'\beta], \tag{4.50}$$

where β and γ are the same population parameters for β and γ, respectively, as for the original population. Note that the original samples of z_{ij} are assumed to satisfy the same NER model (1.17) with X_{ik} (Y_{ik}) replaced by x_{ij} (y_{ij}). Because the original samples are treated as the bootstrap population, following Efron's original idea, the population parameters, β, γ, for the bootstrap samples are the same as those for the original samples.

Nevertheless, the proposed bootstrap procedure is non-parametric in the sense that the assumed model, (1.17), plays no role in drawing the bootstrap samples. In particular, the BPE of β and γ, based on the original samples, are not used anywhere in the bootstrapping; and the population quantities of interest are $\bar{Y}_i$, $1 \le i \le m$, whose bootstrap analogies are $\bar{y}_{i\cdot}, 1 \le i \le m$. This is different from the parametric bootstrap of [126], where the BPE of the model parameters, based on the original samples, are used to draw bootstrap samples under the assumed model.

Also note that, because the $\bar{X}_i$ are known, they are treated as known constants and therefore do not change during the bootstrapping (it does not make sense to "estimate" something that one already knows).

Other than those, the procedure follows closely the standard bootstrap idea (e.g., [54]; also see [32] for an application to SAE). The bootstrap estimator of $\text{MSPE}(\hat{\theta}_i) = \text{E}(\hat{\theta}_i - \bar{Y}_i)^2$ is given by

$$\widehat{\text{MSPE}}(\hat{\theta}_i) = \frac{1}{B}\sum_{b=1}^{B}\{\hat{\theta}_i^{(b)} - \bar{y}_{i\cdot}\}^2, \tag{4.51}$$

where $\hat{\theta}_i^{(b)}$ is (4.50) with β, γ replaced by their BPE based on the bth bootstrapped samples, $z^{(b)}$.

TABLE 4.10
Empirical performance of $\widehat{\text{MSPE}}$.

m	n_i	i	MSPE	$\widehat{\text{MSPE}}$	%RB	RMSE	i	MSPE	$\widehat{\text{MSPE}}$	%RB	RMSE
10	5	1	.041	.042	4.5	.103	6	.034	.043	26.3	.070
10	10	1	.036	.036	-0.4	.068	6	.034	.036	6.4	.070
20	5	1	.031	.032	4.1	.051	6	.028	.031	12.5	.046
10	5	2	.046	.038	-16.1	.078	7	.032	.040	25.4	.078
10	10	2	.035	.033	-4.1	.078	7	.033	.034	2.7	.068
20	5	2	.031	.029	-7.2	.050	7	.030	.031	3.6	.055
10	5	3	.038	.042	10.2	.121	8	.042	.042	-0.4	.150
10	10	3	.037	.036	-1.7	.091	8	.033	.035	7.5	.067
20	5	3	.031	.032	4.4	.052	8	.030	.031	4.1	.058
10	5	4	.056	.052	-7.6	.121	9	.050	.042	-15.0	.074
10	10	4	.037	.040	6.3	.072	9	.034	.034	-1.0	.063
20	5	4	.040	.035	-11.3	.068	9	.034	.030	-11.1	.049
10	5	5	.033	.037	11.8	.066	10	.041	.043	3.1	.082
10	10	5	.032	.033	2.5	.066	10	.034	.033	-2.9	.073
20	5	5	.024	.025	2.9	.052	10	.035	.033	-7.9	.062

One might be concerned that, because the n_i's may be small in typical SAE problems, there may not be many distinct bootstrap samples for each small area. However, the data consist of not just one, but many small areas. When all of the small areas are combined, there are, still, a lot of distinct bootstrap samples, even if the n_i's are small.

Ref. [128] reported results of a simulation study, in which the performance of their proposed bootstrap MSPE estimator is evaluated. The simulation was under a special case of the first simulated example in Section 4.1.4 with $b = 0.5$ but under smaller sample sizes. Specifically, starting with $m = 10$ and $n_i = 5$, we increase either n_i, from 5 to 10, or m, from 10 to 20.

First consider the design-based bias of $\widehat{\text{MSPE}}(\hat{\theta}_i)$. Two finite populations were generated, and then fixed, so that the finite population for $m = 10$ was a sub-population of the finite population for $m = 20$. Table 4.10 reports, for the first 10 small areas (these are all the small areas that are common under different values of m), the simulated true MSPE (MSPE), the simulated mean of $\widehat{\text{MSPE}}(\hat{\theta}_i)$ ($\widehat{\text{MSPE}}$), and the %RB. Also reported is the square root of the mean squared error (RMSE) over the simulations, defined as

$$\sqrt{\frac{1}{K}\sum_{k=1}^{K}(\widehat{\text{MSPE}}_{i,k} - \text{MSPE}_i)^2}$$

for the ith small area, where MSPE_i is the simulated MSPE for the ith small area (which does not depend on k), and $\widehat{\text{MSPE}}_{i,k}$ is the MSPE estimate based on the kth simulated data set. The number of bootstrap samples used to evaluate the MSPE estimator (4.51) is $B = 100$. All results are based on $K = 1,000$ simulation runs.

It is seen that, overall, the results improve when either n_i or m increase, but, in terms of %RB, the improvement is more universal, or effective, when

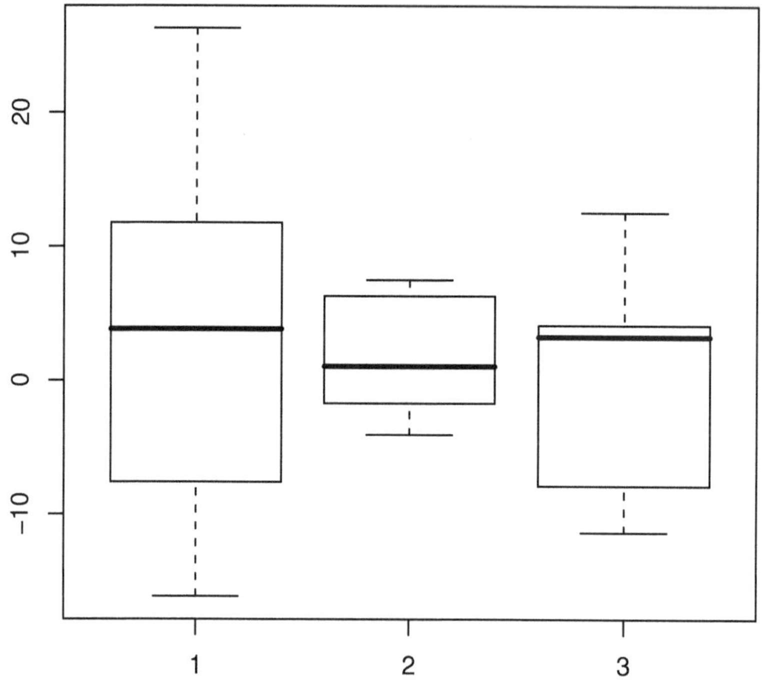

FIGURE 4.6
Boxplots of %*RB*. *1:* $m = 10, n_i = 5$; *2:* $m = 10, n_i = 10$; *3:* $m = 20, n_i = 5$.

n_i increases. This is mainly due to the fact that, as n_i increases, the sample provides a better approximation to the population; hence, the bootstrap distribution better approximates to the population distribution. Also note that, depending on the area, the sign of the RB can be either positive or negative. This is mainly due to the area-to-area difference (recall that the populations are fixed) as well as the bootstrap errors.

To see some overall measures, the mean and s.d. of the %RBs for the 10 small areas are, for $m = 10, n_i = 5$, mean = 4.2%, s.d. = 14.8%; for $m = 10, n_i = 10$, mean = 1.5%, s.d. = 4.2%; and for $m = 20, n_i = 5$, mean = −0.6%, s.d. = 8.1%. The boxplots of the %RBs are presented in Figure 4.6. The plots further illustrate the pattern of improvement.

On the other hand, in terms of RMSE, the improvement is much more significant when m increases than when n_i increases. This is because having a larger m reduces the MSPEs, in general; hence, naturally, the corresponding MSPE estimates also drop. In other words, both the estimator and the parameter (the MSPE) decrease, which typically results in a reduction in RMSE.

4.2 Real-data examples

We discuss four real-data examples to illustrate the OBP, its comparison with the EBLUP, as well as measures of uncertainty for OBP. The first example is used to illustrate the difference between the BPE and MLE, and that between the OBP and EBLUP. The second example is regarding prediction of incubation period for COVID-19. The third example is about prediction of the number of deaths due to esophageal cancer. The last example is related to evaluation of school-wide effectiveness of education programs on tobacco use prevention and cessation.

4.2.1 Hospital data revisited

Recall the hospital data discussed in Example 1.3. Ref. [67] proposed a Fay-Herriot model, which is (1.16) with $x_i'\beta = \beta_0 + \beta_1 x_i$, where x_i is the severity index. An inspection of the scatter plot (see Figure 4.7) suggests some non-linear trend. It appears that a quadratic model would fit the data well except for a potential "outlier" at the upper right corner.

The question is what to do in this situation. One option would be to look for a more complex model that fits the data better. This approach will be explored in the next chapter. Another option is to stay with the relatively simple quadratic model, but take into account the potential model misspecification. This would also avoid over-fitting, especially given the small sample size. The quadratic model can be expressed as

$$y_i = \beta_0 + \beta_1 x_i + \beta_2 x_i^2 + v_i + e_i, \quad i = 1, \dots, 23. \tag{4.52}$$

The BPE for fitting model (4.52) are $\hat{\beta}_0 = 0.280$, $\hat{\beta}_1 = -0.989$, $\hat{\beta}_2 = 3.261$, and $\hat{A} = 2.099 \times 10^{-4}$. Plots of the direct estimates (i.e., the raw data) and the OBPs and EBLUPs are presented in Figure 4.7.

Recall the difference between the BPE and MLE is in terms of different weights assigned to the direct estimates; see (4.9) and (4.11). The effect of differential weighting of hospitals is clearly evident in the rightmost plot which compares the OBPs to the EBLUPs (the line of identity is overlayed). In this case, using the BPE estimates forces the OBPs to be systematically larger in magnitude then their EBLUP counterparts.

4.2.2 Prediction of incubation period for Covid-19

Since the first case reported in Wuhan, China in December 2019, the 2019 coronavirus disease (COVID-19) had overwhelmed the world with its rapid growth in cases and deaths. From a scientific point of view, the incubation period (IP) of the disease plays an important role in prevention and control efforts, as well as mathematical modeling of the coronavirus transmission. The

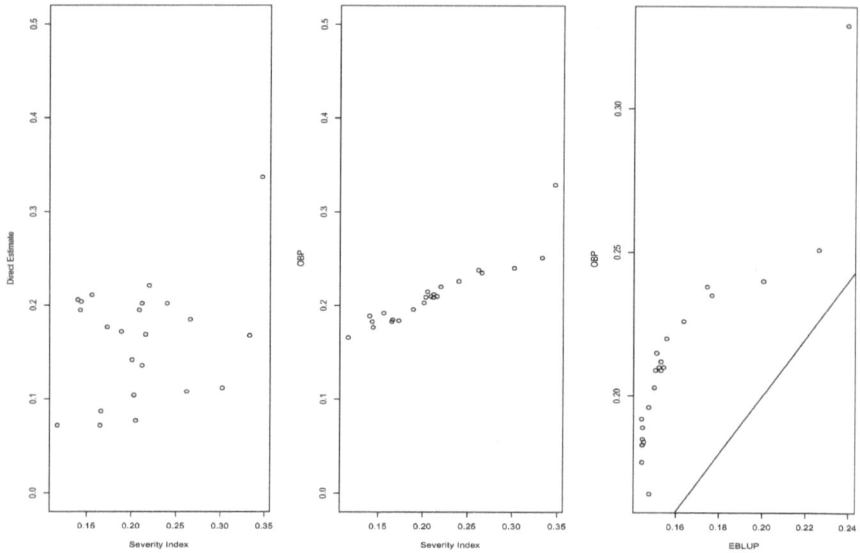

FIGURE 4.7
Trio of plots from the hospital dataset showing Left: the direct estimates vs the severity index; middle: the OBP vs the severity index (notice the smoothing effect produced); and right: the OBP vs the EBLUP estimates. The last plot reflects the effect the differential weighting of hospitals to estimate the BPE vs the MLE of β. The line of identity is overlaid.

initial quarantine period in China was fixed at 14 days, without considering any other auxiliary information; see, for example, [81].

A complication of the existing IP in the human population was that it had to do with both the nature of the disease and the regional efforts used to test and report the disease cases. The city of Wuhan, the center of the disease outbreak, and the entire Hubei province (of which Wuhan is the capital city), was considered different from other provinces in China in terms of the magnitude of the outbreak. In a study reported in [162], the authors were able to obtain data from 1,493 confirmed cases outside Hubei province in China, including histories tracking these cases to the first reported contact with a disease carrier. The cases were reported from 27 provinces of China (all from mainland China except Hubei) crossed by three age groups, namely, less than 23, 23–55, and greater than 55 years old, which were treated as small areas. This leads to a total of 64 small areas from which data were available. The data were reported in the supplement of [162].

Among a limited number of auxiliary variables, age was considered a potential factor that contributed to the IP distribution. Figure 4.8 presents a

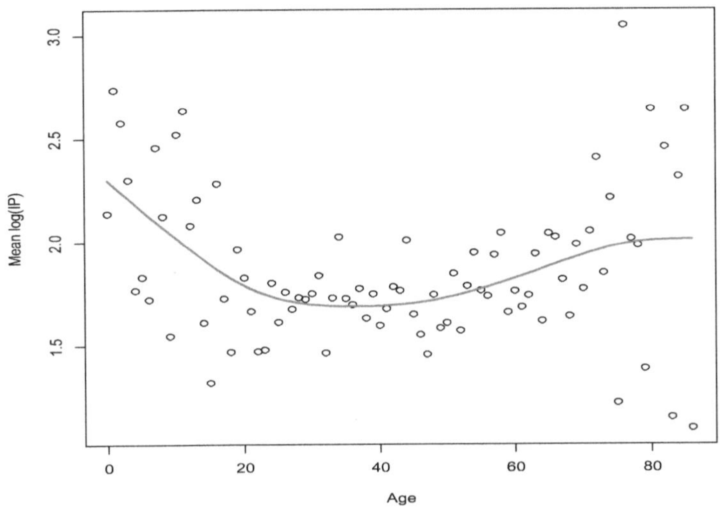

FIGURE 4.8
Scatter plot of mean log(IP) against age.

plot of the reported mean (natural) logarithm of IP against mean age for the 64 areas. A nonlinear (quadratic) pattern appears apparent. Due to such an observation, a quadratic functional relationship with the age was considered.

Furthermore, empirical evidence suggests that there is a potential climate effect. Specifically, provinces under continental climate conditions tend to have a lower IP than those under non-continental climate conditions do. Let I_i be an indicator, which is equal to one if the ith small area is associated with a continental-climate province, and zero otherwise. The proposed model is

$$y_i = \beta_0 + \beta_1 x_i + \beta_2 x_i^2 + \beta_3 I_i + v_i + e_i, \quad i = 1, 2, \cdots, 64, \tag{4.53}$$

where y_i is the observed mean log(IP) for the ith area and x_i is the mean age for the ith area, and v_i and e_i satisfy the conditions of the Fay-Herriot model. In particular, D_i, $1 \le i \le m$ were obtained via approximation to the sampling variance. Specifically, let σ^2 be the population variance of log(IP). Then, we have $D_i = \sigma^2/n_i \approx \hat{\sigma}^2/n_i$, where n_i is the sample size for the ith small area, and $\hat{\sigma}^2$ is the sample variance of log(IP) obtained from the data combining all provinces and age groups. The D_is, along with other information, are provided in Table A.2 of the Supplementary Material of [162].

The BPEs of the model parameters are $\hat{\beta}_0 = 2.518$, $\hat{\beta}_1 = -0.042$, $\hat{\beta}_2 = 0.001$, $\hat{\beta}_3 = -1.365$, and $\hat{A} = 0.062$. The estimate of β_2, the coefficient of x_i^2 in (4.53), is seen to be positive, whereas that of β_3, the coefficient of the continental-climate indicator, is negative. These are consistent with the empirical observations discussed above.

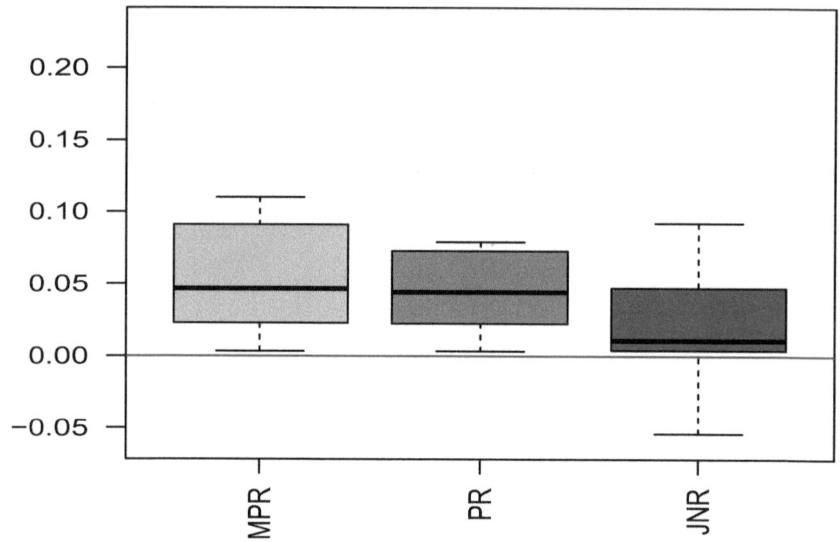

FIGURE 4.9
Boxplots of MSPE estimates.

Furthermore, the OBPs and their corresponding MSPE estimates using the MPR, PR, and JNR methods were obtained. Box plots of the MSPE estimates are presented in Figure 4.9. It is seen that the PR and MPR estimates are all positive; on the other hand, there are some negative values for the JNR estimates. Figure 4.10 presents the OBPs with corresponding margins of error, defined as (plus/minus) the square root of the MSPE estimate, for areas within the second age group (23–55). Note that some margins of error for JNR are not available due to the negative MSPE estimates.

4.2.3 Esophageal cancer data

We illustrate the OBP method for small area counts using a esophageal cancer data set. The data involve 87 counties in Minnesota, USA, originally reported by [135]; also see [225]. The response variable, y, is a count representing the observed number of deaths due to esophageal cancer in a certain county. As noted (see the end of §4.1.3), an offset term is often used in modeling count data. In this case, an apparent offset is the logarithm of the expected number of deaths due to esophageal cancer for the same county.

Ref. [184] considered a special case of the model introduced in Section 4.1.5 [above (4.47)] with $x_i'\beta = \beta$, and λ_i being the expected number of deaths due to esophageal cancer for county i. The BPE of the model parameters are found to be $\hat{\beta} = 0.861$ and $\hat{\phi} = 0.108$. The OBP of μ_i are computed for all

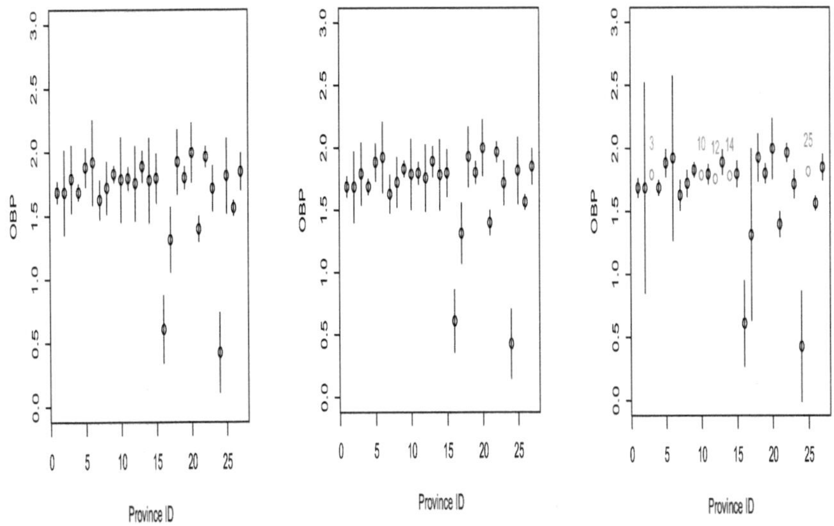

FIGURE 4.10
OBP with margins of error: age group #2 (age 23-55). From left to right: MPR, PR, JNR. Red circles indicate cases with negative MSPE estimates, with the corresponding case numbers showing above.

87 counties. Partial results (for the first 22 counties) are presented in Tables 4.11 and 4.12.

Also presented are square roots of the MSPE (RMSPE) estimates based on the naive estimator ($\widehat{\text{MSPE}}_1$), the second-order unbiased estimator ($\widehat{\text{MSPE}}$), and the CJN bootstrap estimator ($\widetilde{\text{MSPE}}$). Note that the original CJN method did not allow an offset term; in other words, $\lambda_i = 1$ [34]. If the offset is incorporated, the CJN estimator should be computed as

$$\widetilde{\text{MSPE}}(\hat{\mu}_i) = \frac{1}{B}\sum_{b=1}^{B}(\hat{\mu}_{i,b}^* - y_i/\lambda_i)^2,$$

where $\hat{\mu}_{i,b}^*$ is the OBP computed based on the bth bootstrap sample, $b \in \{1, \dots, B\}$, and B is the bootstrap sample size, which is set to be 200.

It is seen that the RMSPE estimates based on $\widehat{\text{MSPE}}$ are consistently larger than those based on $\widehat{\text{MSPE}}_1$. This is consistent with the results of an earlier simulation study (see Table 4.9), which show that $\widehat{\text{MSPE}}_1$ has a significantly larger negative bias compared to $\widehat{\text{MSPE}}$. As for $\widetilde{\text{MSPE}}$, there is no consistent pattern but the estimates seem to be more variable (for example, some value as high as 0.852). This is also consistent with the pattern observed in Table 4.9, which shows that $\widetilde{\text{MSPE}}$ has a much higher CV.

TABLE 4.11
Minnesota cancer data (counties 1–22), OBP and square roots of MSPE estimates.

County #	y_i	λ_i	OBP	$\sqrt{\widehat{\text{MSPE}}_1}$	$\sqrt{\widehat{\text{MSPE}}}$	$\sqrt{\widetilde{\text{MSPE}}}$
1	14	7	1.350	0.236	0.285	0.524
2	52	49	1.029	0.125	0.132	0.141
3	8	11	0.788	0.212	0.247	0.184
4	16	11	1.183	0.212	0.247	0.306
5	9	9	0.929	0.223	0.264	0.259
6	0	3	0.650	0.272	0.344	0.415
7	20	16	1.107	0.190	0.215	0.245
8	12	11	0.985	0.212	0.247	0.255
9	12	11	0.985	0.212	0.247	0.254
10	8	13	0.718	0.202	0.233	0.189
11	10	11	0.887	0.212	0.247	0.217
12	1	6	0.588	0.244	.297	0.252
13	8	10	0.829	0.217	0.255	0.215
14	11	15	0.782	0.193	0.221	0.199
15	1	4	0.677	0.262	0.326	0.280
16	0	2	0.708	0.284	0.364	0.495
17	3	6	0.719	0.244	0.297	0.223
18	14	20	0.751	0.176	0.197	0.172
19	51	59	0.864	0.115	0.121	0.112
20	1	5	0.629	0.252	0.311	0.279
21	14	12	1.033	0.207	0.240	0.256
22	8	8	0.925	0.229	0.274	0.263

Another observation is that, when the values of y_i (and, correspondingly, λ_i) are relatively large, such as for county 27, the RMSPE estimates are relatively small, and the values of the three RMSPE estimates are fairly close. This may be explained by the limiting behavior of a Poisson process; see, for example [113] (Theorem 10.4). In short, note that the CV of a random variable, $Y \sim \text{Poisson}(\lambda)$, is $1/\sqrt{\lambda}$, which goes to zero as λ goes to ∞. As a result, the OBP associate with a large y_i and λ_i is relatively more stable.

4.2.4 The TVSFP study

As an application OBP to unit-level data, we consider data from the Television School and Family Smoking Prevention and Cessation Project (TVSFP). For a complete description of the TVSFP study, see [93]. The original study was designed to test independent and combined effects of a school-based social-resistance curriculum and a television-based program in terms of tobacco use prevention and cessation. The subjects were seventh-grade students from Los Angeles (LA) and San Diego in the State of California in the United States.

TABLE 4.12
Minnesota cancer data (counties 23–44), OBP and square roots of MSPE estimates.

County #	y_i	λ_i	OBP	$\sqrt{\widehat{\text{MSPE}}_1}$	$\sqrt{\widehat{\text{MSPE}}}$	$\sqrt{\widetilde{\text{MSPE}}}$
23	9	9	0.929	0.223	0.264	0.237
24	15	14	0.987	0.198	0.226	0.225
25	10	16	0.712	0.190	0.215	0.166
26	3	3	0.895	0.272	0.344	0.313
27	319	298	1.064	0.054	0.055	0.060
28	4	7	0.736	0.236	0.285	0.232
29	9	7	1.043	0.236	0.285	0.335
30	8	8	0.925	0.229	0.274	0.240
31	25	16	1.305	0.190	0.215	0.298
32	4	5	0.839	0.252	0.311	0.280
33	9	5	1.189	0.252	0.311	0.543
34	12	14	0.859	0.198	0.226	0.217
35	0	3	0.650	0.272	0.344	0.408
36	2	6	0.654	0.244	0.297	0.233
37	6	4	1.053	0.262	0.326	0.471
38	8	5	1.119	0.252	0.311	0.435
39	4	2	1.062	0.284	0.364	0.852
40	18	9	1.421	0.223	0.264	0.477
41	1	4	0.677	0.262	0.326	0.294
42	7	9	0.820	0.223	0.264	0.226
43	6	11	0.690	0.212	0.247	0.179
44	1	2	0.797	0.284	0.364	0.300

The students were pretested in January 1986 in an initial study. The same students completed an immediate post-intervention questionnaire in April 1986, a one-year follow-up in April 1987, and a two-year follow-up in April 1988.

In their analysis, [128] considered a subset of the TVSFP data involving students from 28 LA schools, where the schools were randomized to one of four study conditions: (a) a social-resistance classroom curriculum (CC); (b) a media (television) intervention (TV); (c) a combination of CC and TV; and (d) a no-treatment control. A tobacco and health knowledge scale (THKS) score was one of the primary study outcome variables, and the one used for this analysis. The THKS consisted of seven questionnaire items used to assess student tobacco and health knowledge. A student's THKS score was defined as the sum of the items that the student answered correctly. Only data from the pretest and immediate post-intervention were available for the analysis.

More specifically, the data only involved subjects who had completed the THKS at both of these time points. On the one hand, the complete-record data set up an ideal "before-after" situation; on the other hand, the missing data, that is, those from subjects who had completed the questionnaire at only one time point, might have provided additional useful information. For example, it was possible that a subject did not complete the follow-up because he or she did not find the program helpful. Unfortunately, the incomplete data were not available. As a result, there is a potential risk of selection bias for the complete-record-only analysis. In all, there were 1,600 students from the 28 schools, with the number of students per school ranging from 18 to 137.

Ref. [93] carried out a mixed-model analysis based on a number of NER models to illustrate maximum likelihood estimation for the analysis of clustered data. [128] considered a different problem on estimating the small area means for the difference between the immediate post-intervention and pretest THKS scores (the response). Here, the "small area" is understood as a number of major characteristics (e.g., residential area, teacher/student ratio) that affect the response, but are not captured by the covariates in the model (i.e., linear combination of the CC, TV and CCTV indicators).

Note that, traditionally, the words "small areas" correspond to small geographical regions or sub-populations, for which adequate samples are not available, and such information as residential characteristics or teacher/student ratios would be used as additional covariates. However, such characteristic information are not available. Due to such as reason, [128] defined these unavailable information as "area-specific", so that they can be treated as the (small-area) random effects. This is consistent with the fundamental features of the random effects that are often used to capture unknown effects or information (e.g., [123]), and extends the traditional notion of SAE.

Thus, a small area is the seventh graders in all of the U.S. schools that share the similar major characteristics as a LA school involved in the data over a reasonable period of time (e.g., 5 years) so that these characteristics had not change much during the time, neither had the social/educational relevance of the CC and TV programs. There were 28 LA schools in the TVSFP data that correspond to 28 sets of characteristics, and the data were considered random samples from the 28 small areas defined as above. As such, each small area population is large enough so that $n_i/N_i \approx 0, 1 \le i \le 28$. Recall that the n_i's in the TVSFP sample range from 18 to 137, while the N_i's are expected to be at least tens of thousands. Note that the only place in the OBP where the knowledge of N_i is required is through the ratio n_i/N_i.

The proposed NER model can be expressed as (1.17) with

$$x_{ij}'\beta = \beta_0 + \beta_1 x_{i,1} + \beta_2 x_{i,2} + \beta_3 x_{i,1} x_{i,2},$$

where $x_{i,1} = 1$ if CC, and 0 otherwise; $x_{i,2} = 1$ if TV, and 0 otherwise. It follows that all the auxiliary data x_i are at the area level; as a result, the value of $\bar{X}_i$ is known for every i.

As noted, the sample sizes for some small areas are quite large, but there are also areas with relatively (much) smaller sample sizes. This is quite common in real-life problems. Because the auxiliary data are at area-level, one has $\bar{X}_i'\beta = \bar{x}_i'\beta$. It is easy to show that the BP (4.34) can be expressed as

$$\tilde{\theta}_i = \left\{ r_i + (1 - r_i)\frac{n_i\gamma}{1 + n_i\gamma} \right\} \bar{y}_i + \frac{1 - r_i}{1 + n_i\gamma}\bar{x}_i'\beta.$$

It is seen that, when n_i is large, the BP is approximately equal to $\bar{y}_i$, the design-based estimator, which has nothing to do with the parameter estimation. Thus, when n_i is large, there is not much difference between the OBP and the EBLUP. On the other hand, when n_i is small or moderate, one expects some difference between the OBP and the EBLUP. The simulation results in Section 4.1.4 showed that the difference between OBP and EBLUP in terms of the MSPE depends on to what extent the assumed model is misspecified.

It should be noted that the response, y_{ij}, is the difference in the THKS scores, and possible values of the THKS score are integers between 0 and 7. Clearly, such data is not normal. The potential impact of the non-normality is two-fold. On the one hand, it is likely that the NER model, as proposed by [93], is misspecified, in which case expression (4.34) is no longer the BP, and the Gaussian ML (REML) estimators are no longer the true ML (REML) estimators. On the other hand, even without the normality, (4.34) can still be justified as the best linear predictor (BLP; e.g., [209], sec. 7.3). Furthermore, the Gaussian ML (REML) estimators are consistent and asymptotically normal even without the normality assumption [104]; also see [123], ch. 1.

Other aspects of the NER model include homoscedasticity of the error variance across the small areas. Figure 4.11 shows the histogram of the sample variances of the 28 small areas. The bimodal shape of the histogram suggests potential heteroscedasticity in the error variance, yet another type of possible model misspecification.

Due to these concerns, the OBP method is naturally considered. Results of the OBP analysis for the 28 small areas are presented in Table 4.13. The BPE of the parameters are $\hat{\beta}_0 = 0.206$, $\hat{\beta}_1 = 0.687$, $\hat{\beta}_2 = 0.213$, $\hat{\beta}_3 = -0.288$, and $\hat{\gamma} = 0.003$. Although interpretation may be given for the parameter estimates, there is a concern about possible model misspecification (in which case the interpretation may not be meaningful), as noted earlier.

Regardless, here our main interest is prediction rather than interpretation; thus, we focus on the OBP. In addition to the OBPs, we also computed the corresponding $\widehat{\text{MSPE}}$, and their square roots as the measures of uncertainty. As a comparison, the EBLUPs for the small areas as well as the corresponding square roots of the MSPE estimates $\widetilde{\text{MSPE}}$, obtained using the Prasad-Rao method (P-R; [196]) are also included in the table.

It is seen that the OBPs are all positive, even for the small areas in the control group. As for the statistical significance (here "significance" is defined as that the OBP is greater in absolute value than 2 times the corresponding square root of the MSPE estimate), the small area means are significantly positive for all of the small areas in the (1,1) group. In contrast, none of the small area mean is significantly positive for the small areas in the (0,0) group. As for the other two groups, the small area means are significantly positive

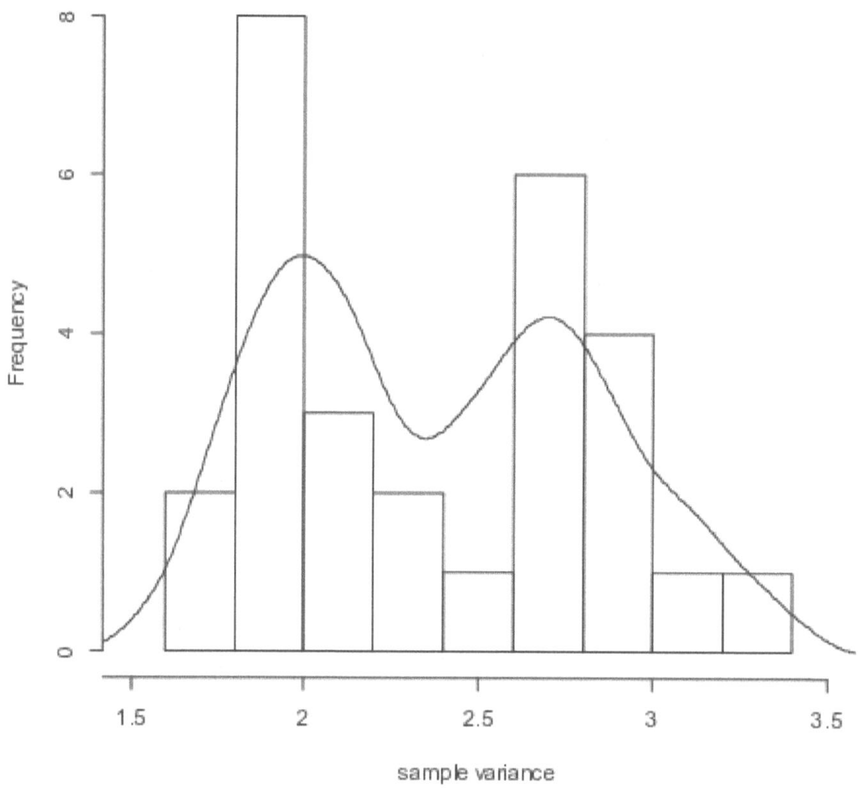

FIGURE 4.11
Histogram of sample variances and a kernel density smoother.

for all the small areas in the (1,0) group, and they are significantly positive for all but two small areas in the (0,1) group. There are 7, 8, 7 and 7 small areas in the (0,0), (0,1), (1,0) and (1,1) groups, respectively.

Comparing the OBP with the EBLUP, the values of the latter are generally higher, and their corresponding MSPE estimates are mostly lower. In terms of statistical significance, the EBLUP results are significant for the (1,1), (1,0) and (0,1) groups, and insignificant for the (0,0) group. It should be noted that the P-R MSPE estimator for the EBLUP is derived under the normality assumption, while in this case the data is clearly not normal, as noted earlier. Thus, the measure of uncertainty for the EBLUP may not be accurate. This is different from the case in Section 4.1.5 (for the Fay-Herriot model), where the model misspecification is only in the mean function. For example, just because the (square roots of the) MSPEs for the EBLUPs are lower, compared to those for the OBPs, it does not mean the corresponding true MSPEs for the EBLUPs are lower than those for the OBPs. It is also observed that the MSPE

TABLE 4.13
OBP, EBLUP, and measures of uncertainty for TVSFP data.

ID	CC	TV	OBP	$\sqrt{\widehat{\text{MSPE}}}$	EBLUP	$\sqrt{\widetilde{\text{MSPE}}}$
403	1	0	.886	.171	.913	.121
404	1	1	.844	.296	.856	.121
193	0	0	.215	.207	.217	.120
194	0	0	.221	.137	.221	.134
196	1	0	.878	.171	.907	.124
197	0	0	.225	.158	.223	.126
198	1	1	.771	.220	.807	.131
199	0	1	.426	.142	.453	.130
401	1	1	.826	.133	.844	.127
402	0	0	.188	.171	.199	.123
405	0	1	.394	.147	.432	.129
407	0	1	.508	.300	.508	.133
408	1	0	.871	.240	.903	.123
409	0	0	.230	.125	.227	.136
410	1	1	.778	.304	.813	.124
411	0	1	.409	.195	.444	.115
412	1	0	.913	.219	.930	.126
414	1	0	.929	.257	.941	.127
415	1	1	.869	.199	.872	.135
505	1	1	.790	.154	.818	.136
506	0	1	.389	.169	.428	.134
507	0	1	.426	.148	.452	.135
508	0	1	.411	.108	.442	.136
509	1	0	.915	.097	.929	.143
510	1	0	.880	.119	.905	.143
513	0	0	.185	.215	.197	.123
514	1	1	.866	.144	.870	.140
515	0	0	.180	.102	.192	.143

estimates for the EBLUPs are more homogeneous cross the small areas. This may be due to the fact that the P-R MSPE estimator for EBLUP is obtained assuming that the NER model is correct, while the proposed MSPE estimator for OBP does not rely on such an assumption.

In terms of the different methods of intervention, in spite of the potential difference in the small area characteristics, the CC and TV programs appears to be successful in terms of improving the students' THKS scores (whether the improved THKS score means improved tobacco use prevention and cessation is a different matter though). It also seems apparent that the CC program was relatively more effective than the TV program. Without the intervention of any of these programs, the THKS score did not seem to improve in terms of the small area means. In terms of the statistically significant results, when CC=0 and TV=0, the THKS score did not seem to improve; when CC=1, the THKS score seemed to improve; and, when CC=0 and TV=1, the improvement of the THKS score was not so convincing.

4.3 Extensions of OBP

Several methods have been developed in recent years that may be viewed as extensions of the OBP. In this section we discuss two of those extensions that are characteristic in some ways.

4.3.1 Observed best selective predictor

[218] proposed a method, called observed best selective prediction (OBSP), that combines covariate selection and estimation of parameters in a single operation, and does so by minimizing the observed MSPE.

The method is developed under the Fay-Herriot model. Suppose there are p candidate covariate, denoted by $1, \dots, p$, and J candidate models such that candidate Model j corresponds to a subset $I_j \subset \{1, \dots, p\}$ with $p_j = |I_j|$ ($|S|$ denotes the cardinality of set S), $j = 1, \dots, J$. To highlight the covariate selection, write the Fay-Herriot model, (1.16), under Model j as

$$y_i = x'_{i(j)}\beta_{(j)} + v_i + e_i, \tag{4.54}$$

$i = 1, \dots, m$, where $x_{i(j)} = (x_{ik})_{k \in I_j}$, $\beta_{(j)} = (\beta_k)_{k \in I_j}$, and everything else is the same as in the Fay-Herriot model.

As in Section 4.1.1, let us first assume that A is known. Let $\theta = (\theta_i)_{1 \le i \le m}$, and $\tilde{\theta}_{(j)} = [\tilde{\theta}_{i(j)}]_{1 \le i \le m}$ be the vector of BPs under Model j, that is, $\tilde{\theta}_{i(j)}$ is the BP of θ_i under Model j. Then, by (4.7), the BPE of $\beta_{(j)}$ is given by

$$\hat{\beta}_{(j)} = \{X'_{(j)}\Gamma^2 X_{(j)}\}^{-1} X'_{(j)}\Gamma^2 y, \tag{4.55}$$

where $\Gamma = \text{diag}\{D_i/(A + D_i), 1 \le i \le m\}$, and $X_{(j)} = [x'_{i(j)}]_{1 \le i \le m}$. The corresponding OBP for θ is given by

$$\hat{\theta}_{(j)} = y - \Gamma\{y - X_{(j)}\hat{\beta}_{(j)}\} = (I_m - \Gamma)y + \Gamma X_{(j)}\hat{\beta}_{(j)}. \tag{4.56}$$

Note that $\hat{\theta}_{(j)}$ is $\tilde{\theta}_{(j)}$ with $\beta_{(j)}$ replaced by $\hat{\beta}_{(j)}$. It is seen that the OBP is a weight average of the direct estimator, y (with weight $I_m - \Gamma$) and the regression estimator, $X_{(j)}\hat{\beta}_{(j)}$ (with weight Γ), under Model j.

To incorporate covariate selection into the OBP, a thurstoning expression can be derived for the MSPE of $\hat{\theta}_{(j)}$, that is,

$$\text{MSPE}\{\hat{\theta}_{(j)}\} = \text{E}\{|\hat{\theta}_{(j)} - \theta|^2\} = \text{E}\{C(j) + \cdots\} \tag{4.57}$$

(see [218]), where the expectation is not model-dependent, and

$$C(j) = \{y - X_{(j)}\hat{\beta}_{(j)}\}'\Gamma^2\{y - X_{(j)}\hat{\beta}_{(j)}\} + 2\mathrm{tr}\{M_{(j)}\} \tag{4.58}$$

with $M_{(j)} = P_{(j)}\Gamma D$, $P_{(j)} = \Gamma X_{(j)}\{X'_{(j)}\Gamma^2 X_{(j)}\}^{-1}X'_{(j)}\Gamma$, $D = \mathrm{diag}(D_i, 1 \le i \le m)$, and the $\cdots$ does not depend on j. Following the OBP idea, the next step is to find j, which corresponds to a subject I_j, that minimizes $C(j)$. Let

$$\hat{j} = \mathrm{argmin}_{1\le j\le J}C(j). \tag{4.59}$$

Let $X_{\mathrm{f}} = (x'_i)_{1\le i\le m}$, where $x_i = (x_{ik})_{1\le k\le p}$, that is, the covariate matrix corresponding to the full set of candidate covariates. For any $1 \le j \le J$, define $\hat{\beta}_{[j]} = (\hat{\beta}_{[j]k})_{1\le k\le p}$, where

$$\hat{\beta}_{[j]k} = \begin{cases} \hat{\beta}_{(j)k}, & \text{if } k \in I_j, \\ 0, & \text{otherwise.} \end{cases}$$

The observed best selective predictor (OBSP) is then given by [see (4.56)]

$$\hat{\theta} = (I_m - \Gamma)y + \Gamma X_{(\hat{j})}\hat{\beta}_{(\hat{j})} = (I_m - \Gamma)y + \Gamma X_{\mathrm{f}}\hat{\beta}_{[\hat{j}]}. \tag{4.60}$$

The estimator $\hat{\beta}_{[\hat{j}]}$ may be called the best selective predictive estimator (BSPE), although [218] has given it a different name.

Note. The covariates selection method, (4.59), is similar to the predictive model selection idea proposed by [122] (sec. 6.4). Also see §6.4 in the sequel.

So far we have assumed that A is known. In practice, A is unknown. Ref. [218] suggests to estimate A using maximum likelihood (ML) or REML under the full model, that is, $X = X_{\mathrm{f}}$. Other methods of estimation for A that the authors suggested include the Prasad-Rao estimator [196], the Fay-Herriot estimator [59], and the BPE [126].

Note. It appears that one could extend the derivation above to estimate β, A and select the covariates altogether using the OBP idea. To do so, note that, by the derivation in [218], one can obtain a full thurstoning expression of (4.57) by bringing back the $\cdots$, that is,

$$\begin{aligned}\mathrm{MSPE}\{\hat{\theta}_{(j)}\} &= \mathrm{E}\{|\hat{\theta}_{(j)} - \theta|^2\} \\ &= \mathrm{E}[\{y - X_{(j)}\hat{\beta}_{(j)}\}'\Gamma^2\{y - X_{(j)}\hat{\beta}_{(j)}\} \\ &\quad +\mathrm{tr}\{P_{(j)}^{\perp}\Gamma D\} + \mathrm{tr}(D)],\end{aligned} \tag{4.61}$$

where $P_{(j)}^{\perp} = I_m - P_{(j)}$. One can estimate A and select j by minimizing the expression inside the expectation on the right side of (4.61), which is equivalent to minimizing

$$C(A, j) = \{y - X_{(j)}\hat{\beta}_{(j)}\}'\Gamma^2\{y - X_{(j)}\hat{\beta}_{(j)}\} + \mathrm{tr}\{P_{(j)}^{\perp}\Gamma D\}. \tag{4.62}$$

This can be done by first minimizing $C(A, j)$ with respect to A for fixed j, leading to $\hat{A}_{(j)}$, then minimizing $C(\hat{A}_{(j)}, j)$ with respect to $1 \le j \le J$.

Again, let the latter minimizer be $\hat{j}$. The BSPE of β is then given by $\hat{\beta}_{(\hat{j})}$, or $\hat{\beta}_{[\hat{j}]}$ according to the definition above, with A replaced by $\hat{A}_{(\hat{j})}$. Denote this estimator of β again by $\hat{\beta}_{(\hat{j})}$ or $\hat{\beta}_{[\hat{j}]}$. The OBSP is then given by (4.60) with the A in Γ replaced by $\hat{A}_{(\hat{j})}$. In [218] (discussion at the end of the paper), the authors suggested that they did not estimate A this way due to a concern that such an approach may be unstable for covariate selection. The issue may need further investigation.

The authors of [218] showed that, when the true underlying model is within the scope of the candidate models, that is, $\mu_i = x'_{i(j)}\beta_{(j)}$ for some $1 \leq j \leq J$ and some $\beta_{(j)}$, the proposed OBSP method is consistent in the sense that

$$\max_{1 \leq k \leq p} \mathrm{E}\left(\hat{\beta}_{[\hat{j}]k} - \beta^*_k\right)^2 \to 0,$$

where β^* is the true β such that $\beta^*_k = 0$ if $k \notin I$ and I is the smallest subset containing all nonzero β coefficients under the true model.

In terms of measure of uncertainty, [218] proposed a parametric bootstrap estimator of the MSPE following [19]. The authors showed that, again, when the true underlying model is within the scope of the candidate models, the parametric bootstrap MSPE estimator is second-order unbiased. It should be noted that assuming that the true underlying model is within the scope is a serious assumption, which may not hold in practice. It would be more desirable to develop an uncertainty measure beyond such an assumption; see §4.1.5.

On the other hand, the assumption that the true model is within the scope of candidate models makes it easier to study theoretical properties, such as consistency and second-order unbiasedness of the MSPE estimator.

Simulation studies carried out in [218] compared the predictive performance of the proposed OBSP with several competitors, including the OBP based on the full model, and OBP based on covariates selected by other model selection procedures, namely the AIC [4] and BIC [208], denoted by OBP-AIC and OBP-BIC, respectively. For the OBSP, the authors considered two ways of estimating A, namely REML or BPE, both based on the full model. The two OBSP methods are denoted by OBSP-REML and OBSP-OBP.

The data were simulated either under an underlying model that is within the scope of the candidate models, or underlying models that are outside the scope of the candidate models, including nonlinear mean functions or non-Gaussian random effects and/or errors. The results showed that, when the underlying model is within the scope of the candidate models, OBP-BIC performed the best; when the underlying model is outside the scope of the candidate models, OBSP-REML performs the best, overall.

The authors of [218] also ran simulation studies to investigate the empirical performance of the proposed parametric bootstrap MSPE estimator under a simulation setting such that that the true model is within the scope of the candidate models. The simulation results showed satisfactory performance of the MSPE estimator in terms of %RB and RMSE. It is unclear, however, how the parametric bootstrap MSPE estimator performs when the true underlying model is outside the scope of the candidate models, such as under some of the scenarios described in the last paragraph.

The method was illustrated using data from a family income and expenditure survey (FIES) in Japan. The data were available for $m = 47$ prefectures based on the FIES in November 2015, which were considered as the small areas. As is well known, income data distribution is typically skewed. Here, the area means of household expenditures on education, with the sample size around 100 per area, was considered, which is more normal-look-like due to the central limit theorem. Still, the logarithm of the mean expenditure on education, in unit of 1,000 Japanese Yen, was used as the outcome variable.

The auxiliary information include data from the National Survey of Family Income and Expenditure (NSFIE) and the population census. The NSFIE areal mean of household expenditure on education is based on larger sample size, and therefore more reliable than FIES. But the NSFIE is conducted only once every five years. Thus, the NSFIE is not accurate enough, but it still provides useful auxiliary information. Specifically, the following variables from NSFIE in 2014 are used: the logarithm of the reported areal mean of household expenditure on education per month based on NSFIE (edc), the logarithm of the population size based on the population census (pop), the proportion of the people under 15 based on the population census (young), and the proportion of the labor force based on the population census (labor). The sampling variances, $D_1, \dots, D_m$, were calculated as the sample variances of the logarithm of the reported areal mean of household expenditure on education in November in FIES for the past 15 years.

The candidate models under consideration were all subsets of the above auxiliary variables including at least an intercept. The authors of [218] applied three methods: OBSP-REML, OBP-AIC, and OBP-BIC to the data. OBP-AIC and OBP-BIC only selected the edc auxiliary variable, while OBSP-REML selected edc plus the young auxiliary variable. This seems to be reasonable, because information-criteria (AIC or BIC) prioritize interpretation in selecting the covariates while the selection criterion embedded in OBSP focuses on prediction. Typically, information criteria tend to select a model that is parsimonious, while a prediction-oriented selection criterion is more in favor of a "larger" model, that is, a model that includes more variables. See, for example, Section 6.4 of [213].

Another issue involved here was prediction after model selection. In a way, the MSPE estimator for the OBSP has taken into consideration of the additional variation due to variable selection. The question is what happened in this real-data application. To see this, [218] compared their bootstrap MSPE estimates for the OBSP with another approach, in which the bootstrap samples were drawn based on the variables selected by OBSP from the original data. In other words, in each bootstrap iteration, the selected model from the original data was simply applied to the bootstrap samples without selecting a suitable model from the bootstrap samples. The latter approach resulted about 35% drop in MSPE estimates. This issue was also noted by [117].

The issue about variable selection, or more generally model selection, will be discussed more systematically in Chapter 6.

4.3.2 Compromised best predictor

Recall Theorem 4.1 in Section 4.1.1, which shows that the MSPE of an EBP can be decomposed as a constant plus two other terms that depend on the specification of the EBP. The best performer for these two terms are the OBP and BLUP, respectively. Inspired by the result, some may conjecture that somewhere in the middle between the OBP and the BLUP lies a predictor who does better, in terms of the MSPE, than both the OBP and the BLUP.

It turns out this conjecture is indeed true. In a recent work, [95] considered a convex combination, or weighted average, of the BLUP and OBP. The authors showed that, by carefully choosing the weights, in a data-driven manner, the resulting predictor, called compromised best predictor (CBP), preserves the robustness of the OBP while retaining the efficiency of the EBLUP when the underlying regression model is correctly specified. More specifically, the construction of the weight can adapt to varying degree of mean function misspecification so that the resulting parameter estimator is automatically closer to the MLE when the model is well specified, but it will more closely resemble the BPE when there is substantial model misspecification.

Ref. [95] develops the CBP method mainly under the Fay-Herriot model, although the authors also consider an extension under a similar framework of Section 4.1.1, and an application of estimating an average of small area means. An extension to the NER model was described in the supplement of [95].

The key idea comes, again, from Theorem 4.1, where W is assumed to be an diagonal matrix. Note that, in (4.13), $\tilde{\zeta}$ depends on the given weights, w_i, $1 \leq i \leq m$, on the diagonal of W, thus denoting $\tilde{\zeta} = \tilde{\zeta}_w$, where $w = (w_i)_{1 \leq i \leq m}$. While OBP uses a thurstoning expression for the MSPE in (4.6), CBP uses a similar thurstoning expression for the MSPE in (4.13). In fact, the authors showed that, for fixed w, A, the MSPE in (4.13) can be expressed as

$$\text{MSPE}(\tilde{\zeta}_w) = \text{E}\{\widetilde{\text{MSPE}}(w, A)\}, \tag{4.63}$$

where, with $D = \text{diag}(D_i, 1 \leq i \leq m)$, $B_A = \text{diag}(B_{A,i}, 1 \leq i \leq m)$, $B_{A,i} = D_i/(A + D_i)$, $1 \leq i \leq m$, and $U_{w,A} = B_A\{X(X'WX)^{-1}X'W - I_m\}$, we have

$$\widetilde{\text{MSPE}}(w, A) = y'U'_{w,A}U_{w,A}y + 2\text{tr}(U_{w,A}D) + \text{tr}(D). \tag{4.64}$$

The expression of $\widetilde{\text{MSPE}}(w, A)$ is similar to the expression inside the expectation on the right side of (4.6)—both are quadratic functions of y. Thus, $\widetilde{\text{MSPE}}(w, A)$ is also an observed MSPE. The difference is that the observed MSPE in (4.6) is a function of β and A, while the observed MSPE in (4.64) is a function of w and A.

Following the OBP idea, and assuming A is known, for now, the goal would be to choose the optimal weights, w, that minimizes the observed MSPE in (4.64). This involves optimization over m different weights, which could be a daunting task. To simplify the mission, [95] considers choosing a single weight, $\lambda \in (0,1)$, for a convex combination of the MLE weights and the BPE weights. The choice is well motivated in view of Theorem 4.1.

Recall the expressions of the BPE, (4.9), and MLE, (4.11), under the Fay-Herriot model, assuming that A is known. Thus, the BPE, MLE weights are

$$w_{\mathrm{BPE},i}(A) = \frac{B_{A,i}^2}{\sum_{i=1}^m B_{A,i}^2}, \quad 1 \le i \le m; \tag{4.65}$$

$$w_{\mathrm{MLE},i}(A) = \frac{(A+D_i)^{-1}}{\sum_{i=1}^m (A+D_i)^{-1}}, \quad 1 \le i \le m, \tag{4.66}$$

respectively, where $B_{A,i}$ is defined below (4.63). Note that both weights are functions of A. The compromise weights are therefore given by

$$\begin{aligned} w_{\mathrm{c},i}(\lambda, A) &= \lambda w_{\mathrm{MLE},i}(A) + (1-\lambda) w_{\mathrm{BPE},i}(A) \\ &= \frac{\lambda B_{A,i}^2}{\sum_{i=1}^m B_{A,i}^2} + \frac{(1-\lambda)(A+D_i)^{-1}}{\sum_{i=1}^m (A+D_i)^{-1}}, \end{aligned} \tag{4.67}$$

$1 \le i \le m$, where the subscript c refers to "compromise".

So, instead of finding the vector of weights, w, that minimizes the observed MSPE, (4.64), over all possible weights $w_i, 1 \le i \le m$, now one only needs to finding a single weight, λ, that minimizes (4.64) with $w = w_{\mathrm{c}}(\lambda, A) = [w_{\mathrm{c},i}(\lambda, A)]_{1 \le i \le m}$. Furthermore, the variance A can also get involved, so the minimization is over $\lambda \in (0,1)$ and $A \ge 0$. Denote the minimizer by $(\hat{\lambda}, \hat{A})$.

Once the optimal weights are determined, the β vector is estimated by the WLS estimator, (4.12), which is denoted by $\hat{\beta}$. The EBP of the small-area mean, $\zeta_i = \mu_i + v_i$, is obtained by replacing β, A in the BP formula, (4.10), by $\hat{\beta}, \hat{A}$, respectively. The result is the CBP, denoted by $\hat{\zeta}_{\mathrm{CBP}}$.

In terms of theoretical properties of the CBP, [95] studied behavior of the CBP in terms of the asymptotic risk. For a predictor, $\tilde{\zeta} = (\tilde{\zeta}_i)_{1 \le i \le m}$, of $\zeta = (\zeta_i)_{1 \le i \le m}$, a loss function is defined as

$$\mathcal{L}_m(\tilde{\zeta}, \zeta) = \frac{1}{m} \sum_{i=1}^m (\tilde{\zeta}_i - \zeta_i)^2. \tag{4.68}$$

The risk is defined as $R_m(\tilde{\zeta}) = \mathrm{E}\{\mathcal{L}_m(\tilde{\zeta}, \zeta)\} = \mathrm{MSPE}(\tilde{\zeta})/m$.

Consider an "oracle predictor" of ζ who actually observed ζ. This person (if you wish) does not need to predict, because he knows it already; nevertheless, he wants to play for fun by considering the BP, (4.10), where β is replaced by (4.12) with $W = \mathrm{diag}(w_i, 1 \le i \le m)$ and w_i given by the right side of (4.67). This way, (4.10) is a function of λ and A, denoted by $\tilde{\zeta}_i(\lambda, A)$, $1 \le i \le m$.

The oracle predictor can choose the optimal λ, A by minimizing the loss of $\tilde{\zeta}(\lambda, A) = [\tilde{\zeta}_i(\lambda, A)]_{1 \le i \le m}$, that is,

$$\frac{1}{m}\sum_{i=1}^{m}\{\tilde{\zeta}_i(\lambda, A) - \zeta_i\}^2 \tag{4.69}$$

over λ, A (he can do so because he observes all the ζ_i's). Clearly, no one beats this oracle because he gets the "insider" of knowing the ζ_i's. However, [95] showed that, in spite of not knowing the ζ_i's, the CBP has asymptotically the same risk as the oracle predictor. More precisely, let $\lambda^{\text{o}}, A^{\text{o}}$ denote the minimizer of (4.69). The oracle predictor is then $\tilde{\zeta}^{\text{o}} = \tilde{\zeta}(\lambda^{\text{o}}, A^{\text{o}})$. Ref. [95] showed that, under certain regularity conditions, one has

$$\lim_{m\to\infty}\{R_m(\hat{\zeta}_{\text{CBP}}) - R_m(\tilde{\zeta}^{\text{o}})\} = 0. \tag{4.70}$$

In addition to the CBP, [95] also proposed a plug-in CBP. This is defined by modifying (simplifying) the CBP procedure. Instead of minimizing (4.64), with w_i given by (4.67), over λ, A, the A in the MLE weight (4.66) is replaced by its REML estimator, and the A in the BPE weight (4.65) is replaced by its BPE. This way, one only needs to minimize (4.64) with respect to λ. Simulation results (see below) showed that the plug-in CBP has better empirical performance than the CBP. However, no theoretical justification similar to (4.70) is established for the plug-in CBP.

Ref. [95] conducted a series of simulation studies, in which the predictive performance of CBP, plug-in CBP was compared with OBP and different versions of EBLUP. Below we summarize the results under different scenarios, in which the underlying model is either correctly specified, or having varying degrees of model misspecification, from minor to moderate and to severe. When the underlying model is correctly specified or only slightly misspecified, the EBLUP methods outperformed the OBP-type methods, including the OBP, CBP and plug-in CBP. On the other hand, when the underlying model is at least moderately misspecified, the OBP-type predictors outperformed the EBLUPs. Within the OBP-type methods, the plug-in CBP uniformly performed the best among the three in all scenarios; the CBP performed better than the OBP except in some cases of severe model misspecification. Overall, the simulation results suggested that the CBP (including the plug-in CBP) methods offered robust compromises between the EBLUP and OBP.

The authors of [95] also discussed an application of CBP and plug-in CBP to estimation of normative gait speed in older adults using data from the National Health and Aging Trends Study (NHATS). The study focused on older people, that is, 65 or older. The gait speed is a simple measure of a person's ability to walk, which, in turn, is viewed as an important measure of the aging process. Specifically, it is measured as the time it took for a person to walk 3 meters at "a usual pace" of the person. Two measures were taken and the average was reported as the gait speed, for each person.

The aim was to estimate the average gait speed within key demographic groups recorded by NHATS. In all, there were 48 small areas formed by the demographic characteristics of sex (male and female), race (white non-Hispanic, black, Hispanic, and other), and age (65–69, 70– 74, 75–79, 80–84, 85–89, 90+). For each small area, the average gait speed of sampled older adults, with the sample size n_i for the ith small area, was used as the direct estimate of the mean gait speed, y_i, $i = 1, \dots, 48$. The associated sample variance, D_i, was computed as s_i^2/n_i, where s_i is the sample s.d. for the i small area.

The authors also considered a smaller subset of the small areas by excluding those that contained the "other" race category and the 4 areas where the race was "Hispanic" and the age categories were either 85–89 or 90+. There were 32 small areas left after the exclusion.

The only available auxiliary information, suggested by the existing literature [14], was the person's height. This was defined as $x_i = u_i - 65$, where u_i was the average height for the ith small area (in inches). It is unlikely that the mean gait speed can be modeled linearly by the single auxiliary variable. Thus, naturally, the CBP method was considered due to robustness to model misspecification. In addition to the CBP and plug-in CBP, the OBP, EBLUP-REML, regression-REML, and direct estimates were also computed for comparison. Here, regression-REML only uses $x_i'\hat{\beta}$ as the predictor, where $\hat{\beta}$ is the REML estimator of β, that is, (4.12) with $W = \hat{V}^{-1}$ and $\hat{V} = \hat{A}I_m + D$, $\hat{A}$ being the REML estimator.

The 48-area analysis did not show difference between the CBP, plug-in CBP and EBLUP-REML. In fact, in this case, the estimated weight, $\hat{\lambda}$, for CBP is 1 and that for the plug-in CBP is 0.999. As a result, there was no "compromise" because (almost) all of the weight was given to the MLE [see (4.67)]. On the other hand, the 32-area analysis showed more interesting differences. In the case, the $\hat{\lambda}$ for the CBP is 0.635 and that for the plug-in CBP is 0.531. Thus, in the latter case, the compromise was working.

What was apparently lacking was measures of uncertainty for the CBP, or plug-in CBP. Statistics, or statistical data science as it is often called these days, is not just for producing a bunch of numbers as estimates. It is also about providing measures of uncertainty. Here, similar to Section 4.1.5, the uncertainty measure needs to be developed under the prospect of model misspecification, which makes it a more challenging problem. Given the work reported in Section 4.1.5 (under the Fay-Herriot model), we expect follow-up studies to fill this gap.

5

More Flexible Models

It is not uncommon that the investigator has the flexibility of choosing the assumed model for the real-life application. Here, the flexibility includes using a model that links the outcome and predictors in a more complex way. The more complex model may be treated as an extension of a simpler model, such as a linear model, in the sense that, in case the simple model fails, the more complex model is still valid. An SAE method developed under such a complex model is less likely to suffer from model failure; in this way, robustness against model misspecification is achieved, at least to some extent. In this chapter, we discuss several types of the more complex models, including semi-parametric, non-parametric and functional models.

5.1 Preliminaries

1. Approximating functions. A subfield of mathematics is called approximation theory. Basically, it is about how to approximate a "well behaved" function by a class of functions that are linear combinations of some simpler functions, called base functions.

For example, a well-known theorem in measure theory states that a function that is Lebesgue measurable can be approximated by a sequence of simple functions. Here, a simple function is a linear combination of indicator functions of sets that are Borel measurable.

Another example of the approximation theory is the polynomial approximation. It states that any continuous function, $f(x)$, can be approximated by a sequence of polynomials, $P_s(x)$. A polynomial is a linear combinations of simple functions like 1 (a constant is also viewed as a function), x, x^2, etc. such that $P_s(x) = a_0 + a_1 x + \cdots + a_s x^s$. Here, s is the degree of the polynomial and $a_j, 0 \leq j \leq s$ are coefficients of the polynomial such that $a_s \neq 0$. The approximation becomes accurate when $s \to \infty$.

Such an approximation can be rephrased in a way of infinite series of simple functions. For example, $f(x) = P_\infty(x) = \sum_{j=0}^{\infty} a_j x^j$, where $a_0, a_1, \ldots$ are coefficients of the infinite series. With this expression, even though $f(x)$ is not a polynomial, it is "reachable" within the scope of polynomial approximations in the sense that it may be viewed as a polynomial whose order is infinity.

DOI: 10.1201/9781003395171-5

The latter form of the approximation is closely related to a more advanced topic called functional analysis, in which the simple functions, such as those in the polynomial approximation, are bases of a Hilbert space (e.g.,[197]).

Much of the approximation theory has its multivariate extension. For example, a bi-variate polynomial of second degree may be expressed as

$$P_2(x_1, x_2) = a_{0,0} + a_{1,0}x_1 + a_{0,1}x_2 + a_{2,0}x_1^2 + a_{0,2}x_2^2 + a_{1,1}x_1x_2,$$

where $a_{0,0}, a_{1,0}$, etc. are coefficients.

One particular type of approximation, which arguably has made more impact in applied statistics than any other type of approximation, is spline approximation. See, for example, [237] (ch. 5). Again, let us consider the univariate case for illustration. An approximating spline is in the form of

$$\tilde{f}(x) = \beta_0 + \beta_1 x + \cdots + \beta_p x^p + \gamma_1(x - \kappa_1)_+^p + \cdots + \gamma_q(x - \kappa_q)_+^p, \quad (5.1)$$

where p is the degree of the spline, q is the number of knots, κ_j, $1 \le j \le q$ are the knots, and $x_+ = x1_{(x>0)}$. Graphically, a P-spline is pieces of (pth degree) polynomials smoothly connected at the knots. The degree of the spline, p, is typically not high. For example, linear spline ($p = 1$) and cubic spline ($p = 3$) are most widely used.

On the other hand, an important issue in spline approximation is how to control the number of knots involved. Theoretically, with a sufficiently large number of knots, one can approximate a function with any degree of accuracy. But approximation accuracy is not the only thing one cares about.

Often times, one would like the approximating function to have a certain degree of smoothness; otherwise, the approximating function is too wiggly that it loses interpretability, which is sometimes desirable. Having too many knots does not help in this regard; it can also cause over-fitting the data. Note that, even if one fits today's data perfectly, tomorrow is a different day, and one may have a different set of data that defeats the "perfect fit".

One way to maintain the smoothness is via a smoothing penalty. Let (x_i, y_i), $i = 1, \dots, n$ be a set of paired data, and one is looking for a function, $f(x)$, that fits the data by minimizing

$$\sum_{i=1}^{n} \{y_i - f(x_i)\}^2 + \lambda \int \{f''(x)\}^2 dx. \quad (5.2)$$

Here, $f(\cdot)$ is assumed to be twice differentiable, and $\lambda \ge 0$ is a smoothing parameter that controls the trade-off between fitting the data and roughness of the function. It is known that the minimizer to (5.2) is a cubic spline (e.g., [80]). In fact, (5.2) has motivated practical implementation of spline approximations; see below.

It is known that all spline functions can be expressed as linear combinations of the so-called basic splines, or B-splines. For a given set of knots, say, $\kappa_1 < \cdots < \kappa_q$, a B-spline of order p is a function that is non-zero over an interval, say, $[\kappa_j, \kappa_{j+p+1})$, and zero elsewhere. Specifically, let $B_{j,p}(x)$ denote the j B-spline of order p. Then, we have $B_{j,0}(x) = 1_{(\kappa_j \le x < \kappa_{j+1})}$, $1 \le j \le q - 1$. The higher-order B-Splines can be defined recursively as

$$B_{j,p}(x) = \frac{x - \kappa_j}{\kappa_{j+p} - \kappa_j} B_{j,p-1}(x) + \frac{\kappa_{j+p+1} - x}{\kappa_{j+p+1} - \kappa_{j+1}} B_{j+1,p-1}(x),$$

$1 \le j \le q - p - 1$.

2. Gaussian processes. A stochastic process, $X(t), t \in T$, where T is a continuous subset of R, the real line, is an extension of time series, in which case T is a subset of discrete times. In other words, a stochastic process is a random function that the curve of the function depends on a chance process. A stochastic process, $G(t), t \in T$ is called a Gaussian process if for any collections of points in T, $t_1 < \cdots < t_k$, $[G(t_1), \dots, G(t_k)]'$ has a multivariate normal distribution.

A Wiener process is a Gaussian process such that for any $0 \le t_1 < \cdots < t_k \le 1$, the random variables $W(t_2) - W(t_1), \dots, W(t_k) - W(t_{k-1})$ are independent (this property is called independent increments) and distributed as $N(0, t_2 - t_1), \dots, N(0, t_k - t_{k-1})$, respectively.

The Wiener process is named after Norbert Wiener, who received his Ph.D. at the age of 18. In a series of papers published in 1918, Wiener defined what is called the Brownian motion, $B(t), t \ge 0$, that is, a stochastic process that satisfies the following conditions:
(i) $B(t)$ has independent and stationary increments [i.e., the increments are independent (see above) and the joint distribution of the increments does not change under a shift of time];
(ii) $B(t) \sim N(0, \sigma^2 t)$ for any $t > 0$, where σ^2 is a positive constant;
(iii) with probability one, $B(t)$ is a continuous function of t.
It follows that a Brownian motion is a Gaussian process. It is, however, more than just a Gaussian process, because it is a continuous function of t with probability one. A Gaussian process, in general, is not guaranteed continuous with probability one (e.g., [3]). The conditional stochastic process $B(t), 0 \le t \le 1$, given $B(1) = 0$ is called the *Brownian bridge* (or tied-down Brownian motion). Another way of defining the Brownian bridge is by the process $U(t) = B(t) - tB(1), 0 \le t \le 1$. It can be shown that the Brownian bridge is a Gaussian process with mean 0 and covariances $\text{cov}\{U(s), U(t)\} = s(1 - t)$, $s \le t$.

Gaussian distribution, or normal distribution, is the limit of convergence in distribution of sequence of random variables in many cases, according to the central limit theorem. Likewise, Gaussian process, especially the Brownian motion, is often associated with the limiting process of a sequence of stochastic processes, such as the empirical processes, in terms of convergence in distribution in a functional space. See, for example, Chapter 7 of [113].

5.2 Non-parametric methods for SAE

Non-parametric/semi-parametric models in SAE has received much attention since the paper of Opsomer et al. [188]. The latter proposed non-parametric

SAE models for both area and unit-level data. For simplicity, we mainly focus on area-level models for illustration, which can be expressed as

$$y_i = f(x_i) + v_i + e_i, \quad i = 1, \dots, n, \tag{5.3}$$

where $f(\cdot)$ is an unknown function, and everything else is the same as the standard Fay-Herriot model, (1.16). Clearly, the function $f(\cdot)$ is what distinguishes (5.3) from the Fay-Herriot model, which may be viewed as a special case with $f(x) = x'\beta$ for some vector of unknown parameters, β. In the latter case, f is unknown only up to β. In other words, the form of f is known. It is clear that a non-parametric model offers more flexibility than the Fay-Herriot model in modeling the mean function of the outcome.

On the other hand, it is more difficult to make inference about the model if f is completely unknown. A common strategy in practice is to approximate f by a function that can be chosen from a rich family of parametric functions. One such approach is called P-spline. For simplicity, consider the case of univariate x. Then, a P-spline approximation of $f(x)$ is given by (5.1). One then focuses on inference about the approximating model, which is (5.3) with the $f(x_i)$ replaced by its spline approximation evaluated at x_i.

To explain how the approximating model is fitted, let us first consider a simpler case, that is, (5.3) without v_i, the area-specific random effect. In this case, the model can be fitted by penalized least squares (PLS)—that is, by minimizing

$$|y - X\beta - Z\gamma|^2 + \lambda|\gamma|^2, \tag{5.4}$$

with respect to β and γ, where $y = (y_i)_{1\le i\le n}$, the ith row of X is $(1, x_i, \dots, x_i^p)$, the ith row of Z is $[(x_i - \kappa_1)_+^p, \dots, (x_i - \kappa_q)_+^p]$, $1 \le i \le n$, and λ is a penalty, or smoothing, parameter. The PLS is motivated by the smoothing spline penalizer (5.2). The question is how to choose λ.

To determine λ, [232] used the following interesting connection to a linear mixed model (LMM). Suppose that the e_i's are distributed as $N(0, \tau^2)$. Then if the γ's are treated as independent random effects with the distribution $N(0, \sigma^2)$, the minimizer of (5.4) is the same as the best linear unbiased estimator (BLUE) of β and BLUP of γ, provided that λ is identical to the ratio τ^2/σ^2. Thus, the P-spline model is fitted the same way as fitting the following LMM (e.g., [123])

$$y = X\beta + Z\gamma + e. \tag{5.5}$$

It should be pointed out that the connection is asymptotically valid only if the true underlying function f is not a P-spline. To see this, consider the following example.

Example 5.1. Suppose that the true underlying function is a quadratic spline with two knots, expressed as

$$f(x) = 1 - x + x^2 - 2(x-1)_+^2 + 2(x-2)_+^2, \quad 0 \le x \le 3. \tag{5.6}$$

The shape of the function is a half-circle between 0 and 1 facing up, a half-circle between 1 and 2 facing down, and a half-circle between 2 and 3 facing up. Note that this function is smooth in that it has a continuous derivative. Obviously, the best approximating spline is f itself, for which $q = 2$. However, if one uses the above LMM connection, the ML (or REML) estimator of σ^2 is consistent only if $q \to \infty$, that is, the number of appearances of the spline random effects goes to infinity. This can be seen from the asymptotic theory (e.g., [104]), but, intuitively, it also makes sense even without the theory.

Nevertheless, in most applications of P-splines, the unknown function f is unlikely to be a P-spline, so, in a way, (5.1) is only used as an approximation, with the approximation error vanishing as $q \to \infty$. We shall focus on this more practical situation from now on.

A good thing about the LMM connection is that it can easily incorporate random effects already existing in the model, such as the v_i in (5.3). More generally, one may replace the v_i in (5.3) by $w_i'v$, where w_i is a known vector and $v = (v_i)_{1\le i\le m}$. Then, with the spline approximation described above, the LMM (5.5) is extended to

$$y = X\beta + Z\gamma + Wv + e, \tag{5.7}$$

where the ith row of W is w_i', $1 \le i \le m$. Using the spline–mixed-model connection, one looks to fit the LMM (5.7) under the assumption that γ, v, and ϵ are uncorrelated with means 0 and covariance matrices Σ_γ, Σ_v, and Σ_ϵ, respectively. The BLUE and BLUP are given by

$$\begin{aligned}
\tilde{\beta} &= (X'V^{-1}X)^{-1}X'V^{-1}y, \\
\tilde{\gamma} &= \Sigma_\gamma Z'V^{-1}(y - X\tilde{\beta}), \\
\tilde{v} &= \Sigma_v W'V^{-1}(y - X\tilde{\beta}),
\end{aligned}$$

where $V = \mathrm{Var}(y) = Z\Sigma_\gamma Z' + W\Sigma_v W' + \Sigma_\epsilon$ (e.g., [123], §2.3). The EBLUE and EBLUP, denoted by $\hat{\beta}$, $\hat{\gamma}$, and $\hat{v}$, are obtained by replacing V by $\hat{V}$ in the corresponding expressions. Here, we assume that $V = V(\theta)$, where θ is a vector of variance components, so that $\hat{V} = V(\hat{\theta})$. The REML estimator is used for $\hat{\theta}$, as suggested by [188].

The non-parametric method can be extended to the NER model for unit-level data. The model can be extended to the finite populations of small areas under a super-population model, which treats the finite population as a random sample from a super-population that satisfies

$$Y_{ij} = X_{ij}'\beta + X_{ij}'\gamma + v_i + \epsilon_{ij}, \tag{5.8}$$

$j = 1, \dots, N_i$, where N_i is the population size for the ith small area or sub-population. The term $X_{ij}'\beta + Z_{ij}'\gamma$ arises as the result of a similar spline approximation to $f(x_{ij})$, with $X_{ij} = (1, x_{ij}, \dots, x_{ij}^p)'$ and

$$Z_{ij} = [(x_{ij} - \kappa_1)_+^p, \dots, (x_{ij} - \kappa_q)_+^p]'.$$

Again, the v_i in (5.8) can be extended to $W_{ij}'v$, where W_{ij}' is a known vector, although in most SAE applications one has $W_{ij}'v = v_i$; we shall focus on this case in the sequel. The population mean for the ith small area is approximately equal to the mixed effect

$$\theta_i = \bar{X}_{i,\mathrm{P}}'\beta + \bar{Z}_{i,\mathrm{P}}'\gamma + v_i, \tag{5.9}$$

where $\bar{X}_{i,\mathrm{P}}$ and $\bar{Z}_{i,\mathrm{P}}$ are the population means of X_{ij} and Z_{ij}, respectively, over $j = 1, \dots, N_i$, and the approximation error is $O_{\mathrm{P}}(N_i^{-1/2})$. Thus, by ignoring the approximation error, we may treat θ_i as the small-area mean. The EBLUP of θ_i is then given by

$$\hat{\theta}_i = \bar{X}_{i,\mathrm{P}}'\hat{\beta} + \bar{Z}_{i,\mathrm{P}}'\hat{\gamma} + \hat{v}_i, \tag{5.10}$$

where $\hat{v}_i$ is the ith component of $\hat{v}$.

The MSPE of the EBLUP is a standard measure of uncertainty. Here, however, the results of Prasad–Rao [196] may not apply. The reason is that the original Prasad-Rao estimator apply only to the case where the covariance matrix of y, V, is block-diagonal. It is easy to see that V is not block-diagonal in this case. Ref. [39] extended the Prasad–Rao method to general linear mixed models, which may not have a block-diagonal covariance structure. Their results are applicable to the current situation. Ref. [188] applied the result of [39] to obtain a second-order unbiased estimator of the MSPE of EBLUP. See [113], sec. 13.4 for detail.

[200] extended the robust EBLUP method of [215] to the non-parametric NER model described above. The extension was not straightforward due to a similar difficulty as mentioned in the last paragraph. Recall, in Section 3.2, the robust EBLUP was developed based on the residual vector $r = U^{-1/2}(y - X\beta)$, where U is the diagonal matrix with the diagonal elements equal to the those of $V = \mathrm{Var}(y)$. The role of $U^{-1/2}$ was supposed to be standardization. This worked in the previous case of §3.2, because, back then, V is block-diagonal. However, in the current case, V is not a block-diagonal matrix, as noted above. As a result, the elements of r were large in absolute values, as reported in [200]. The latter authors took a different route to robustizing the EBLUP in the current case.

Recall Henderson's MME (3.10). The procedure can be combined with estimation of the variance components in an iterative manner; see [60]. Given the current values of the EBLUPs, the original procedure updated the estimates of the variance components via the following formulae:

$$\begin{aligned}
\tilde{\sigma}_\gamma^2 &= \frac{1}{q - t_2}\sum_{k=1}^{q}\tilde{\gamma}_k^2, \\
\tilde{\sigma}_v^2 &= \frac{1}{m - t_3}\sum_{i=1}^{m}\tilde{v}_i^2, \\
\tilde{\sigma}_e^2 &= \frac{1}{n - t_4}\sum_{i=1}^{m}\sum_{j=1}^{n_i}\tilde{e}_{ij}^2,
\end{aligned} \tag{5.11}$$

where $\tilde{u}$, $\tilde{v}$ are the EBLUPs of u, v, respectively, in the current iteration, and $\tilde{e} = y - X\tilde{\beta} - Z\tilde{\gamma} - W\tilde{v}$, with $W = \text{diag}(1_{m_i}, 1 \le i \le m)$, is the "EBLUP" of e. Furthermore, $t_2 = \text{tr}(T_{22})/\sigma_\gamma^2$, $t_3 = \text{tr}(T_{33})/\sigma_v^2$, T_{22}, T_{33} being some matrices associated with Henderson's MME, and $t_4 = p + q + m - t_2 - t_3$. In (5.11), the variance components on the right sides are replaced by their current values, and this is how the iterative procedure is run.

To obtain robust versions of the variance components, the $\tilde{v}_i$ and $\tilde{e}_{ij}$ in the last two equations in (5.11) are replaced by their robust versions obtained via Huber's functions; see Section 3.2. The first equation in (5.11) is not modified because $\tilde{\gamma}$ is not modified by outliers, as noted by [200]. The next step is to modify Henderson's mixed model equation, (3.10), applied to this case, by replacing the y by its pseudo-values,

$$\tilde{y} = X\beta + Z\gamma + Wv + \sigma_e \Psi(\sigma_e^{-1} e),$$

where γ, v and σ_e are evaluated at their current values. The rest of the procedure is very much the same as Henderson's MME; see [200] for further details. The latter authors called their method robust EBLUP under semi-parametric mixed models (robust semi-parametric EBLUP).

For measure of uncertainty, [200] developed a similar bootstrap MSPE estimator to that of [215] under the spline LMM.

Empirical results from a simulation study showed the proposed robust semi-parametric EBLUP performed well in terms the MSPE when the true mean function is not nearly linear and the spline model has "enough" knots (e.g., $K = 20$). The proposed bootstrap MSPE estimator also seemed to track the true MSPE well.

The bootstrap MSPE estimator mentioned in the last sentence may be viewed non-parametric in some sense, because it is based on a spline approximation to a non-parametric function. See [6] for a similar approach to estimating the MSPE of benchmarked OBP, a topic to be discussed in the sequel (see Chapter 7). In an earlier work, [88] proposed a non-parametric bootstrap method for estimating the MSPE of EBLUP under an extended version of the NER model. It was not clear how to bootstrap non-parametrically in a mixed-model situations. It seems obvious that Efron's i.i.d. bootstrap cannot be applied directly to the data. In fact, a mixed effects model, linear or generalized linear, is conditional on the covariates, X. Under this setting, the vector y of all of the observations is viewed as a single n-dimensional observation; there is no other replication of this observation (in other words, one does not have an i.i.d. sample of size greater than 1).

On the other hand, there are some i.i.d. random variables "inside" the model, at least for most mixed models that are practically used. For example, in most cases, the random effects are assumed to be i.i.d. However, one cannot bootstrap the random effects directly, because they are not observed. Ref. [88] had a clever idea on how to bootstrap the random effects non-parametrically under an NER model.

Suppose that the random effects are i.i.d. but not necessarily normally distributed. In fact, one does not know the distribution of the random effects up to some unknown parameter(s). Then, the random effects cannot be bootstrapped parametrically. On the other hand, as noted, one cannot use Efron's [51] bootstrap because the random effects are not observed. A solution by [88]: Depending on what one wants.

In many cases, the quantity of interest does not involve every piece of information about the distribution of the random effects. For example, [88] observed that, up to a term of order $o(m^{-1})$, the MSPE of EBLUP involves only the second and fourth moments of the random effects and errors. This means that for random effects and errors from different distributions with the same second and fourth moments, the corresponding MSPEs of the EBLUPs are different only by a term of $o(m^{-1})$. This observation leads to the following strategy: (i) estimate the second and fourth moments of the random effects and errors; (ii) draw bootstrap samples of the random effects and errors from distributions that match the estimated second and fourth moments; (iii) use the NER model to generate the bootstrap data, and estimate the MSPE based on the bootstrapped data.

Let us be more precise on how to execute (i)–(iii). For any $\mu_2, \mu_4 \geq 0$ such that $\mu_2^2 \leq \mu_4$, let $D(\mu_2, \mu_4)$ denote the distribution of a random variable, ξ, such that $\mathrm{E}(\xi) = 0$ and $\mathrm{E}(\xi^j) = \mu_j, j = 2, 4$. Ref. [88] suggests using moment estimators, $\hat{\sigma}_v^2$, $\hat{\sigma}_e^2$, $\hat{\gamma}_v$, and $\hat{\gamma}_e$, of σ_v^2, σ_e^2, γ_v, and γ_e, where γ_v, γ_e are the fourth moments of v_i and e_{ij}, respectively, with the following constraints:

$$\hat{\sigma}_v^4 \leq \hat{\gamma}_v, \quad \hat{\sigma}_e^4 \leq \hat{\gamma}_e.$$

Suppose that we wish to predict the mixed effects

$$\zeta_i = X_i'\beta + v_i, \tag{5.12}$$

where X_i is a known vector of covariates, such as the population mean of the x_{ij}'s. Denote the EBLUP of ζ_i by $\hat{\zeta}_i$. To obtain a bootstrap estimator of the MSPE of $\hat{\zeta}_i$, $\mathrm{MSPE}_i = \mathrm{E}(\hat{\zeta}_i - \zeta_i)^2$, one draws samples $v_i^*, i = 1, \dots, m$, independently from $D(\hat{\sigma}_v^2, \hat{\gamma}_v)$, and $e_{ij}^*, i = 1, \dots, m, j = 1, \dots, n_i$, independently, from $D(\hat{\sigma}_e^2, \hat{\gamma}_e)$. Then, mimicking the NER model, define

$$y_{ij}^* = x_{ij}'\hat{\beta} + v_i^* + e_{ij}^*, \quad i = 1, \dots, m, j = 1, \dots, n_i, \tag{5.13}$$

where $\hat{\beta}$ is the EBLUE of β, whose formula is the same as (1.37) but with $\hat{V}$ being the estimated covariance matrix of y under the NER model. Let $\hat{\zeta}_i^*$ be the bootstrap version of $\hat{\zeta}_i$, obtained the same way as $\hat{\zeta}_i$ except using y^* instead of y. The bootstrap MSPE estimator is given by

$$\widetilde{\mathrm{MSPE}}_i = \mathrm{E}\{(\hat{\zeta}_i^* - \zeta_i^*)^2 | y\}, \tag{5.14}$$

where $\zeta_i^* = X_i'\hat{\beta} + v_i^*$ [see (5.12)], and the conditional expectation is evaluated by replications of y^*. Ref. [88] showed that (5.14) produces a first-order unbiased MSPE estimator, that is, its bias is $O(m^{-1})$.

To obtain a second-order unbiased MSPE estimator, whose bias is $o(m^{-1})$, [88] proposes to bias-correct $\widetilde{\text{MSPE}}_i$ using double-bootstrap. The first bootstrap is done as above. Then, conditional on v^* and e^*, draw samples $v_i^{**}, i = 1, \ldots, m$ and $e_{ij}^{**}, i = 1, \ldots, m, j = 1, \ldots, n_i$ independently from $D(\hat{\sigma}_v^{*2}, \hat{\gamma}_v^*)$ and $D(\hat{\sigma}_e^{*2}, \hat{\gamma}_e^*)$, respectively, where $\hat{\sigma}_v^{*2}$, and so on, are computed the same way as $\hat{\sigma}_v^2$, and so on, except using y^* instead of y. Similarly, define

$$y_{ij}^{**} = x_{ij}'\hat{\beta}^* + v_i^{**} + e_{ij}^{**}, \quad i = 1, \ldots, m, j = 1, \ldots, n_i,$$

and compute the double-bootstrap version of $\hat{\zeta}_i$, $\hat{\zeta}_i^{**}$. Then, compute

$$\widetilde{\text{MSPE}}_i^* = \text{E}\{(\hat{\zeta}_i^{**} - \zeta_i^{**})^2 | y^*\}, \tag{5.15}$$

where $\zeta_i^{**} = X_i'\hat{\beta}^* + v_i^{**}$ and the conditional expectation is evaluated by replications of y^{**}. One then estimates the bias of $\widetilde{\text{MSPE}}_i$ by

$$\widetilde{\text{bias}}_i = \text{E}(\widetilde{\text{MSPE}}_i^* | y) - \widetilde{\text{MSPE}}_i,$$

where the conditional expectation is evaluated, again, by replications of y^*. This leads to the bias-corrected MSPE estimator

$$\widehat{\text{MSPE}}_i = \widetilde{\text{MSPE}}_i - \widetilde{\text{bias}}_i = 2\widetilde{\text{MSPE}}_i - \text{E}(\widetilde{\text{MSPE}}_i^* | y). \tag{5.16}$$

[88] showed that, under some regularity conditions, $\widehat{\text{MSPE}}_i$ is a second-order unbiased MSPE estimator.

Regarding (ii), in particular, the choice of $D(\mu_2, \mu_4)$ is clearly not unique. There were a few suggestions from [88]. The simplest is a three-point distribution depending on a $p \in (0, 1)$ such that $\text{P}(\xi = 0) = 1 - p$ and

$$\text{P}\left(\xi = \frac{1}{\sqrt{p}}\right) = \text{P}\left(\xi = -\frac{1}{\sqrt{p}}\right) = \frac{p}{2}.$$

It is easily verified that $\text{E}(\xi) = 0, \text{E}(\xi^2) = 1$, and $\text{E}(\xi^4) = 1/p$. Thus, by letting $p = \mu_2^2/\mu_4$, the distribution of $\sqrt{\mu_2}\xi$ is $D(\mu_2, \mu_4)$. Note that the inequality $\mu_2^2 \leq \mu_4$ always hold with the equality if and only if ξ^2 is degenerate, that is, a constant with probability one. Another choice would be the re-scaled Student t-distribution, whose degrees of freedom $\nu > 4$ and is not necessarily an integer. The distribution has first and third moments 0, $\mu_2 = 1$, and $\mu_4 = 3(\nu - 2)/(\nu - 4)$, which is always greater than 3, meaning that the tails of the distribution are heavier than those of the normal distribution.

The double-bootstrap procedure is computationally intensive. What is more, so far, there have not been published comparisons of the method with other existing methods that also produce second-order unbiased MSPE estimators, such as the Prasad–Rao [196], jackknife [118], parametric bootstrap [19], and Sumca [133].

It should be noted that there are recent development in non-parametric bootstrapping in LMM. For example, [203] considered bootstrapping data under a general LMM (4.3) with $\mu = X\beta$, that is, $y = X\beta + Zv + e$. The bootstrapped data are supposed to be generated under a model like $y^* = X\hat{\beta} + Zv^* + e^*$, where $\hat{\beta}$ is the EBLUE, given by (1.28) with the variance components in V replaced by their, say, REML estimators. The question is how to generate v^* and e^*. The traditional method is to generate those under the normal distributions with estimated variance components. The authors of [203] proposed to generate v^* and e^* in a "semi-parametric" fashion. Recall the BLUE, $\tilde{\beta}$, in (1.28) and BLUP, $\tilde{v}$, in (1.29). One defines the BLUP of e as

$$\tilde{e} = y - X\tilde{\beta} - Z\tilde{v}.$$

Because $\tilde{v}$ and $\tilde{e}$ are linear functions of y, analytic expressions of their covariance matrices can be derived, which are matrix expressions involving the variance components. One then use the inverse square roots of those covariance matrices to standardize $\tilde{v}$ and $\tilde{e}$. In case the variance components are unknown, they are replaced by their REML (or ML) estimators. Let $\hat{v}_{\rm s}$ and $\hat{e}_{\rm s}$ denote the standardized versions of $\tilde{v}$ and $\tilde{e}$. One then treats the components of $\hat{v}$ and $\hat{e}$ as i.i.d., and bootstrap (sampling with replacement) as in the original bootstrap of Efron [51]. Once $\hat{v}_{\rm s}$ and $\hat{e}_{\rm s}$ are bootstrap, the bootstrap samples, v^* and e^*, are obtained by multiplying the corresponding square roots of the estimated covariance matrices. However, when the bootstrap method is applied to inference about the small-area means, the asymptotic theory that ensures convergence in distribution of the inference about the small area means requires that the sample sizes for all of the small areas go to infinity, which is not really an SAE situation (however, see discussion in Section 7.6).

5.3 Semi-parametric models

Several authors have proposed semi-parametric models for SAE under the framework of generalized linear models.

Ref. [28] proposed a robust semi-parametric GLMM for SAE with discrete outcomes, which may be viewed as an extension of an earlier work of M-quantiles [29]. To our understanding, the method is called semi-parametric because, as noted by the authors, the approach follows "the tradition of M-estimation" [99] that does not require explicit distributional assumptions of GLMM. Recall the qth M-quantile defined in Section 3.3. Here, for a random variable Y, whose distribution is associated with covariates x and parameter vector ψ, the authors considered a standardized version, $Q = Q_q(x; \psi)$, defined as the solution to

$$\int \psi_q \left(\frac{y - Q_q}{\sigma_q} \right) dF(y) = 0,$$

where $\psi_q(t) = 2\psi(t)\{q1_{(t>0)}+(1-q)1_{(t\leq 0)}\}$, $\psi(\cdot)$ being a user-defined influence function, and σ_q is a scale parameter. The M-quantile predictor of a small-area mean, $\bar{Y}_d$, where d corresponds to the small area or domain, is defined as

$$\widehat{\bar{Y}}^{\text{MQ}} = \frac{1}{N_d}\left\{\sum_{j\in s_d} y_{dj} + \sum_{j\notin s_d} \hat{Q}_{\hat{\theta}_d}(x_{dj})\right\}, \tag{5.17}$$

where N_d is the domain population size (assumed known), s_d is the subset of sampled indexes, x_{dj}, y_{dj} are the sampled covariates and outcome for the jth unit in the sample, $\hat{Q}(x_{dj})$ is the estimated M-quantile and $\hat{\theta}_d$ the estimated quantile coefficient for domain q. The authors discussed different ways of defining the estimated M-quantile coefficients for binary data and recommended their Definition C; see [28] for details.

Ref. [28] compared the proposed M-quantile predictor with a GLMM predictor, which replaces the Q terms inside the $\{\cdots\}$ in (5.17) by the conditional expectation of the non-sampled y_{dj} under the assumed GLMM, and then replace the fixed and random effects in the conditional expectation by their standard estimators under the GLMM (see [123], sec. 3.6.1) in a simulation study. The simulation settings included a case where the assumed GLMM holds and a case involving measurement errors in x and misclassifications (of binary outcome) in y. The result showed that, in the case of GLMM, the GLMM predictor performed better than the M-quantile predictor in terms of the bias; the trend is reversed in the case of measurement errors and misclassifications. As another performance measure, the authors also considered the root MSE (RMSE). The simulation result showed that, in terms of RMSE, the GLMM predictor and M-quantile predictor performed similarly in the case of GLMM, and the M-quantile predictor had advantage in the case of measurement errors and misclassifications. The simulation study was carried out as a "sensitivity analysis" of an application to unemployment estimation using data from the UK Labour Force Survey.

Two methods were developed to produce uncertainty measures for the M-quantile predictors in terms of the MSPE. The first is based on an approximation method that ignored variation in estimating the M-quantile coefficient. The second is a random-effects block bootstrap (REB) method based on [25]. A simulation study was carried out that compared the two approaches of MSPE estimation. The result showed that the REB method performed better. The REB method was based on its stability and simplicity.

Ref. [234] extended the mixed logistic model (see Section 1.5.3) to incorporate auxiliary variables whose association with the logit probability on the left side of (1.18) may be complex. The extended model is expressed as

$$\text{logit}(\pi_{ijk}) = x'_{ijk}\beta + m(z_{ij}) + v_i, \tag{5.18}$$

where i represents the survey strata (e.g., county), j the census tract within the strata, and k the individual within the census tract; π_{ijk} is the conditional probability of the binary outcome, y_{ijk}, equal to 1 given the strata-level random effect, v_i; x_{ijk} is an observed individual-level predictors; and β is an unknown vector of fixed effects. The main difference between (5.18) and (1.18) is the term $m(z_{ij})$. Here, z_{ij} is a known vector of auxiliary variables at the census-tract level, and $m(\cdot)$ is an unknown, non-parametric function.

The y_{ijk} and x_{ijk} are assumed to be available; the z_{ij} is available from other sources, such as population data or larger-scale survey data, not necessarily at the same time of the surveys. In an application study discussed in [234], y_{ijk}, x_{ijk} were obtained from the California Health Interview Survey (CHIS), 2011–2012; z_{ij} included the Nielsen–Claritas population data and Census-tract summary tables from the American Community Survey (ACS).

Much of the procedure followed the lines of [188]. Namely, the non-parametric function is approximated via the P-splines, whose connection to the LMM is utilized for model fitting. It should be noted that the spline-coefficient random effects are crossed over all areas. As a result, the data cannot be divided into independent groups when the spline coefficients are treated as random effects (because the same spline coefficients are related to each data point). Under the LMM, this is not a problem computationally, because the full data follow a multivariate normal distribution anyway. Under a GLMM, however, this can be very challenging computationally. The authors of [234] indicated that they used the GLIMMIX procedure in SAS, version 9.3 to fit the mode. It is believed that the procedure is Laplace-approximation based, which produces inconsistent estimators of the model parameters (e.g., [105], [16]). See below for the use of a more advanced computing method, called data-cloning, in computing the MLE under a GLMM.

Ref. [234] also discussed selection of the number of knots for the splines. The authors proposed using the coefficient of variation (CV) with a 30% threshold as cut-off. In a related work, [125] proposed to use the fence methods (e.g., [122]) to select the number of knots with a data-driven cut-off. See the next chapter for details.

A main objective of [234] is to generate health estimates by zip code. It was anticipated that one could borrow strength spatially if the geographic location is built into the model. The authors incorporated such spatial information using the neighborhood-level auxiliary variables in z and the non-parametric function $m(\cdot)$. It should be noted that incorporating spatial information is not just in terms of the mean function. Spatial correlation should also be taken into consideration, which, for example, can have impact on assessing the uncertainty in estimation. Specifically, the random effects involved in the approximating LMM (see Section 5.2) may be correlated spatially, and this was not taken into consideration. See, for example, [226], [227].

The latter concern was taken into consideration in [96] under an area-level model. The area-level model can be expressed as

$$f(y_i|\eta_i, \phi) = \exp\left\{\frac{y_i\eta_i - b(\eta_i)}{\phi} + c(y_i, \phi)\right\}, \tag{5.19}$$

$$\eta_i = x_i'\beta + f(t_i) + v_i, \tag{5.20}$$

$i = 1, \ldots, m$, where y_i is the direct estimate from area i, whose conditional distribution given η_i follows that of an exponential-family with the specified functions $b(\cdot)$ and $c(\cdot, \cdot)$, and ϕ is a dispersion parameter, which in some cases is known. Furthermore, x_i is an observed area-level covariates, t_i is another area-level auxiliary variable, $f(\cdot)$ is an unknown smooth function, which is assumed nonlinear, and v_i is an area-specific random effect. Again, $f(t)$ is approximated via the splines with knots at $\kappa_j, j = 1, \ldots, q$ and coefficients $\gamma_k, k = 1, \ldots, d+q+1$ (d is the degree of polynomial). As a result, (5.20) can be approximated by

$$\tilde{\eta}_i = x_i'\beta + z_i'\gamma + v_i = w_i'\alpha + v_i, \tag{5.21}$$

where $z_i = (1, t_i, \ldots, t_i^d, (t_i - \kappa_1)_+^d, \ldots, (t_i - \kappa_q)_+^d)'$, $\gamma = (\eta_k)_{1 \le k \le d+q+1}$, $w_i' = (x_i', z_i')$, and $\alpha = (\beta', \gamma')'$. Note that, unlike the approaches previously described, here the spline coefficients γ are treated as fixed effects.

Importantly, the area-level random effects, $v_i, i = 1, \ldots, m$ are not assumed to be independent. Following [226], let $v = (v_i)_{1 \le i \le m}$. It is assumed that

$$v \sim N(0, \Sigma), \tag{5.22}$$

where $\Sigma = \Sigma(\psi)$ is an $m \times m$ covariance matrix depending on a vector of dispersion parameters, ψ, that incorporates the spatial correlation among the area-specific random effects. See Section 1.6.

Due to the spatial correlation among the random effects, fitting the GLMM is computationally challenging. The main difficult comes from the fact that, under (5.22), the likelihood function cannot be simplified as a product of low-dimensional integrals. Note that, in the case of independent area-specific random effects, the likelihood function under the GLMM is a product of m integrals, each of the dimension of v_i (for example, if v_i is a scalar random effect, the dimension of the integral is 1). The low-dimensional integrals can be numerically evaluated via numerical integration or simple Monte-Carlo methods; see, for example, sec. 4.1 of [123].

To facilitate computation of the MLE of the parameters involved in the spatial GLMM, [96] followed a data-cloning (DC) approach of [226] (also see [225]). The method was originally proposed by [157] and was further developed in [156]. In a way, it uses the Bayesian computational approach for a frequentist purpose, namely, the computation of the MLE. Let π denote the prior density function of θ. Then, the posterior is given by

$$\pi(\theta|y) = \frac{p(y|\theta)\pi(\theta)}{p(y)}, \tag{5.23}$$

where $p(y)$ is the integral of the numerator with respect to θ, which does not depend on θ. Note that $p(y|\theta)$ corresponds to the likelihood function. There are computational tools using Markov-chain Monte-Carlo (MCMC; e.g., [69], [160]) for posterior simulation that generate random variables from the posterior without having to compute the numerator or denominator of (5.23).

Thus, we can assume that one can generate random variables from the posterior. The key idea of DC is the following: If the observations y were repeated independently from K different individuals such that all of these individuals result in exactly the same data, y, denoted by $y^{[K]} = (y, \dots, y)$, then the posterior based on $y^{[K]}$ is given by

$$\pi_K(\theta|y^{[K]}) = \frac{\{p(y|\theta)\}^K \pi(\theta)}{\int \{p(y|\theta)\}^K \pi(\theta) d\theta}. \tag{5.24}$$

[156] showed that, as K increases, the right side of (5.24) converges to a multivariate normal distribution whose mean vector is equal to the MLE, $\hat{\theta}$, and whose covariance matrix is approximately equal to $K^{-1} I_{\mathrm{f}}^{-1}(\hat{\theta})$, where $I_{\mathrm{f}}(\theta)$ denotes the Fisher information matrix evaluated at θ.

Thus, for large K, one can approximate the MLE by the sample mean of $\theta^{(1)}, \dots, \theta^{(B)}$, generated from the posterior distribution (5.24). Similarly, for large K, $I_{\mathrm{f}}^{-1}(\hat{\theta})$ can be approximated by K times the sample covariance matrix of $\theta^{(1)}, \dots, \theta^{(B)}$.

Importantly, because B, K are, at least in principle, up to one's choice; one can make sure that they are large enough so that the approximation is accurate enough. In this regard, a reasonable choice of B would reduce the sampling variation and therefore improve the DC approximation and, at the same time, preserve the computational efficiency. See [156] for discussion on how to choose B and K from practical points of view.

As for the prior π, [156] suggests that it be chosen according to computational convenience and be proper (to avoid improper posterior). In many cases, there is a computationally feasible method that produces a consistent and asymptotically normal estimator of θ, say, $\tilde{\theta}$. See, for example, Section 4.2 of [123]. Let $V_{\tilde{\theta}}$ be the asymptotic covariance matrix of $\tilde{\theta}$. Then, one possible choice of the prior would be a multivariate normal distribution with mean vector $\tilde{\theta}$ and covariance matrix $V_{\tilde{\theta}}$. Such a choice of prior may be regarded as empirical Bayes.

Once the MLE is obtained, the predictor of η_i is based on Laplace approximation to the BP of η_i, that is, $\tilde{\eta}_i = \mathrm{E}(\eta_i|y, \alpha_0, \psi_0)$, where α_0, ψ_0 are the true α, ψ, respectively. The BP is intractable due to similar difficulty as for computing the MLE. The Laplace approximation is based on the second-order Taylor expansion,

$$y_i\eta_i - b(\eta_i) \approx y_i\eta_i^0 - b'(\eta_i^0) + \{y_i - b'(\eta_i^0)\}(\eta_i - \eta_i^0) - \frac{1}{2}b''(\eta_i^0)(\eta_i - \eta_i^0)^2,$$

where η_i^0 is the maximizer of $f(\eta_i|y, \alpha_0, \psi_0)$, the conditional pdf of η_i given y, evaluated under the true parameter vectors, which are then replaced by their MLEs, leading to the EBP. It should be noted that [96] considered $\mathrm{E}(\eta_i|y_i, \alpha_0, \psi_0)$, which is not the BP. The BP should be $\mathrm{E}(\eta_i|y, \alpha_0, \psi_0)$, which is not equal to $\mathrm{E}(\eta_i|y_i, \alpha_0, \psi_0)$ given the spatial correlations (as a result, $y_1, \dots, y_m$ are not independent; see (A.16) of [113]). An estimator of the MSPE of the EBP is also derived using the Laplace approximation.

The mixed effect η_i is of direct practical interest in some cases. For example, [96] considered an application of their semi-parametric spatial GLMM to modeling the number of life insurance contracts, in which case y_i is assumed to have a Poisson distribution with mean η_i, for the ith province in Iran, and t_i is the mean work experience of the insurance sales branch for the ith province.

The authors also carried out a simulation study under the same simulation setting of [226], who considered a parametric spatial Poisson mixed model. The data were simulation with the true $f(t) = 1 + 2t - 3\sin(t)$ and the random effects v generated under a conditional auto-regressive (CAR) model that induces spatial correlation. The simulation results showed that the small-area predictors based on the semi-parametric spatial GLMM (i.e., Poisson mixed model) performs better than that based on the parametric spatial GLMM in terms of the MSPE. However, performance of the proposed MSPE estimator is not evaluated.

Also using the spline approximation to the non-parametric component, and the DC method for inference, [214] considered a semi-parametric GLMM that combines time-series and cross-sectional data. The model is similar to (5.19) and (5.20) with the index i replaced by it, and

$$\eta_{it} = f(x_{it}) + v_i + e_{it}, \tag{5.25}$$

$i = 1, \dots, m$, $t = 1, \dots, T$. Here, y_{it} is the outcome of area i at time t, x_{it} is a corresponding vector of auxiliary variables, $f(\cdot)$ is an unknown function, v_i is an area-specific random effects, and e_{it} is an within-area and time-dependent error. The random effects v_i are assumed to be independent and distributed as $N(0, \sigma_v^2)$. The errors e_{it} are assumed to follow a first-order auto-regressive, or AR(1), process, that is, it satisfies

$$e_{it} = \rho e_{i\underline{t-1}} + \epsilon_{it}, \tag{5.26}$$

where ρ is an unknown coefficient satisfying $|\rho| < 1$, and ϵ_{it} are independent and distributed as $N(0, \sigma_\epsilon^2)$, and are independent with the random effects v_i. To our understanding, the model is called semi-parametric, instead of non-parametric, due to the parametric (AR) model for the errors. The model may be viewed as an extension of LMM in SAE for data that are both time-series and cross-sectional (e.g., [47]).

The non-parametric function, $f(\cdot)$, in (5.25) is then approximated via the splines as in [234]. Recall the latter used a Laplace-approximation based method to estimate the parameters, which is computationally attractive but results in inconsistent estimators of the model parameters. Here, the authors of [214] used the DC method (see above) to carry out a likelihood-based inference. The MLE is known to be consistent in cases of GLMM with clustered (e.g., [107]) and crossed random effects [110]. However, the GLMM with the random spline coefficients does not belong to one of these two types. Perhaps, a suitable use of the subset argument, introduced in [110], can be used to establish the consistency of the MLE in this case?

Based on the DC theory [156], [214] proposed to predict the small area mean, $\theta_{it} = \mathrm{E}(y_{it}|\eta_{it})$, as the posterior mean of η_{it} under the prior of a multivariate normal distribution, whose mean is the MLE (computed via the DC) and whose covariance matrix is K times the sample covariance matrix of $\psi^{(1)}, \dots, \psi^{(B)}$, where ψ denotes the vector of all unknown parameters under the assumed model [see the notes below (5.24) and the following paragraph]. The posterior mean is evaluated via the MCMC method. The posterior also provides a measure of uncertainty as the posterior variance, also evaluated from the MCMC samples. See [214] for details.

Simulation results from [214] showed improvement of the proposed method over the parametric method, in which case $f(x)$ is assumed to be a linear function of x, when $f(x)$ is truly nonlinear. The improvement is in terms of the average (over the small areas) MSPE of the predictor of the small area mean and the percentage average (again, over the small areas) absolute relative bias of the MSPE estimator, which is the empirical (or estimated) posterior variance, as well as the coverage probability and length of the prediction intervals, constructed from the MCMC samples.

In one of the applications discussed in [214], the authors considered estimation of the rate of physician visits related to total respiratory morbidity (TRM) conditions. Here, y_{it} is the total number of TRM physician visits of the ith Regional Health Authorities District (RHAD) and year t, and n_{it} is the total population at risk for the ith RHAD and year t. There are 217 RHADs in the province of Manitoba, Canada, which were considered as the small areas, and 10 years from 2000 to 2010. A binomial mixed logistic model was proposed, that is, $y_{it}|\theta_{it} \sim \mathrm{Binomial}(n_{it}, \theta_{it})$, such that

$$\mathrm{logit}(\theta_{it}) = \beta_0 + \beta_1 z_{it} + f(x_{it}) + v_i + e_{it}, \tag{5.27}$$

where z_{it}, x_{it} are the percentage of female and average age of the ith RHAD and year t. It was noted that the relationship of the observed visit rate, at the logit scale, and the average age, was not linear. Therefore, a linear spline (i.e., the degree of polynomial, p, for the spline is equal to 1) with $L = 40$ knots was used to approximate $f(\cdot)$. The DC method with $K = 20$ was used to fit the model. The authors reported the values of the MLEs and associated standard errors for the model parameters. The DC method was also used to approximate the posterior distribution of θ_{it} (via MCMC), including the posterior means and associated measures of uncertainty.

There are different approaches to semi-parametric SAE models. For example, [164] proposed a kernel-based approach to estimating a non-parametric component in a semi-parametric GLMM, in which the conditional mean of a unit-level outcome given the random effects can be expressed as

$$\begin{aligned} \mu_{ij} &= \mathrm{E}(y_{ij}|v_i, x_{ij}, t_{ij}) \\ &= g(x_{ij}'\beta + \lambda(t_{ij}) + z_{ij}'v_i), \end{aligned} \tag{5.28}$$

where $g(\cdot)$ is a known link function, v_i is a vector-valued, area-specific random effect, x_{ij}, t_{ij} are observed vectors of auxiliary variables, and z_{ij} is a sub-vector

of $(1, x'_{ij})'$. Furthermore, $\lambda(\cdot)$ is an unknown function. It is assumed that
(i) (y_{ij}, x_{ij}, t_{ij}) are i.i.d. given the random effects;
(ii) the random effects v_i are i.i.d. $N(0, \sigma^2)$; and
(iii) the (x_{ij}, t_{ij})'s are independent with the random effects.
The authors proposed a procedure that combines the idea of maximum a posterior (MAP; see below) and kernel estimation.

Ref. [114] took another look at Henderson's method of jointly estimating the fixed and random effects in LMM (e.g., [123], p. 90). Let y be a vector of responses, and θ a vector of dispersion parameters. Let $f(y, v|\beta, \theta)$ be the joint probability density, with respect to a certain measure, of y and v given β and θ, where v is the vector of random effects, and β the vector of fixed effects. Because $\max_{v,\beta} f(y, v|\beta, \theta) = \max_\beta \max_v f(y, v|\beta, \theta)$, the maximization can be done in two steps. In the first step one finds $\tilde{v} = \tilde{v}(\beta; y, \theta)$ that maximizes $f(y, v|\beta, \theta)$ for fixed β. In the second step, one finds $\hat{\beta} = \hat{\beta}(y, \theta)$ that maximizes $f(y, \tilde{v}|\beta, \theta)$, and computes $\hat{v} = \hat{v}(y, \theta) = \tilde{v}(\hat{\beta}; y, \theta)$.

Let us now focus on the first step. Observe that

$$f(y, v|\beta, \theta) = f(y|\beta, \theta) f(v|y, \beta, \theta). \tag{5.29}$$

The first factor on the right side of (5.29) corresponds to the likelihood function for estimating β and θ, and the second factor the posterior density of v given y (if one would like, assuming that a non-informative prior has been assigned to v). Henderson's idea was to find v and β that jointly maximize $f(y, v|\beta, \theta)$. Because the first factor does not depend on v, maximizing $f(y, v|\beta, \theta)$ is equivalent to maximizing the posterior. Note that, although, in LMM, the maximizers $\hat{v}$ and $\hat{\beta}$ correspond to the BLUP and BLUE, respectively, they are no longer such predictor and estimator in GLMM. Still, the method is intuitive in the sense that $\hat{v}$ maximizes the posterior. For such a reason, the method is called MAP, and $\hat{v}$ and $\hat{\beta}$ maximum posterior estimators, or MPE. See Section 3.6.1 of [123] for more details, including how to estimate the dispersion parameters, θ.

For a fixed value of t, and given values of ψ, v (note that σ^2 is part of ψ), [164] proposes to find the value $\lambda(t) = \tilde{\lambda}(t)$ that maximizes a kernel estimator of the conditional expectation

$$\mathrm{E}\{\log f(Y|X, T, v; \psi, \lambda(t)) + \log f(v|\sigma^2)\},$$

where (Y, X, T, v) is an i.i.d. copy of $(y_{ij}, x_{ij}, t_{ij}, v_i)$. The kernel estimator is

$$\sum_{i=1}^{m} \sum_{j=1}^{n_i} K_h(t - t_{ij}) \log f(y_{ij}|x_{ij}, t_{ij}, v_i; \psi, \lambda(t)) + \sum_{i=1}^{n} \log f(v_i|\sigma^2), \tag{5.30}$$

where $K_h(\cdot)$ is a q-dimensional $[q = \dim(t)]$ kernel depending on a vector $h = (h_k)_{1 \le k \le q}$ of bandwidths. Note that there is no kernel on $f(v|\sigma^2)$ because it does not depend on t. The maximization of (5.30) is over $\lambda \in \Lambda$, a space of candidate functions for λ, chosen by the user. In particular, if Λ is a class of splines with given knots, the maximization is with respect to a set of spline coefficients, which one has been familiar with by now.

The maximization of (5.30) produces an estimator $\tilde{\lambda}(t)$, which is a function of ψ and v. This is called the least favorable curve [164]. One then maximize the objective function (5.30), with $\lambda(t)$ replaced by $\tilde{\lambda}(t)$, with respect to ψ, v in a way similar to the MAP.

As for measure of uncertainty, the authors proposed a bootstrap procedure and provided a theoretical justification, also in terms of asymptotic approximation of the distribution of (Y, X, T) by the bootstrap distribution.

In addition to estimation, [164] also considered testing the following hypothesis, $H_0 : \lambda(t) = c + \gamma' t$ vs $H_1 : \lambda(t) \neq c + \gamma' t$ for some constant c and constant vector $\gamma \in R^q$. Note that, under the null hypothesis, the semi-parametric GLMM reduces to the standard (parametric) GLMM, and the estimation procedure is equivalent to the MAP. Asymptotic analysis of MAP has been previously studied (see [114]). A (new) asymptotic normality result was established in [164], under suitable conditions, to derive the asymptotic null distribution.

One concern with the MAP-type methods is that they do not produce consistent estimators of the model parameters, namely, the fixed effects and variance components under the standard asymptotic framework for SAE, that is, the number of areas grows while the within-area sample sizes remain bounded. In such a case, it is not possible to consistently estimate the area-specific random effects, no matter what method is used, but one should be able to obtain consistent estimators of the fixed parameters. The MLE is consistent under such an asymptotic framework (e.g., [107]) but not the MPE, neither is the estimators of [164] described above.

5.4 Functional mixed effects models

In earlier discussions, we have encountered SAE problems with cross-sectional and time-series data (e.g., [47], [214]). There is another way to incorporate this types of problems under the framework of functional mixed effects models. Here, the small-area means at different times are viewed as realizations of a functions of time, t. This way, there is one small-area mean which is a function of t, instead of multiple small-area means, one for each time.

There have been studies in functional mixed effects models. These include models for functional data, in which the random effects are Euclidean (i.e., not functional), and models in which both the data and the random effects are both functional. Ref. [82] noted that data in many cases arise as curves, such as growth curves and economical data collected over time. The author proposed a functional mixed effects model (FMEM), in which the random effects are modeled as realizations of a Gaussian process. Suppose that there is a response curve associated with each of the m subjects (e.g., patients, firms, small areas). Let y_{ij} be the observed value of the ith curve at time t_{ij} satisfying

$$y_{ij} = x_{ij}'\beta(t_{ij}) + z_{ij}'\alpha_i(t_{ij}) + \epsilon_{ij},$$

$i = 1, \dots, m, j = 1, \dots, n_i$, where $\beta(t) = [\beta_k(t)]_{1 \le k \le p}$ is a $p \times 1$ vector of fixed functions, $\alpha_i(t) = [\alpha_{ik}(t)]_{1 \le k \le q}$ is a $q \times 1$ vector of random functions that are modeled as realizations of Gaussian processes, $\alpha(t) = [\alpha_k(t)]_{1 \le k \le q}$, with zero means, $x_{ij} = (x_{ijk})_{1 \le k \le p}$ and $z_{ij} = (z_{ijk})_{1 \le k \le q}$ are known covariate and design matrices, respectively, and ϵ_{ij} are independent $N(0, \sigma_\epsilon^2)$ and are independent with the α process.

In fitting the FMEM, [82] proposed to first approximated the fixed and random functions by smoothing splines. The author then used a connection built by Wahba [231] to model $\beta(t)$ and $\alpha(t)$ as

$$\begin{aligned} \beta_k(t) &= b_{1k} + b_{2k}t + \lambda_{b,k}^{-1/2} \int_0^t W_{b,k}(s)ds, \quad k = 1, \dots, p; \\ \alpha_k(t) &= a_{1k} + a_{2k}t + \lambda_{a,k}^{-1/2} \int_0^t W_{a,k}(s)ds, \quad k = 1, \dots, q, \end{aligned}$$

where $(b_{1k}, b_{2k})' \sim N(0, \tau I_2)$ with $\tau \to \infty$ (diffuse prior), and $(a_{1k}, a_{2k})' \sim N(0, \sigma_k^2 D)$, D being an unknown covariance matrix, and $W_{b,k}(s), W_{a,k}(s)$ are Weiner processes (see Section 5.1). Here, a_{1k} and a_{2k} are considered random intercept and random slope, respectively.

In terms of statistical inference, [82] was mainly interested in testing linearity of $\beta_k(t)$ and $\alpha_k(t)$, that is, $\lambda_{b,k}^{-1} = 0$, or $\lambda_{a,k}^{-1} = 0$. Using a result from [210] on likelihood ratio test for parameters on the boundary of the parameter space, [82] showed that, under the null hypothesis of $\lambda_{b,k}^{-1} = 0$, the asymptotic distribution of the likelihood ratio statistic (LRS) is a mixture of 0 and χ_1^2 with equal weight (1/2 for both). As for testing for a single component of $\beta(t)$, that is, $\beta_k(t) = 0$, it is equivalent to testing for $b_{1k} = b_{2k} = \lambda_{b,k}^{-1} = 0$. The asymptotic null distribution of the LRS is a mixture of χ_2^2 and χ_3^2 with equal weight. Below we illustrate the fitting of the functional mixed effects model using a simulated example.

Example 5.2. The data were generated under the model

$$y_i(t) = 5x_{i1}\sin(2\pi t) + 5x_{i2}\sin(\pi t) + r_i(t) + e_i, \quad i = 1, \dots, 10, t \in \{t_1, \dots, t_{30}\}.$$

For each subject i, x_{i1} is generated from Uniform$[-0.5, 0.5]$ and x_{i2} is binary with equal number of 0s and 1s. Thus, we have $\beta_1(t) = 2\sin(2\pi t)$ and $\beta_2(t) = 5\sin(\pi t)$. Also, we have $r_i(t) \sim N(0, \sigma_1^2\Sigma(t))$ with $\sigma_1^2 = 90$ and $\Sigma(t)$ is simulated by the reproducing kernel $R^1(t_i, t_i)$ [e.g., [82]]. Finally, $e_i = (e_{ij})_{1 \le j \le 30}$ and $e_{ij} \overset{\text{i.i.d.}}{\sim} N(0, 1)$. Furthermore, the generated data are

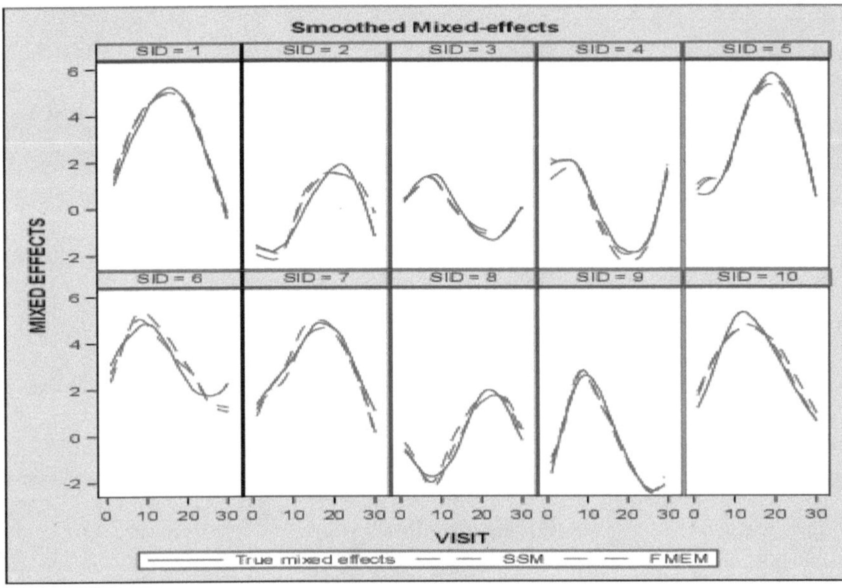

FIGURE 5.1
Fitted curves via SSM and FMEM.

subject to random missing with probability 0.1. The model was fitted using two methods proposed in [82], namely, the state space model (SSM) and functional mixed effects model (FMEM). There are software packages in SAS that implement these methods, PROC SSM for SSM and PROC MIXED for FMEM. See Figure 5.1 for illustration.

Ref. [167] noted that the main interest in SAE are prediction of mixed effects associated with the small area means, and developed the FMEM for such an interest. Consider the case of area-level cross-sectional and time-series data. One may initially begin with a functional regression model,

$$y_i(t) = \beta_{i0}(t) + \sum_{k=1}^{p} \beta_{ik}(t)x_{it}(t) + e_i(t), \tag{5.31}$$

where $y_i(t)$ is the outcome function for the ith small area, of which, the observed outcome data at different time points are realizations; $x_{ik}(t)$, $k = 1, \dots, p$ are the corresponding covariate functions; $\beta_{ik}(t)$, $k = 0, \dots, p$ are area-specific coefficient functions; and $\epsilon_i(t)$ is the error function. Note that (5.31) is similar to the linear regression model but with two main differences: (i) the coefficients are functions of t rather than constants; and (ii) the coefficients are also area-dependent.

The next step is to decompose each area-specific function into two parts, a fixed function that is common for all areas, and a random function, that is, Gaussian process that is area-specific. This decomposition is similar to the standard approach in the analysis of longitudinal data (e.g., [50]). We have

$$\beta_{ik}(t) = \beta_k(t) + v_{ik}(t), \quad k = 0, 1, \dots, p, \tag{5.32}$$

where $\beta_k(t)$ is an unknown function and $v_{ik}(t)$ an area-specific Gaussian process, which is assumed to have mean zero and covariance function

$$\text{cov}[v_{ik}(s), v_{ik}(t)] = \sigma_k(|s-t|, \psi_k),$$

where $\sigma(\cdot, \psi_k)$ is a covariance function that depends on a vector ψ_k of dispersion parameters. An example of such a function is $\sigma_k(u, \psi_k) = \tau^2 e^{-\rho u}$ for $u \geq 0$, where τ^2 is an unknown variance and $\rho > 0$ is another unknown parameter. In this case, $\psi_k = (\sigma^2, \rho)'$, which does not depend on k.

Now comes the approximation step. The functions $\beta_k(t)$ are approximated by B-splines (see Section 5.1),

$$\beta_k(t) \approx \sum_{l=1}^{L} B_{l,d}(t) b_{kl}, \quad k = 0, 1, \dots, p, \tag{5.33}$$

where $B_{l,d}(t), 1 \leq l \leq L$ are, respectively, the B-splines of order d with knots $\kappa_1 < \cdots < \kappa_L$ and $b_{kl}, 1 \leq l \leq L$ are the spline coefficients.

By combining (5.31)–(5.33), one can write the outcome function as

$$y_i(t) \approx \sum_{k=0}^{p} b_k' z_{ik}(t) + v_{i0}(t) + \sum_{k=1}^{p} x_{ik}(t) v_{ik}(t) + e_i(t), \tag{5.34}$$

where $z_{ik}(t) = B_d(t) x_{ik}(t)$ with $B_d(t) = [B_{l,d}(t)]_{1 \leq l \leq L}$.

The approach to inference about the model is similar to that described in Section 5.2 and subsequent contexts. Namely, after treating the spline coefficients as random effects, and discretize the Gaussian processes at their realized values, treated as another random effect factor, one can express the right side of (5.34) in the form of a LMM. The variance-covariance structure of the LMM is not block-diagonal, and it incorporates the B-splines as well as the covariance function of the Gaussian processes.

As noted, the main goal is to produce estimates of the small area means as well as the associated measures of uncertainty. The small area means are estimated via the EBLUP method. For the measure of uncertainty, a Prasad-Rao (P-R; [196]) type MSPE estimator is developed. The authors noted that some conditions of [39], which establishes the second-order unbiasedness of the P-R MSPE estimator under LMM without the block-diagonal covariance structure, are not satisfied in the current situation. The authors carried out their asymptotic analysis to derive their MSPE estimator.

An advantage of the FMEM approach is seen as its ability to incorporate time-dependent coefficient in both fixed and random slopes using the method of the analysis of longitudinal data (e.g., [50]). A simulation study carried out in [167] showed this. Specifically, the latter authors compared their FMEM method with some existing methods based on LMM for cross-sectional and time-series data, such as [47], as well as the direct estimates based on the survey data. The LMM methods are based on non-time-dependent fixed effects, a random intercept, a serial AR(1) error, plus a noise, which is a simple case of the LMM for longitudinal data without random slope (e.g., Chapter 5 of [50]). The simulation results showed that, under the scenarios of time-dependent coefficients, the FMEM method outperformed the LMM methods, which is what one would expect. Under a scenario of no time-dependent coefficients, the FMEM and LMM methods performed similarly. The proposed MSPE estimator also performed satisfactorily in the simulation study.

The FMEM method was illustrated using a monthly unemployment data from the $m = 51$ states of the U.S., including the District of Columbia (D.C.). The authors considered the logarithm of the unemployment rate as the response, which was averaged over the three months of each quarter. This produced $T = 44$ quarterly outcomes, from 2001 to 2010, for each state and the D.C. The authors used B-splines with equidistant knots at

$$\kappa = 4, 9, 14, 19, 24, 29, 34, 39$$

for the FMEM method, and compared the results with those of the LMM methods, namely, [47] and [201]. The methods seemed to produce similar results both in terms of the estimated quarterly mean outcomes and in terms of the estimated MSPEs of the estimated mean outcomes.

5.5 Concluding remarks

As noted in [113] (sec. 13.4), there is a cap between the computational and theoretical developments in the spline-based non-parametric/semi-parametric methods, in particular when applied to SAE problems.

The majority of the methods, if not all of the methods, assume that (a) there is an unknown, non-parametric function that is approximated by a spline function. The attention then shifts to (b) how to fit the spline-approximated model, which is parametric. The spline-approximated model is (c) then fitted using a spline-LMM connection. There are two issues associated with the theoretical justification of the spline approximation and its LMM connection.

The first question is regarding how well the splines approximate the true underlying function. This is about the difference between (a) and (b). From a mathematical point of view, the problem has been well studied (e.g., [87]); but the problem is not even clear once it is connected to the LMM. Recall the unknown function, $f(x)$, is approximated by the spline given by (5.1). Here, x is a fixed value associated with the mean of y, and the knots $\kappa_j, 1 \le j \le q$ are usually treated as deterministic (e.g., equidistant). If the spline coefficients, $\gamma_j, 1 \le j \le q$ are treated as random effects, the standard assumption is that

$\mathrm{E}(\gamma_j) = 0, 1 \le j \le q$. It then follows that

$$\mathrm{E}\{\tilde{f}(x)\} = \beta_0 + \beta_1 x + \cdots + \beta_p x_p. \tag{5.35}$$

Typically in the spline approximation, the degree of polynomial, p, is chosen to be fairly low; in fact, in most cases p is 1 (linear spline) or 3 (cubic spline). If p does not increase, (5.35) cannot be a good approximation to $f(x)$ unless $f(x)$ itself is a pth order polynomial. If $\mathrm{E}\{\tilde{f}(x)\}$ is not a good approximation to $f(x)$, what can we expect from $\tilde{f}(x)$ as an approximation to $f(x)$?

One way to step out of this confusion is to stay focused on the "small-area mean". Recall the basic Fay-Herriot model, (1.16). The unconditional mean is $x_i'\beta$, but this is not what we are interested. We are interested in the small-area mean, which is $x_i'\beta + v_i$. Here, similarly, the unconditional mean is (5.35), but this is not what we are interested. We are interested in (5.35) plus $\sum_{j=1}^q \gamma_j (x - \kappa_j)_+^p$, that is, (5.1), which is analogous to the small-area mean.

The second question has to do with the asymptotic behavior of the spline-LMM connection. Let us, once again, consider (5.1). If the spline coefficients γ_j are treated as random effects, the standard assumption is that they are i.i.d. with mean zero and constant variance. Then, as q increases, the right side of (5.1) cannot be convergent. To see this, let $p = 1$, that is, we are considering linear splines. Let $A < \kappa_1 < \cdots < \kappa_q < B$, where $[A, B]$ is the interval range of x, and $1 \le r \le q - 1$ be such that $\kappa_r \le x < \kappa_{r+1}$. Then, the right side of (5.1) is equal to

$$\beta_0 + \beta_1 x + \sum_{j=1}^{r} (x - \kappa_j)\gamma_j. \tag{5.36}$$

As q increases, one is filling more knots to the range $[A, B]$; as a result, r is also increasing. However, the mean of (5.36) is $\beta_0 + \beta_1 x$, and its variance is $\tau^2 \sum_{j=1}^r (x - \kappa_j)^2$, where $\tau^2 = \mathrm{var}(\gamma_j)$, which goes to ∞ as r increases, provided that $\tau^2 > 0$.

The discussion above suggests that, if the γ_j's are treated as random effects, they need to be suitably standardized so that the total variance remains bounded. Such kind of standardization has been considered, for example, in genome-wide association study (GWAS; e.g., [120]). For example, if the γ_j in (5.1) is replaced by $\gamma_j/\sqrt{q}$, one can write

$$\tilde{f}(x) = \sum_{k=0}^{p} \beta_k x^k + \sum_{j=1}^{q} w_{q,j}(x)\gamma_j, \tag{5.37}$$

with $w_{q,j}(x) = (x - \kappa_j)_+^p/\sqrt{q}$. Assuming, again, that $\gamma_j, 1 \le j \le q$ are i.i.d. with mean 0 and variance τ^2, the variance of (5.37) is $\tau^2 q^{-1} \sum_{j=1}^q (x - \kappa_j)_+^{2p}$, which can well be bounded, or even convergent. Some asymptotic framework could be developed from here to justify the spline-LMM connection.

Similarly, a way to get out of the seemingly confusing circle about the spline-LMM connection is to focus on the small area means. Forget about the spine-LMM assumptions. The most important thing is the computational procedure built on the LMM, that produces predictors of the mixed effects of interest, such as the EBLUP. Of course, the latter is a product of fitting the LMM, but it can perform well even without the LMM assumption, or, perhaps, under a different assumption. This certainly deserves some serious research, both theoretically and empirically.

6

Model Selection and Diagnostics

Carefully choosing the model for SAE; checking its appropriateness before using it are important procedures not just in SAE but in the practice of statistics, in general, to reduce the possibility, and consequences, of model misspecification. The first procedure is called model selection; the second is called model diagnostics.

Model selection has a long and celebrated history in statistics and, more recently, in data science. Traditional model selection strategies include the information criteria and shrinkage selection/estimation methods; see, for example, [182], [122] (ch. 1). Below we first provide a brief literature review of frequently used methods, discussing their applications and limitations in SAE.

6.1 Model selection: Overview

The importance of model selection in SAE was noted earlier by prominent researchers; see, for example, [8] and [77]. More recently, Lahiri [149] edited a collection of papers on topics of model selection, some covering survey topics. In this collection, [46] discussed a model selection method based on computation of the frequentist's Bayes factor in choosing between a fixed effects model and a random effects model. The authors focused on a one-way random effects model and noted a connection between the choice of fixed or random effects models and the test of hypothesis that the variance of the random effects is zero. Note that, however, not all model selection problems can be formulated as hypothesis testing problems.

Ref. [175] demonstrated the limitations of Mallows' Cp statistic in selecting the fixed covariates in an NER model (see Section 1.5.2), which is a special case of the LMM (e.g., [123]). The authors showed with simulation results that the Cp method without modification does not work well when the variance of the random effects is large. The authors proposed a modified Cp criterion by adjusting the intra-cluster correlations and showed that it performed similarly as the Cp in the regression settings.

In a survey-related context, [55] noted that model selection, especially the well-known likelihood-based methods such as the BIC [208], had to date received little attention in survey research literature. The authors

DOI: 10.1201/9781003395171-6

further developed a robust model selection method in the context of complex surveys. Their method is based on an approximation to the Bayes factor, leading to what is called a finite population BIC.

Ref. [192] presents a variety of important developments in SAE, including methods of model selection, noting that mixed effects models, such as LMM and GLMM (e.g., [123]) are extensively used in SAE. The author pointed out that applying classical model selection criteria, such as the AIC [4], is not straightforward because the likelihood requires specification of the distribution of the random effects, and it is difficult to determine the effective number of parameters (see below for further discussion). The author further discussed extensions of the information criteria, including the conditional AIC [228] and the fence methods [132]; see below.

All of the information criteria can be expressed in the form of the generalized information criterion (GIC) ([186], [212]), namely,

$$\text{GIC}(M) = \hat{Q}_M + \lambda_n |M|, \tag{6.1}$$

where $\hat{Q}_M$ is a measure of lack-of-fit by the model, M (see below for more discussion); $|M|$ is the dimension of M, defined as the number of free parameters under M; and λ_n is a penalty for complexity of the model, which may depend on the effective sample size, n. Let $\mathcal{M}$ denote the space of candidate models. In principle, the GIC needs to be evaluated for all $M \in \mathcal{M}$, and the model with the minimum GIC is selected.

Ref. [132] noted a number of difficulties with the information criteria when applied to non-conventional situations, including the following.

1. The effective sample size. The penalty, λ_n, may depend on n. For example, if the data are i.i.d., the effective sample size should be the same as the actual sample size, because every new observation provides the same amount of new information. On the other hand, if all the data points are identical, the effective sample size should be 1, regardless of the number of observations, because every new observation provides no additional information. Of course, the latter case is extreme, but there are many practical situations where the observations are correlated, even though they are not identical. In such a case, the effective sample size can be expected to be somewhere between 1 and n, but one may not know where.

2. The dimension of a model, $|M|$. In some cases, such as the ordinary linear regression, the dimension is simply the number of parameters in the model. In other situations where nonlinear, adaptive models are fitted, the dimension of the model can be substantially different from the number of parameters. Ref. [241] developed the concept of generalized degrees of freedom (GDF) to track model complexity. For example, in the case of multivariate adaptive regression splines [64], k nonlinear terms can have an effect of approximately $3k$ degrees of freedom. While a general algorithm in its essence, the GDF approach requires significant computations. It is not at all clear how a plug-in of GDF for $|M|$ affects the selection performance of the criterion.

3. Unknown distribution. As also noted in [192], in many cases, the distribution of the data is not fully specified (up to a number of unknown parameters); as a result, the likelihood function is not available. For example, suppose that normality is not assumed in an LMM. Now suppose that one wishes to select the fixed covariates using AIC or BIC. It is not clear how to do this because $\hat{Q}(M)$, which is supposed to be minus two times the maximized log-likelihood in such a case, is not available.

4. Finite-sample performance, and the effect of a constant. Even in standard situation such as regression variable selection, there are practical issues. For example, the BIC is known to have the tendency to overly penalize "bigger" models. Here, a bigger model is more complex in the sense of having more unknown parameters than a smaller model; for example, in regression variable selection, a bigger model involves more predictors, hence more unknown regression coefficients, than a smaller model. In other words, the penalizer, $\lambda_n = \log n$, may be a little too much in some cases. In such a case, one may wish to replace the penalizer by $c\log(n)$, where c is a constant less than one. What c should be used? Asymptotically, the choice of c does not make a difference in terms of consistency of model selection, so long as $c > 0$. However, practically, it does. In fact, there have been several modifications of the BIC aiming at improving the finite-sample performance. For example, [18] proposed a δ-BIC method by replacing the $\lambda_n = \log n$ in BIC by $\lambda_n = \delta \log n$, where δ is a constant carefully chosen to optimize the finite-sample performance. However, the choice of δ relies on extensive Monte-Carlo simulations and is determined on a case-by-case basis. It is not easy to generalize the δ-BIC method.

The information criteria may also run into some computational difficult, as noted below.

5. Computational burden. The computational burden of GIC depends on $\mathcal{M}$, the space of candidate models. Consider, for example, a problem of regression variable selection. Let $x_1, \dots, x_p$ be all the candidate predictors of an outcome variable, y. A subset of $\mathcal{P} = \{1, \dots, p\}$ is to be selected via the GIC. If the selection is over all possible subset of $\mathcal{P}$, which is called all-subset selection, there are 2^p different subsets. If p is large, this can be computationally very challenging. For example, high-dimensional variable selection problems are increasingly encountered in modern data science. This means that p can be in the scale of thousands or even higher. The total number of different subsets is then an astronomical number. It is simply not possible to do all-subset selection. As a partial solution, one may implement the procedure in a step-wise, in which the predictors are selected sequentially. Standard step-wise selection methods include forward selection and backward elimination (e.g., [148], sec. 9.4).

Moreover, there are some more important, practical concerns. Broadly speaking, model selection is not a statistical problem. This is because one does not do model selection for fun (although it could be fun at times). There is usually a real-life problem that drives the need for model selection. If so, why is a criterion for model selection, such as the AIC, solely based on statistical consideration? Practical considerations, specific to the real-life problem that one is dealing with, must be allowed to play a role. The question is how.

Ref. [132] (also see [182], [192],[122]) proposed a class of model selection strategies, known as the fence methods, that are particularly suitable for non-conventional model selection problems. What is more, the fence can (potentially) take practical considerations into consideration. The basic idea of the fence may be described as three steps:

1. Build a "statistical fence" to ensure that the model fits the data reasonably well. The fence isolates a subset of candidate models that meet the model-fitting threshold.

2. Within the fence, incorporate other considerations to identify the optimal model. For example, parsimony is often used, with the objective of selecting the simplest eligible model (within the fence), typically characterized as having the minimum dimension (equal to the number of free parameters under the model). More importantly, practical considerations can be incorporated when searching for the optimal model within the fence. Such practical considerations can be scientific, economical, legal, or political (e.g., the model must not require privacy-protected information to "train").

3. Finally, the threshold of the fence may be determined by "letting the data speak".

The building of the fence in step 1 is via the following inequality:

$$Q(M) - Q(M_*) \leq c, \tag{6.2}$$

where M denotes a candidate model; $Q(\cdot)$ is a measure of lack-of-fit, that is, a function of M such that M such that $Q(M_1) \geq Q(M_2)$ if M_1 is a sub-model (i.e., a special case) of M_2; M_* is a baseline model, that is, a model with minimum Q; and c is a threshold. For example, in regression model selection, let M_{f} denote the full model, that is, the model that includes all the candidate predictors. If Q is chosen as the sum of squares of residuals (SSR), then we have $M_* = M_{\mathrm{f}}$. Another popular choice of Q is the negative log-likelihood. See [122] for other examples of Q.

A challenging problem is how to choose the threshold c of the fence. Here, we explain the idea of step 3 by considering a special case.

Let parsimony be the criterion for selecting the optimal model within the fence. For simplicity, consider a "classical" setting, in which the space of candidate models contains a true model, i.e., a model under which the data are generated. In this setting, an "over-fitting model" that includes the true model as a special case is also a correct model. However, an over-fitting model may not be optimal because it may be simplified. In particular, M_{f} is a correct model. Let M_c denotes the model selected using the fence procedure with the given threshold c. Ideally, one would like to choose c such that

$$P_c = \mathrm{P}(M_c = M_{\mathrm{opt}}) \tag{6.3}$$

is maximized, where M_{opt} denotes the optimal model. However, two things are unknown:

(i) What is the probability distribution (P), in (6.3)? Typically, the probability distribution depends on the model. Here, we do not know what the model is because that is what is subject to selection.

(ii) What is M_{opt} in (6.3)? If this were known, our job would have been done already.

Ref. [132] (also see [124]) proposed strategies to solve (i) and (ii). For (i), the authors suggested that, because M_{f} is at least a correct model, one can bootstrap under M_{f} and approximate the probability P by the empirical (i.e., bootstrap) probability $\text{P}^{?}$.

For (ii), the authors suggested using the idea of maximum likelihood. For a given candidate model, M, let $p^*(M) = \text{P}^*(M_c = M)$, that is, the empirical probability that M is selected. Let p^* denote the maximum empirical probability over all candidate models. To illustrate with an example, suppose that 100 bootstrap samples are drawn. Model 1 is selected 10 times; Model 12 is selected 85 times; Model 17 is selected 4 times; Model 32 is selected once, and no other models are selected. Then, we have $p^*(M_1) = 0.1$, $p^*(M_{12}) = 0.85$, $p^*(M_{17}) = 0.04$, $p^*(M_{32}) = 0.01$, and $p^*(M) = 0$, $M \notin \{M_1, M_{12}, M_{17}, M_{32}\}$. It follows that $p^* = 0.85$. Note that p^* depends on the threshold c, that is, $p^* = p_c^*$. That is exactly the point! The idea is to choose c that maximizes p_c^*. The fence with the threshold c chosen this way is called adaptive fence, or AF [132].

It should be noted that such a plot of p_c^* vs c is typically W-shaped, instead of Λ-shaped. This is because, if M_* is unique, then for $c = 0$, by (1), the only model within the fence is M_*; therefore, $p_0^* = 1$ because M_* must be selected every time. To see what happens when c is very large, assume that the parsimony criterion is used to selection the optimal model within the fence and that there is a unique model, M_0, with the minimum dimension among the candidate models. Then if c is sufficiently large, every candidate model is within the fence; thus, M_0 must be selected every time, hence $p_c^* = 1$. Typically, one would look for a peak in the middle of the p_c^* vs c plot. For example, in regression variable selection, this means that one believes the optimal model is neither M_{f} nor M_0; in other words, some, but not all, of the candidate predictors are associated with the outcome variable.

A real-data example is provided in the next section to illustrate the fence method.

Regarding the computational aspect, there is a fence algorithm that allows the fence to proceed more efficiently computationally. Suppose, once again, that parsimony is the criterion for selecting the optimal model within the fence (otherwise, the algorithm may be modified accordingly). Then, in short, the algorithm may be described as follows: Check the candidate models, from the simplest to the most complex. Once one has discovered a model that falls within the fence and checked all of the other models of the same simplicity for their membership within the fence, one stops.

An alternative strategy for model selection, which is computationally more attractive than the information criteria, especially for high-dimensional selection problems, is the shrinkage selection/estimation method, which has become popular since the Lasso [224]. The method applies mainly to the variable selection in regression or generalized linear models. It is a type of penalized estimation procedure with a suitably chosen penalty function, such as the L^1 penalty. Depending on the choice of a penalty parameter, or tuning parameter, only a subset of the estimated coefficients are nonzero; the rest are (exactly) zero. This way, the method does variable selection and parameter estimation in a single operation, because the selected subset of variables correspond to those whose estimated coefficients are nonzero. See, for example, [182], [122] (ch. 7). We return to this topic in Section 6.4.

6.2 Example

Recall the Iowa crops data discussed in Example 2.3, which involved 12 Iowa counties obtained from the 1978 June Enumerative Survey of the U.S. Department of Agriculture presented in [8](hereafter, BHF, in this section). Ref. [124] used the same data to illustrate the adaptive fence method. The objective of the study was to provide estimates of the land area (in hectares) planted to different types of crops, namely, corn and soybeans. The land area could be measured by sending technicians to the field; this is accurate but too costly. Data from land observatory satellites on crop areas are available; this is much cheaper and more convenient to use, but less accurate. It would be nice to improve the accuracy of the satellite data for estimating the land areas. The SAE method offers a way to do so.

The idea is to borrow strength via a statistical model. The model is carefully chosen (see below) and trained. Here, training is a machine-learning term for fitting a model. To train a model, we need the field measurements and the satellite images for the same fields. This is called training data, another machine-learning term. We specify a relationship between these two measures by fitting the model; in other words, we estimate the unknown parameters involved in the model using the training data. Once the model is trained, we can use it for estimating any specific field's crop area using the satellite data and should do this more accurately than we would have if using only the field's satellite data.

We can do better because we have borrowed strength from the model, which allows us to know more than we would have if we only utilized the raw satellite data. It is quite likely that the model we are using is wrong; in that sense, there is no free lunch. However, this model may still be useful for our objective, depending on how wrong it is.

There are two types of crops of interest: corn or soybeans. For a given type of crops, BHF proposed a super-population NER model, which is a special

TABLE 6.1
Area sample sizes for Iowa crops data.

i	1	2	3	4	5	6	7	8	9	10	11	12
n_i	1	1	1	2	3	3	3	3	4	5	5	6

case of LMM. For example, the corn model is a special case of the NER model (1.17) repeated here as

$$y_{ij} = x_{ij}'\beta + v_i + e_{ij}, \tag{6.4}$$

$i = 1, \ldots, m$, $j = 1, \ldots, N_i$. Here, y_{ij} is the number of hectares of the jth corn area in the ith county; $m = 12$ corresponding to the 12 Iowa counties; and N_i is the population size (the total number of crop areas planted for corn in the county) for the ith county. Furthermore, v_i is an area-specific random effects corresponding the between-area variation, and e_{ij} is an additional error associated with the within-county variation. It is assumed that the v_i's and e_{ij}'s are independent and normally distributed with means zero and different, unknown variances for v and e. As for the mean function of y_{ij}, the BHF model assumes

$$x_{ij}'\beta = \beta_0 + \beta_1 x_{ij1} + \beta_2 x_{ij2}, \tag{6.5}$$

where x_{ij1} and x_{ij2} are the numbers of satellite image pixels classified as corn and soybeans, respectively. A similar model was proposed for the soybeans data, in which case the outcome on the left side of (6.4) is for the soybeans.

The observed data, or training data, are also assumed to satisfy (6.4), with N_i replaced by n_i, $1 \leq i \leq 12$, where the n_i's are area sample sizes given in Table 6.1. The training data are presented in Table 1 of BHF. It is seen that the sample sizes for all of the counties are very small.

Even though model (6.5) was proposed for the mean function, there were concerns on why the numbers of pixels for both types of crops, corn and soybeans, are needed for a single type of outcome, say, corn. One explanation (Fuller; personal communication) is that, for instance, for the estimation of corn areas, although the soybeans pixels are of wrong type, they, at least, suggested that there was something out there, and knowing something is better than knowing nothing. In fact, BHF discussed possibilities of including quadratic terms of $x_{ijk}, k = 1, 2$ on the right side of (4). More generally, we may treat this as a problem of variable selection, where the space of candidate predictors is

$$\{1, x_{ij1}, x_{ij2}, x_{ij1}^2, x_{ij2}^2, x_{ij1}x_{ij2}\}.$$

We applied the fence method to this model selection problem, with the measure of lack-of- fit (Q) being the negative log-likelihood. Table 6.2 presents the model selection results when parsimony is used as the criterion of selecting the optimal model within the fence, and the adaptive fence (AF) method is used to choose the threshold c, with the BHF models obtained with (6.5) included for comparison. Figure 6.1 presents the AF plots of p_c^* vs c for corn

TABLE 6.2
Model selection results and comparison.

Outcome	Predictors	
	BHF Model	AF
Corn	Corn, Soybeans	Corn
Soybeans	Corn, Soybeans	Soybeans

(left) and soybeans (right). The selected model by AF in Table 6.2 correspond to the peak-in-the-middle in each case.

A remaining question is whether the models selected by AF are reasonable. Recall, here, parsimony is the criterion for selecting the model within the fence. For the most part, the criterion prefers a simpler model provided that the model fits the data sufficiently well. This leads to the optimal selection that says model corn from corn; soybeans from soybeans (rather than corn and soybeans for corn, and corn and soybeans for soybeans). Thus, if model simplicity is the only concern besides model fit, the AF selection seems to follow the common sense.

As noted, one does not do model selection for fun; there is usually a real-life problem that drives the need for the model selection. Here, the main interest is estimation of the crop areas. Simplicity, or parsimony, is not necessarily a primary concern. Perhaps some other consideration more directly related to the crop-area estimation should be incorporated into the fence. This has not been done. In Section 6.4, we explore how the main interest of SAE can be incorporated into model selection.

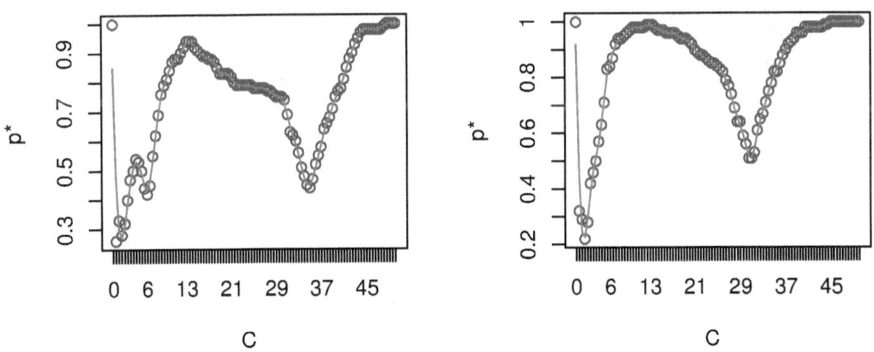

FIGURE 6.1
AF plots: corn (left) and soybeans (right). (Source: [111].)

6.3 Selecting a non-parametric model for SAE

Recall the spline-based model for SAE introduced in Section 5.2. Clearly, a spline is characterized by p, the degree of polynomial, q, the number of knots, and the location of the knots. Note that, given p, q, the location of the knots are typically selected either as equidistant, or by the space-filling algorithm implemented in R [**cover.design()**]. But the question how to choose p and q remains. The general "rule-of-thumb" is that p is 1, 2 or 3, and q proportional to the sample size, n, with 4 or 5 observations per knot. But there are still different choices given the rule-of-thumb. For example, if $n = 200$, the possible choices for q range from 40 to 50, which, combined with the choice of $1, 2, 3$ for p, gives a total of 33 different choices for the spline.

The fence method offers a natural approach to choosing the degree of the spline, p, the number of knots, q, and the smoothing parameter, λ [see (5.4)] at the same time. Note that, however, here the true underlying model is not among the class of candidate models, that is, the true mean function $f(x)$ is not a spline. This is different from what is assumed in the "classical model selection" setting (e.g., [132]).

Furthermore, the role of λ in the model should be made clear: λ controls the degree of smoothness of the underlying model. In this context, a natural measure of lack-of-fit is

$$Q(M, \theta_M; y) = |y - X\beta - Z\gamma|^2.$$

However, $Q(M)$ is not obtained by minimizing $Q(M, \theta_M; y)$ over β and γ without constraint. Instead, we have

$$Q(M) = |y - X\hat{\beta} - Z\hat{\gamma}|^2,$$

where $\hat{\beta}$ and $\hat{\gamma}$ minimize (5.4, hence depend on λ.

Another issue has to do with how to implement the adaptive fence (AF). Recall, in Section 6.1, the AF is carried out by bootstrapping under the full model, $M_{\rm f}$. Here, there may not be a full model among the candidate models. Thus, the M_* in the fence inequality, (6.2) is chosen as the candidate model that has the minimum $Q(M)$.

As for the criterion for selecting the model within the fence, [125] suggests following criterion that combines model simplicity and smoothness. For the models within the fence, choose the one with the smallest q; if there are more than one such model, choose the model with the smallest p. This gives the best choice of p and q. Once p, q are chosen, we choose the model within the fence with the largest λ. The threshold c in (6.2) is chosen adaptively using the AF idea, where parametric bootstrap is used for computing p^*, namely, M_* and the ML estimators under M_* are used for the bootstrapping.

The following theorem is proved in [125]. For simplicity, assume that the matrix $W = (X\ Z)$ is of full rank. Let $P_{W^\perp} = I_n - P_W$, where $n = \sum_{i=1}^m n_i$ and $P_W = W(W'W)^{-1}W'$.

Theorem 6.1. Computationally, the above fence procedure is equivalent to the following: (i) first use the AF to select p and q using (6.2) with $\lambda = 0$ and $Q(M) = y'P_{W^\perp}y$ (see Lemma 6.1 below), and same criterion as above for choosing p, q within the fence; (ii) let M_0^* denotes the model corresponding to the selected p and q, find the maximum λ such that

$$Q(M_0^*, \lambda) - Q(\tilde{M}) \le c^*, \tag{6.6}$$

where for any model M with the corresponding X and Z, we have

$$\begin{aligned} Q(M, \lambda) &= |y - X\hat{\beta}_\lambda - Z\hat{\gamma}_\lambda|^2, \\ \hat{\beta}_\lambda &= (X'V_\lambda^{-1}X)^{-1}X'V_\lambda^{-1}y, \\ \hat{\gamma}_\lambda &= \lambda^{-1}(I_q + \lambda^{-1}Z'Z)^{-1}Z'(y - X\hat{\beta}_\lambda), \\ X'V_\lambda^{-1}X &= X'X - \lambda^{-1}X'Z(I_q + \lambda^{-1}Z'Z)^{-1}Z'X, \\ X'V_\lambda^{-1}y &= X'y - \lambda^{-1}X'Z(I_q + \lambda^{-1}Z'Z)^{-1}Z'y, \end{aligned}$$

and c^* is chosen by the AF (V_λ is defined below but not directly needed here for the computation because of the last two equations).

Note that in step (i) of the Theorem one does not need to deal with λ. The motivation for (6.6) is that this inequality is satisfied when $\lambda = 0$, so one would like to see how far λ can go. In fact, the maximum λ is a solution to the equation

$$Q(M_0^*, \lambda) - Q(\tilde{M}) = c^*.$$

The purpose of the last two equations is to avoid direct inversion of $V_\lambda = I_n + \lambda^{-1}ZZ'$, whose dimension is equal to n, the total sample size. Note that V_λ does not have a block diagonal structure because of ZZ', so if n is large direct inversion of V_λ may be computationally burdensome.

Lemma 6.1. For any M and y, $Q(M, \lambda)$ is an increasing function of λ with $\inf_{\lambda>0} Q(M, \lambda) = Q(M)$.

Ref. [125] carried out a simulation study designed to evaluate performance of the proposed fence method. The authors considered a non-parametric regression, which can be expressed as

$$y_i = f(x_i) + \epsilon_i, \quad i = 1, \ldots, m, \tag{6.7}$$

where $\epsilon_i \sim N(0, \sigma^2)$ and σ^2 is unknown.

Three different cases that cover various situations and aspects were considered. In the first case, Case 1, the true underlying function is a linear function, $f(x) = 1 - x$, $0 \le x \le 1$, hence the model reduces to a linear regression model. The goal is to find out if fence can validate the linear regression in the case that it is valid. In the second case, Case 2, the true underlying function is the quadratic spline with two knots, given by (5.6). Here, one intends to investigate whether the fence can identify the true underlying function in the "perfect" situation, that is, when $f(x)$ itself is a spline. The last case, Case 3,

TABLE 6.3
Nonparametric model selection – Case 1 and Case 2. Reported are empirical probabilities, in terms of percentage, based on 100 simulations that the optimal model is selected.

	Case 1			Case 2		
Sample size	$m = 10$	$m = 15$	$m = 20$	$m = 30$	$m = 40$	$m = 50$
Highest Peak	62	91	97	71	83	97
Confidence L.B.	73	90	97	73	80	96

is perhaps the most practical situation, in which no spline can provide a perfect approximation to $f(x)$. In other words, the true underlying function is not among the candidates. In this case $f(x)$ is chosen as $0.5\sin(2\pi x)$, $0 \leq x \leq 1$.

Several cases of small or median sample size were considered; namely, $m = 10$, 15 or 20 for Case I, $m = 30$, 40 or 50 for Case 2, and $m = 10$, 30 or 50 for Case 3. The covariate x_i were generated from the Uniform$[0, 1]$ distribution in Case 1, and from Uniform$[0, 3]$ in Case 2; then fixed throughout the simulations. As for Case 3, the x_i were chosen to be equidistant. The error standard deviation σ in (6.7) is chosen as 0.2 in Case 1 and Case 2. The value was chosen such that the signal standard deviation in each case is about the same as the error standard deviation. As for Case 3, three different values for σ were considered, 0.2, 0.5 and 1.0. These values were also of the same order as the signal standard deviation in this case.

The candidate approximating splines for Case 1 and Case 2 are the following: $p = 0, 1, 2, 3$, $q = 0$ and $p = 1, 2, 3$, $q = 2, 5$ (so there are a total of 10 candidates). As for Case 3, [125] consider only linear splines (i.e., $p = 1$); furthermore, the number of knots followed the "rule-of-thumb" (i.e., roughly 4 or 5 observations per knot; see above), plus the intercept model ($p = q = 0$), and the linear model ($p = 1$, $q = 0$). Thus, for $m = 10$, $q = 0, 2, 3$; for $m = 30$, $q = 0, 6, 7, 8$; and for $m = 50$, $q = 0, 10, 11, 12, 13$.

Table 6.3, taken from [125], shows the results based on 100 simulations under Case 1 and Case 2. As in [124], two strategies were considered in choosing the threshold c in the fence. The first was the highest peak, that is, choosing c with the highest p^* (i.e., AF); the second was the 95% lower bound, that is, choosing a smaller c corresponding to a peak of p^* in order to be conservative, if the corresponding p^* is greater than the 95% lower bound of the p^* for any larger c that corresponds to a peak of p^*. The 95% lower bound may be viewed as a variation of the AF. It is seen that performance of the AF is satisfactory even with the small sample size. Also, it appears that the confidence lower bound method works better in smaller sample, but makes almost no difference in larger sample. These are consistent with the findings of [124].

Table 6.4, also taken from [125], shows the results for Case 3. Note that, unlike Case 1 and Case 2, here there is no optimal model (an optimal model must be a true model, which does not exist). So, instead of giving the empirical

TABLE 6.4
Nonparametric model selection – case 3. Reported are empirical distributions, in terms of percentage, of the selected models.

	Sample Size # of Knots	$m=10$ 0,2,3		$m=30$ 0,6,7,8		$m=50$ 0,10,11,12,13	
		(p,q)	%	(p,q)	%	(p,q)	%
$\sigma=.2$	Highest Peak	(0,0)	1	(1,0)	9	(1,10)	100
		(1,0)	31	(1,6)	91		
		(1,2)	68				
	Confidence L.B.	(1,0)	24	(1,0)	9	(1,10)	100
		(1,2)	76	(1,6)	91		
$\sigma=.5$	Highest Peak	(0,0)	14	(1,0)	21	(1,0)	13
		(1,0)	27	(1,6)	77	(1,10)	84
		(1,2)	56	(1,7)	2	(1,11)	2
		(1,3)	3			(1,12)	1
	Confidence L.B.	(0,0)	8	(1,0)	8	(1,0)	2
		(1,0)	23	(1,6)	89	(1,10)	94
		(1,2)	65	(1,7)	3	(1,11)	2
		(1,3)	4			(1,12)	2
$\sigma=1$	Highest Peak	(0,0)	27	(0,0)	15	(0,0)	10
		(1,0)	20	(1,0)	18	(1,0)	26
		(1,2)	49	(1,6)	63	(1,10)	60
		(1,3)	4	(1,7)	4	(1,11)	2
						(1,12)	2
	Confidence L.B.	(0,0)	20	(0,0)	1	(0,0)	2
		(1,0)	13	(1,0)	13	(1,0)	13
		(1,2)	59	(1,6)	82	(1,10)	80
		(1,3)	8	(1,7)	4	(1,11)	2
						(1,12)	3

probabilities of selecting the optimal model, the empirical distribution of the selected models is presented in each case. It is apparent that, as σ increases, the distribution of the models selected becomes more spread out. A reverse pattern is observed as m increases. The confidence lower bound method appears to perform better in picking up a model with splines. Within the models with splines, the fence seems to overwhelmingly prefer fewer knots than more knots.

Note that the fence procedure allows one to choose not only p and q but also λ. In each simulation, the values of $\hat{\beta} = \hat{\beta}_\lambda$ and $\hat{\gamma} = \hat{\gamma}_\lambda$, given below (6.6), were computed based on the λ chosen by the AF. The fitted values are calculated by (5.1) with β and γ replaced by $\hat{\beta}$ and $\hat{\gamma}$, respectively. One then took the average of the fitted values over the 100 simulations. Figure 6.2 shows the average fitted values for the three cases ($m = 10, 30, 50$) with $\sigma = 0.2$ under Case 3. The true underlying function values, $f(x_i) = 0.5\sin(2\pi x_i)$, $i = 1, \dots, m$ are also plotted for comparison.

We conclude this section with an example of case study [125].

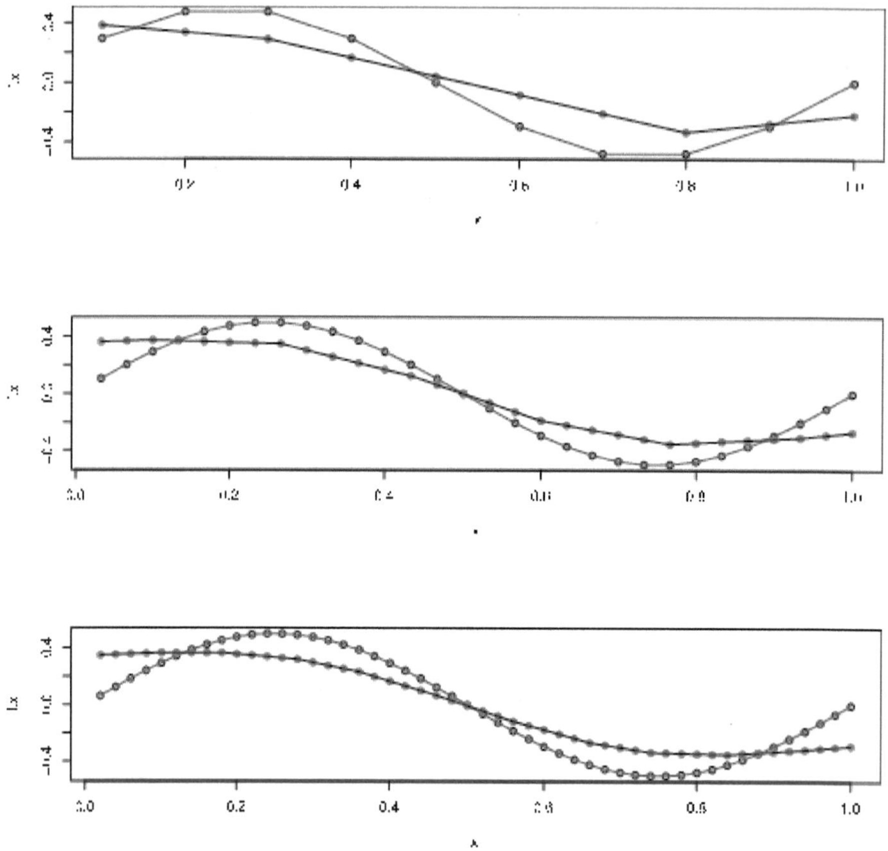

FIGURE 6.2
Case 3 Simulation. Top figure: Average fitted values for $m = 10$. Middle figure: Average fitted values for $m = 30$. Bottom figure: Average fitted values for $m = 50$. In all cases, the red dots represent the fitted values while the blue circles correspond to the true underlying function. (Source: [125].)

Example 6.1 (Case study). Recall the hospital data of Section 4.2.1. The data are presented in Table 6.5 for convenience.

Ref. [67] proposed a Fay-Herriot model for the graft failure rates. The model can be expressed as $y_i = \beta_0 + \beta_1 x_i + v_i + e_i$, where the v_i's are hospital-specific random effects and e_i's are sampling errors. It is assumed that v_i, e_i are independent with $v_i \sim N(0, A)$ and $e_i \sim N(0, D_i)$. Here the variance A is unknown. Based on the model, the author obtained credible intervals for selected contrasts. However, inspections of the raw data (see the right plot of Figure 6.3) suggest some nonlinear trends, which raises the question on whether the fixed effects part of the model can be made more flexible in its functional form.

TABLE 6.5
Hospital data from Morris and Christiansen (1995).

Area	y_i	x_i	$\sqrt{D_i}$
1	.302	.112	.055
2	.140	.206	.053
3	.203	.104	.052
4	.333	.168	.052
5	.347	.337	.047
6	.216	.169	.046
7	.156	.211	.046
8	.143	.195	.046
9	.220	.221	.044
10	.205	.077	.044
11	.209	.195	.042
12	.266	.185	.041
13	.240	.202	.041
14	.262	.108	.036
15	.144	.204	.036
16	.116	.072	.035
17	.201	.142	.033
18	.212	.136	.032
19	.189	.172	.031
20	.212	.202	.029
21	.166	.087	.029
22	.173	.177	.027
23	.165	.072	.025

To answer this question, we fit the Fay-Herriot model as a special member of a class of approximating spline models discussed in this section. More specifically, we assume the model (5.3) with x_i being the severity index. We then consider the approximating spline (5.1) with $p = 0, 1, 2, 3$ and $q = 0, 1, \ldots, 6$ ($p = 0$ is only for $q = 0$). Here the upper bound 6 is chosen according to the "rule-of-thumb" (because $m = 23$, so $m/4 = 5.75$). Note that the Fay-Herriot model corresponds to the case $p = 1$ and $q = 0$. The question is then to find the optimal model, in terms of p and q, from this class.

We apply the AF method described above to this case. Here to obtain the bootstrap samples needed for obtaining c^*, we first compute the ML estimator under the model $\tilde{M}$, which minimizes $Q(M) = y' P_{W^\perp} y$ among the candidate models (Theorem 6.1), then draw parametric bootstrap samples under M_* with the ML estimators treated as the true parameters. This is reasonable because M_* is the best approximating model in terms of the fit, even though under model (5.3) there may not be a true model among the candidate models. The bootstrap sample size is chosen as 100.

The AF has selected the model $p = 3$ and $q = 0$, that is, a cubic function with no knots, as the optimal model. We repeated the analysis 100 times, each time using different bootstrap samples. All results led to the same model: a

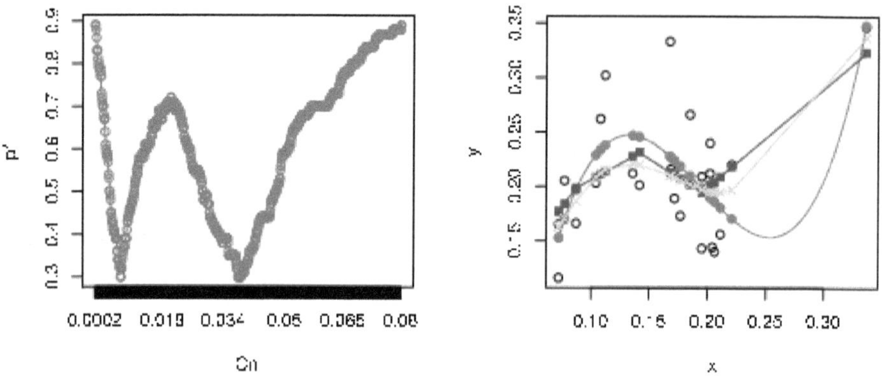

FIGURE 6.3
Left: A plot of p^* against c_n from the search over the full model space. Right: The raw data and the fitted values and curves; red dots and curve correspond to the cubic function resulted from the full model search; blue squares and lines correspond to the linear spline with 4 knots resulted from the restricted model search; green X's and lines represent the GAM fits. (Source: [125].)

cubic function with no knots. The left figure of Figure 6.3 shows the plot of p^* against $c = c_n$ in the AF model selection.

A few comparisons are always helpful. Our first comparison is to fence itself but with a more restricted space of candidate models. More specifically, we consider (5.1) with the restriction to linear splines only, i.e., $p = 1$, and knots in the range of the "rule-of-thumb", i.e., $q = 4, 5, 6$, plus the intercept model ($p = q = 0$) and the linear model ($p = 1$, $q = 0$). In this case, the fence method selected a linear spline with four knots (i.e., $p = 1$, $q = 4$) as the optimal model. The value of λ corresponding to this model is approximately equal to 0.001. The plot of p^* against c_n for this model selection is very similar to the left figure of Figure 6.3 (not presented). In addition, the right plot of Figure 6.3 shows the fitted values and curves under the two models selected by the fence within the different model spaces as well as the original data.

A further comparison can be made by treating (5.3) as a generalized additive model (GAM; e.g., [91]) with heteroscedastic errors. A weighted fit can be obtained with the amount of smoothing optimized by using a generalized cross-validation (GCV) criterion. Here the weights used are $w_i = 1/(A + D_i)$ where the MLE for A is used as a plug-in estimate. Recall that the D_i's are known. This fitted function is also overlaid in the right plot of Figure 6.3. Note how closely the fitted curve resembles that of the restricted space fence.

To expand the class of models under consideration by GCV-based smoothing, we used the BRUTO procedure [[91]] which augments the class of models to look at a null fit and a linear fit for the spline function; and embeds the resulting model selection (i.e., null, linear or smooth fits) into a weighted back-fitting algorithm using GCV for computational efficiency. Interestingly here, BRUTO finds simply an overall linear fit for the fixed effects functional form. While certainly an interesting comparison, BRUTO's theoretical properties for models like (5.3) have not really been studied in depth.

Finally, as mentioned in Section 5.2, by using the connection between P-spline and linear mixed model one can formulate (5.3) with the spline approximation as a LMM, where the spline coefficients are treated as random effects. The problem then becomes a (parametric) mixed model selection problem, and hence, the method of [132] can be directly applied.

6.4 Shrinkage and predictive model selection

We now return to the shrinkage selection methods briefly discussed at the end of Section 6.1. Let us first consider variable selection for a linear regression model: $y = X\beta + \epsilon$, where $y = (y_1, \ldots, y_n)'$ is an $n \times 1$ vector of response, $X = (x_1, \ldots, x_p)$ is an $n \times p$ matrix of predictors, $\beta = (\beta_1, \ldots, \beta_p)'$ is the vector of regression coefficients, and ϵ is a vector of independent errors with mean 0 and variance σ^2. In a high-dimensional selection problem, β is supposed to be sparse, which means that most of the predictors are not associated with the response. A penalized likelihood is used to identify which components of β are to be shrunk to zero and estimate the rest. The penalized likelihood estimator is given by

$$\hat{\beta} \quad = \quad \underset{\beta}{\operatorname{argmin}} \left\{ \left| y - \sum_{j=1}^{p} x_j \beta_j \right|^2 + \sum_{i=1}^{p} \rho(\lambda_j, |\beta_j|) \right\}, \tag{6.8}$$

where $|\cdot|$ is the Euclidean norm, and $\rho(\lambda_j, .)$ is a given non-negative penalty function that depends on the regularization parameter, λ_j. Ref. [224] proposed the Lasso ℓ_1 penalty, with $\rho(\lambda_j, |\beta_j|) = \lambda|\beta_j|$. Ref. [57] proposed a continuous differentiable penalty function (SCAD) whose derivative is defined by

$$(\partial/\partial u)\rho(\lambda_j, u) = \lambda\{1_{(u \leq \lambda)} + \{(a-1)\lambda\}^{-1}(a\lambda - u)_+ 1_{(u > \lambda)}\}$$

for $a > 2$ and $u > 0$. The parameter a is shown to be fairly stable for the SCAD, and is usually set as 3.7 [[57]].

It should be noted that the choice of λ is critically important because, for example, depending on the value of λ, it is possible to have none of the β's shrunk to zero, meaning that all of the predictors are selected, or all of the β shrunk to zero, meaning that none of the predictors are selected. One approach to selecting regularization parameter is to use the information criteria such as the AIC [4] and BIC [208]. Some other data-driven approaches, such as the delete-d cross-validation (CV), generalized CV have also been proposed. See

[122], sec. 7.1 for an overview. Following the AF idea (see Section 6.1), [189] developed a bootstrap approach to choosing the regularization parameter.

Although the shrinkage selection/estimation methods were originally proposed under the setting of regression or generalized linear models, in which cases the data are assumed to be independent, there has been extension of the shrinkage methods to SAE under a LMM, in which case the data are correlated; however, such correlations were ignored in the extension. See, for example, [204]. The latter authors considered model selection for SAE of the proficiency of adult competency in the United States. The goal was to identify and select the best subset of covariates that are related to the outcome variable of adult competency. Here, the outcome variables were measures of two cognitive skills, literacy and numeracy. The data were collected from a total of 185 counties of the United States some of which had small sample sizes. The data sources involved more than 70 county-level variables across five major variable types as potential covariates. The authors proposed a LMM involving three levels of random effects, that is, random effects at the county, state and division level. Due to the relatively large number of potential variables, the authors used a two-phase selection process. In the first phase, a direct application of the Lasso (i.e., treating the LMM as a linear regression model) was used to identify a small set of variables that were "consistently highly correlated with the outcomes". In the second phase, a cross-validation method was used to select the final set of variables. The selection was restricted to the fixed covariates only, that is, the random effect factors (there were 3 of them in this case) were not subject to selection.

There has been extension of the shrinkage selection/estimation methods to joint selection of the fixed and random effects in mixed effects models. Ref. [15] considered such a selection problem in a certain type of LMM, which can be expressed as

$$y_i = X_i\beta + Z_i v_i + e_i, \quad i = 1, \ldots, m, \tag{6.9}$$

where y_i is an $n_i \times 1$ vector of responses for subject i, X_i is an $n_i \times p$ matrix of explanatory variables, β is a $p \times 1$ vector of regression coefficients (the fixed effects), Z_i is an $n_i \times q$ known design matrix, v_i is a $q \times 1$ vector of subject-specific random effects, e_i is an $n_i \times 1$ vector of errors, and m is the number of subjects. It is assumed that the $v_i, e_i, i = 1, \ldots, m$ are independent with $v_i \sim N(0, \sigma^2\Psi)$ and $e_i \sim N(0, \sigma^2 I_{n_i})$, where Ψ is an unknown covariance matrix. The problem of interest, using the terms of shrinkage model selection, is to identify the nonzero components of β and $v_i, 1 \le i \le m$. For example, the components of v_i may include a random intercept and some random slopes.

To take advantage of the idea of shrinkage variable selection, [15] adopted a modified Cholesky decomposition. Note that the covariance matrix, Ψ, can be expressed as $\Psi = D\Omega\Omega' D$, where $D = \text{diag}(d_1, \ldots, d_q)$ and $\Omega = (\omega_{kj})_{1 \le k,j \le q}$ is a lower triangular matrix with 1's on the diagonal. Thus, one can express (6.9) as

$$y_i = X_i\beta + Z_i D\Omega\xi_i + e_i, \quad i = 1, \ldots, m, \tag{6.10}$$

where the ξ_i's are independent $N(0, \sigma^2)$ random variables. The idea is to apply shrinkage estimation to both $\beta_j, 1 \leq j \leq p$ and $d_k, 1 \leq k \leq q$. Note that setting $d_k = 0$ is equivalent to setting all of the elements in the kth column and kth row of Ψ to zero, and thus creating a new sub-matrix by deleting the corresponding row and column, or the exclusion of the kth component of α_i.

However, direct implementation of this idea is difficult, because the ξ_i's are unobserved, even though their distribution is much simpler. To overcome this difficulty, [15] used the E-M algorithm [48]. By treating the ξ_i's as observed, the complete-data log-likelihood can be expressed as

$$l_c = c_0 - \frac{N + mq}{2} \log \sigma^2 - \frac{1}{2\sigma^2}(|y - X\beta - Z\tilde{D}\tilde{\Omega}\xi|^2 + |\xi|^2), \tag{6.11}$$

where c_0 is a constant, $N = \sum_{i=1}^{m} n_i$, $X = (X_i)_{1 \leq i \leq m}$, $Z = \text{diag}(Z_1, \ldots, Z_m)$, $\tilde{D} = I_m \otimes D$, $\tilde{\Omega} = I_m \otimes \Omega$ ($\otimes$ means Kronecker product), $\xi = (\xi_i)_{1 \leq i \leq m}$, and $|\cdot|$ denotes the Euclidean norm. (6.11) leads to the shrinkage estimation by minimizing

$$P_c(\phi|y, \xi) = |y - X\beta - Z\tilde{D}\tilde{\Omega}\xi|^2 + \lambda \left(\sum_{j=1}^{p} \frac{|\beta_j|}{|\tilde{\beta}_j|} + \sum_{k=1}^{q} \frac{|d_j|}{|\tilde{d}_j|} \right), \tag{6.12}$$

where ϕ represents all of the parameters, including the β's, the d's, and the ω's, $\tilde{\beta} = (\tilde{\beta}_j)_{1 \leq j \leq p}$ is given by the right side of (1.28) with the variance components involved in $V = \text{Var}(y)$ replaced by their REML estimators (e.g., [123], sec. 1.3.2), $\tilde{d}_k, 1 \leq k \leq q$ are obtained by decomposition of the estimated Ψ via the REML, and λ is the regularization parameter.

Ref. [15] proposed to use the BIC in choosing the regularization parameter. Here the form L^1 penalty in (6.12) is in terms of the adaptive Lasso [245]. To incorporate with the E-M algorithm, one replaces (6.12) by its conditional expectation given y, and the current estimate of ϕ. Note that only the first term on the right side of (6.12) involves ξ, with respect to which the conditional expectation is taken. The conditional expectation is then minimized with respect to ϕ to obtain the updated (shrinkage) estimate of ϕ.

A similar approach was proposed by [101] for joint selection of the fixed and random effects in GLMMs (e.g., [123]), although the performance of the proposed method was studied only for the special case of LMM.

The primary interest in SAE is estimation, or prediction under a mixed effects model framework, of small area means. It is desirable to incorporate such a practical interest in model selection, as noted in Sections 6.1 and 6.2. So far, this has not been done.

More generally, problems of practical interest often arise in the context of mixed model prediction, in which the quantity of interest can be expressed in terms of a mixed effect. See, for example, [206], [116], for reviews on prediction of mixed effects and its applications. As noted, the fence is flexible in choosing the criterion of optimality in selecting the model within the fence to incorporate consideration of practical interest. Another potentially useful feature of the fence, which so far has not been seriously explored, is that the fence is also flexible in choosing the measure of lack-of-fit, Q, in (6.2), to incorporate the practical interest. Because prediction is currently of main interest, it makes sense to use a predictive measure, rather than an estimation-based measure. Examples of the latter include the negative log-likelihood and the residual sum of squares. See [122] for discussions.

We now describe a general approach to deriving a predictive measure of lack-of-fit using the idea of the OBP. Consider the general setting in Section 4.1.1. It is assumed that $\mu = X\beta$, where the columns of X correspond to a subset of the candidate predictors, or covariates, and β is vector of corresponding fixed effects.

Suppose that G and Σ are known. Then, under the assumed model, the best predictor (BP) of θ, in the sense of minimum MSPE, is the conditional expectation,

$$\begin{aligned} \mathrm{E}_M(\theta|y) &= F'\mu + R'\mathrm{E}_M(v|y) \\ &= F'X\beta + R'GZ'V^{-1}(y - X\beta), \end{aligned} \tag{6.13}$$

where $V = \Sigma + ZGZ'$ and β is the true vector of fixed effects (e.g., [123], p. 88). The E_M in (6.13) denotes conditional expectation under the assumed model, M, rather than the true model (4.3). Although the LMM, that is, (4.3) with $\mu = X\beta$, may be subject to model misspecification, it utilizes the available covariates, X, in a simple way. On the other hand, even if model (4.3) is correct, or close to be correct, it is too broad to be useful; furthermore, it does not make use of any of the available covariates, which is often practically unacceptable. For these reasons, the assumed LMM is often the one of main interest. In other words, one cannot abandon the assumed model; all one could do is to try to do the best under the assumed model, that is, to find the best way to estimate the parameters, in this case β, under the assumed model.

Note that, here, the true model is used in evaluating the predictive performance of the BP, (6.13), to make sure that the evaluation is fair and not model-dependent. This is why, intuitively, the resulting predictor is more robust to model misspecification. The idea introduced in this context is also important when dealing with model selection, because, obviously, the measure of lack-of-fit has to be objective, or "fair", to all of the candidate models.

To derive the best estimator of β, write $B = R'GZ'V^{-1}$ and $\Gamma = F' - B$. Let $\tilde{\theta}$ denote the right side of (6.13), where β is understood as a parameter vector to be determined. The predictive performance of $\tilde{\theta}$ is typically measured by the MSPE, defined as $\mathrm{MSPE}(\tilde{\theta}) = \mathrm{E}(|\tilde{\theta} - \theta|^2)$. Here, E denotes expectation under the true model. It was shown (see Section 4.1.1) that the MSPE can be expressed, alternatively, as

$$\mathrm{MSPE}(\tilde{\theta}) = \mathrm{E}\{(y - X\beta)'\Gamma'\Gamma(y - X\beta) + \cdots\}, \tag{6.14}$$

where $\cdots$ does not depend on β. Note that, unlike the expectation on the left side, the expression inside the expectation on the right side of (6.14) involves a function of the observed data, β, and something unrelated to β. The OBP idea is to estimate β by minimizing the expression inside the expectation on the right side of (6.14), which is equivalent to minimizing the expression without $\cdots$. This leads to the BPE of β, given by (4.7), assuming that $\Gamma'\Gamma$ is nonsingular and X is full rank. See Example 4.1 for an illustration.

As far as this section is concerned, the development leads to a predictive measure of lack-of-fit. Namely, one can define the $Q(M)$ in (6.2) as the minimizer, over β, of the expression without $\cdots$ inside the expectation on the right side of (6.14). Clearly, this measure is designed specifically for the mixed model prediction problem. Also, when it comes to model selection, it is important that the measure of lack-of-fit is "fair" to every candidate model. The above measure $Q(M)$ has this feature, because the expectation in (6.14) is under an objective true model.

The idea of using the above $Q(M)$ (or a similar measure) for variable selection, a special case of model selection, has already been explored in SAE; see Section 4.3.1 for the description of the OBSP method.

The assumption that G and Σ are known can be relaxed, to some extent, especially for G. In fact, it can be shown that, when G is unknown, essentially the same derivation of goes through, and the resulting measure of lack-of-fit, $Q(M)$, is the minimizer of

$$(y - X\beta)'\Gamma'\Gamma(y - X\beta) - 2\mathrm{tr}(\Gamma'\Sigma),$$

assuming that Σ is known.

As for Σ, it may be replaced by its REML estimator (e.g., [123], sec. 1.3.2), assuming that $r = \mathrm{rank}((X_{\mathrm{f}}, Z)) < n$, where X_{f} denote the full X, that is, the X involving all of the candidate predictors. Let T_{f} be an $n \times (n-p)$ matrix such that $\mathrm{rank}(T_{\mathrm{f}}) = n - r$, $T'X_{\mathrm{f}} = 0$ and $T_{\mathrm{f}}'Z = 0$. Suppose that (4.3) with $\mu = X_{\mathrm{f}}\beta_{\mathrm{f}}$ is a correct model for some vector β_{f}. Then, we have $z = T_{\mathrm{f}}'y = T_{\mathrm{f}}'e$, whose distribution depends on Σ only. One can then estimate Σ by its MLE based on z.

Suppose that the purpose of the joint selection is for predicting some mixed effects. We may combine the predictive measure of lack-of-fit with the shrinkage mixed model selection idea discussed above. Namely, we replace the first term on the right side of (6.12) by the following predictive measure,

$$Q(\phi|y) = (y - X\beta)'\Gamma\Gamma'(y - X\beta) - 2\mathrm{tr}(\Gamma'\Sigma). \tag{6.15}$$

We refer to this method as predictive shrinkage selection, or PSS. The idea has been explored by [97]. The authors found that PSS performs better than the shrinkage selection method based on (6.12) not only in terms of the predictive performance, but also in terms of parsimony. The authors also extended the PSS to Poisson mixed models, a special case of GLMM, and obtained similar results.

There is another advantage of the PSS in terms of computation. Denote the first term on the right side of (6.12) by $Q_{\mathrm{c}}(\phi|y, \xi)$. Note that, unlike $Q_{\mathrm{c}}(\phi|y, \xi)$, the unobserved ξ is not involved in $Q(\psi|y)$ defined by (6.15), which is what the PSS is based on. This means that, unlike [15] and [101], PSS does not need to run the E-M algorithm, and thus is computationally (much) more efficient. We illustrate the method in the next section with an example of case study.

6.5 Case study: Prediction for high-speed network

Efficient data access is essential for sharing massive amounts of data among geographically distributed research collaborators. The analysis of network traffic is getting more important in the era of Big Data for utilizing limited resources offered by the network infrastructures and planning wisely large data transfers. The efficiency of data transfer can be improved by learning the current conditions and accurately predicting future network performance. Specifically, short-term prediction of network traffic guides the immediate scientific data placements for network users; long-term forecast of the network traffic enables capacity-planning of the network infrastructure needs for network designers.

Such predictions become non-trivial when the amount of network data grows in unprecedented speed and volumes. One such available data source is NetFlow [103]. The NetFlow measurements provide high volume, abundant specific information for each data flow; some sample records are shown in Table 6.6, where each record contains the following list of variables (the IP addresses are masked due to privacy concerns):

Start, End The start and end time of the recorded data transfer.

Sif, Dif The source and destination interface assigned automatically for the transfer.

SrcIPaddress, DstIPaddress The source and destination IP addresses of the transfer.

SrcP, DstP The source and destination Port chosen based on the transfer type such as email, FTP, SSH, etc.

P The protocol chosen based on the general transfer type such as TCP, UDP, etc.

Fl The flags measured the transfer error caused by the congestion in the network.

Pkts The number of packets of the recorded data transfer.

Octets The Octets measures the size of the transfer in bytes.

TABLE 6.6
Sample NetFlow records.

Start	End	Sif	SrcIPaddress(masked)	SrcP	Dif
DstIPaddress(masked)	DstP	P	Fl	Pkts	Octets
0930.23:59:37.920	0930.23:59:37.925	179	xxx.xxx.xxx.xxx	62362	175
xxx.xxx.xxx.xxx	22364	6	0	1	52
0930.23:59:38.345	0930.23:59:39.051	179	xxx.xxx.xxx.xxx	62362	175
xxx.xxx.xxx.xxx	28335	6	0	4	208
1001.00:00:00.372	1001.00:00:00.372	179	xxx.xxx.xxx.xxx	62362	175
xxx.xxx.xxx.xxx	20492	6	0	2	104
0930.23:59:59.443	0930.23:59:59.443	179	xxx.xxx.xxx.xxx	62362	175
xxx.xxx.xxx.xxx	26649	6	0	1	52
1001.00:00:00.372	1001.00:00:00.372	179	xxx.xxx.xxx.xxx	62362	175
xxx.xxx.xxx.xxx	26915	6	0	1	52
1001.00:00:00.372	1001.00:00:00.372	179	xxx.xxx.xxx.xxx	62362	175
xxx.xxx.xxx.xxx	20886	6	0	2	104

Features of NetFlow data have led to consideration of GLMMs (e.g., [123], ch. 3) for predicting the network performance. There are several reasons for such a consideration. First, the NetFlow records consisted of multiple time series with unevenly collected time stamps. Each of the these time series may be regarded as a group, or "small area", but the grouping structure is not yet clear until the relevant groups are identified. This is different from a traditional SAE problem, where the grouping structure is known in advance.

Furthermore, traditional time series methods such as ARIMA model, wavelet analysis, and exponential smoothing (e.g., [58], sec. 1.3.5) also encounter difficulties, because these methods are mainly designed for evenly collected time stamps and dealing with a single time series. Thus, there is a need for modeling a large number of time series without constraints on even collection of time stamps. GLMM is able to fully utilize all of the variables involved in the data set without requiring evenly-spaced time variable.

Secondly, there is empirical evidence of associations, as well as heteroscedasticity, found in the NetFlow records. For example, it takes longer, in general, to complete the transfer with increasing number of packets in a data transfer. This suggests that the number of packets may be considered as a (fixed) predictor for the duration of data transfer. Also, there appear to be fluctuation among network paths in terms of slope and range in the plots of duration against the number of packets. This suggests that a network path may be associated with a random effect to explain the duration under varying conditions. Thus, again, a mixed effects model seems to be plausible. Moreover, GLMM is more flexible than LMM in terms of the mean-variance association.

Finally, the NetFlow measurements are Big Data. The NetFlow data used for the current analysis were provided by ESnet for the duration from May 1, 2013 to June 30, 2013, with millions of observation for a single router within a day and 14 variables in each record with 30s or 40s interaction terms as candidates. The large volume and complexity of the data require efficient modeling. This is difficult to do with fixed effects modeling. For example, traditional hierarchical modeling requires dividing the data into groups, but the grouping is not clear and requires investigation to identify the variable that classify the observed data. Exploratory data analysis shows that the potential grouping factor could be the path of the data transfer, the delivering time of the day, the transfer protocol used, or the combination of some or all of these. This naturally leads to selection of random effect factors, a topic discussed in the previous selection.

There were different, practical interest that needed to address. For a network user, the established model should be able to predict the duration of a data transfer so that the user can expect how long it would take to complete the data transfer, given the information such as the size of the data, the time of the transfer, the selected path and protocols. On the other hand, for a network designer, the established model should be able to predict the long-time usage of the network so that the designer would know which link in the network is usually congested and requires more bandwidth or rerouting of the path.

These practical interests were addressed in [97], who built a LMM (and a GLMM) for these interests, and compared its prediction accuracy with two traditional selection procedures: Backward-Forward selection (B-F) and Estimation-based Lasso (E-Lasso). The latter refers to the shrinkage selection method, corresponding to (6.12). Here, we focus on the LMM for its simplicity.

The full model involved fixed effects including transfer start time, transfer size (Octets and Packets) and random effects including network transfer conditions such as Flag and Protocol, source and destination Port numbers, and transfer path such as source and destination IP addresses and Interfaces. The PSS procedure, introduced at the end of the previous section, with the Lasso penalty, selected the model

$$y = \beta_{\text{start}} s(x_{\text{start}}) + \beta_{\text{pkt}} x_{\text{pkt}} + Z_{\text{ip-path}} v_{\text{ip-path}} + e, \tag{6.16}$$

where $s(\cdot)$ is a fitted smoothing spline implemented to taken into account that the mean response is usually non-linearly associated with the time variable, x_{start}, with the parameters of the smoothing spline chosen automatically by cross-validation. The parameter estimates and their corresponding P-values for model (6.16) are given in Table 6.7.

The fitted smoothing spline is plotted in Figure 6.4, which shows how the transfer duration varies with start time. Furthermore, in (6.16), $Z_{\text{ip-path}}$ is

TABLE 6.7
Estimates of non-zero fixed effects.

Fixed Effects	Estimates	Standard Error	P-value
Intercept	-13.809	0.914	<2e-16
Start Time	0.574	0.0169	<2e-16
Packets	1.115	0.035	<2e-16

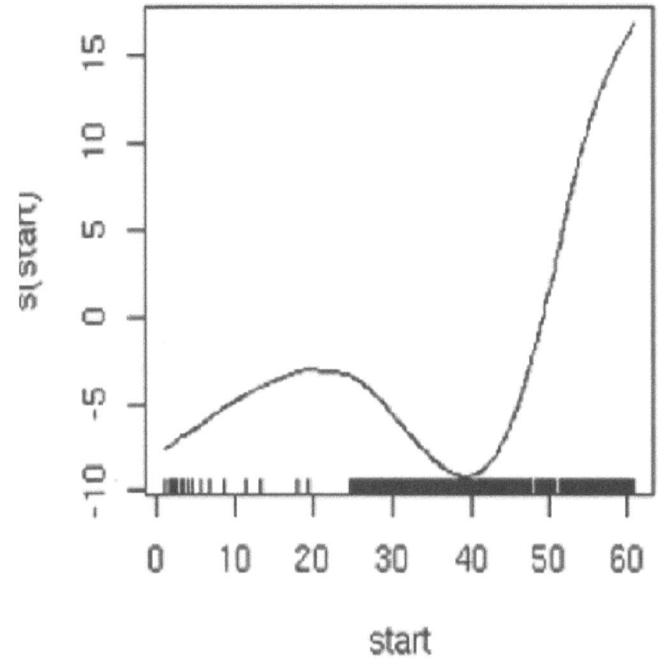

FIGURE 6.4
Fitted smoothing spline: transfer duration vs start time.

the design matrix whose columns correspond to the ip-paths, and $v_{\text{ip-path}}$ is a vector-valued random effect whose components correspond to the ip-paths. Finally, e is an additional error corresponding to the background noise.

In terms of the selection of random effect factors, the PSS has identified six paths with non-zero random-effect standard deviations, indexed as 14, 16, 38, 41, 61, and 83. Among those paths, all except path 61 have the estimated standard deviation of at least 10, while the estimated standard deviation for path 61 is almost zero. A plot is shown in Figure 6.5. The estimated standard deviation for the background noise is 11.239.

As for the comparison of PSS with B-F and E-Lasso, the overall MSPE as well as computation speed for the comparing methods are presented in Table 6.8. Note that, in this case, one actually knew the truth for the prediction, and therefore is able to compute the (exact) MSPE, and record the total computational time, of course. The results show that, in terms of prediction accuracy (MSPE), PSS is about 18 times better than E-Lasso, and 330 times better than B-F. In terms of computational time (using the unit of second), PSS is about 4×10^5 times less than E-Lasso and 3.8×10^8 times less than B-F.

In conclusion, at least for this application, the predictive shrinkage selection method, PSS, greatly improves the prediction accuracy that meets the

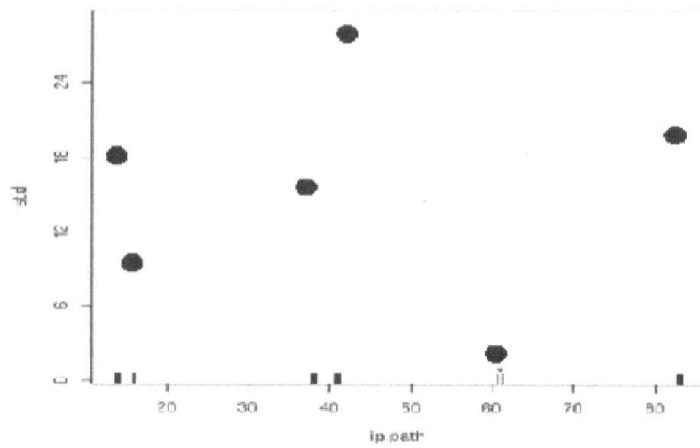

FIGURE 6.5
Estimated non-zero standard deviation.

interests of modeling noted earlier. At the same time, PSS, provides efficient fast algorithm compared to the E-M based E-Lasso and regression based B-F procedure.

TABLE 6.8
Comparison of PSS, B-F, E-lasso in term of MSPE and computing time.

	E-Lasso	B-F	PSS
MSPE	2306	42230	127
Time (in seconds)	6.26×10^{7}	5.43×10^{10}	142

6.6 Variable selection for area and subarea level SAE models

In this section, we focus on variable selection under area level models and subarea level SAE models which are extensively used in practice. A basic area level model, also called the Fay-Herriot model (FH; [59]), uses direct estimators $\hat{\theta}_i$ of area means $\theta_i (i = 1, \ldots, m)$ and associated area level covariates. Direct estimators are obtained from area-specific unit level data, taking survey design into account. Area level covariates are used to link the area means. This leads to a sampling model and a linking model given by $\hat{\theta}_i = \theta_i + e_i$ and $\theta_i = x_i'\beta + v_i$ respectively, where e_i is the sampling error, β is the vector of model parameter, x_i is the $p \times 1$ vector of area level covariates and v_i is a random area effect. Further, e_i has mean 0 and known variance ψ_i, and the sampling errors are assumed to be independent. In practice, the sampling variances ψ_i are obtained by smoothing their direct estimators using generalized variance functions. The area effect v_i has mean 0 and variance σ_v^2, and the area effects are assumed to be independent. Combining the sampling model with the linking model leads to the FH model $\hat{\theta}_i = x_i'\beta + v_i + e_i$ which is then used for variable selection. Note that the linking model alone cannot be used for variable selection because the area means θ_i are not known.

Because of the sampling errors in the FH model, standard methods for linear regression models, such as the AIC, BIC and Mallows' C_p used for variable selection, can lead to biased variable selection when applied to the FH model. Ref. [90] used a conditional AIC method for variable selection that accounts for the sampling errors in the FH model. This method is fairly complex, and practitioners might prefer simple modifications to standard methods that can account for sampling errors in the FH model. We give a brief description of a simple method of estimating the ideal variable selection criteria under the linking model that accounts for the sampling error [151]. The resulting estimation error is shown to converge to 0 in probability as the number of areas increases, unlike the estimation error in the naïve criteria ignoring the sampling errors. The proposed method performed well in simulations unlike the naïve method that ignores the sampling errors in the FH model.

Let $MSE_\theta = (m-p)^{-1}\theta'(I_m - P)\theta$ denote the ideal mean error sum of squares, where $\theta = (\theta_1, \ldots, \theta_m)'$, I_m is the identity matrix of order m, and $P = X(X'X)^{-1}X$ is the standard projection matrix based on the linking model. Then the estimator of MSE_θ is given by $mse_\theta = MSE_{\hat{\theta}} - \bar{\psi}_w$, where $MSE_{\hat{\theta}}$ is obtained by replacing θ by its direct estimator $\hat{\theta}$, and

$$\bar{\psi}_w = \frac{1}{m-p}\sum_{i=1}^{m}(1-h_{ii})\psi_i$$

with $h_{ii} = x_i'(X'X)^{-1}x_i$. We simply replace MSE_θ by mse_θ in the ideal AIC, BIC and C_p which are functions of MSE_θ. For example, the resulting

$$AIC = m\log\left(\frac{m-p}{m}mse_\theta\right) + 2p.$$

In the case of small m, the estimator mse_θ could take a negative value and [151] suggested a simple modification that leads to strictly positive estimator of MSE_θ.

Two-fold subarea models are also often used in practice to estimate subarea and area means. For example, [176] studied adult literacy for counties (subareas) sampled from states (areas) in the United States, using data from the 2003 U. S. National Assessment of Adult Literacy (NAAL). We have areas i and subareas j are sampled from area i. Direct estimators of subarea means $\theta_{ij}(j = 1, \ldots, n_i; i = 1, \ldots, m)$ and associated subarea level covariate vector are denoted as $\hat{\theta}_{ij}$ and x_{ij} respectively. A two-fold subarea model consists of a sampling model $\hat{\theta}_{ij} = \theta_{ij} + e_{ij}$ and a linking model $\theta_{ij} = x_{ij}'\beta + b_{ij}$ respectively, where e_{ij} is the sampling error and $b_{ij} = v_i + u_{ij}$ is the sum of the random area effect v_i and subarea effect u_{ij}. The sampling errors e_{ij} are assumed to be independent with zero means and known variances. Further, the area effect is independent of the subarea effect, and the v_i and u_{ij} are independent and identically distributed with zero means and variances σ_v^2 and σ_u^2 respectively. Under the assumptions, the composite random effects b_{ij} are correlated for each area i with covariance matrix $\Sigma_i = \sigma_v^2 1_i 1_i' + \sigma_u^2 I_i$ where 1_i is the unit vector of length n_i and I_i is the identity matrix of dimension n_i.

We cannot treat the linking model for the two-fold case as a FH-type model on the subarea means because the composite random effects b_{ij} are correlated. It is necessary to transform the covariance matrix Σ_i to a diagonal covariance matrix with equal diagonal elements across areas i, and then apply the variable selection method to the transformed linking model to get the ideal error mean sum of squares. Ref. [21] obtained a parameter-free transformation matrix A_i of order $(n_i - 1) \times n_i$ and full rank that makes the covariance matrix of $A_i b_i$ diagonal with equal diagonal elements across $i = 1, \ldots, m$, where $b_i = (b_{i1}, \ldots, b_{im})'$ [158] used a similar transformation in the context of unit level models. The transformed linking model is then used to get the ideal mean square error sum of squares MSE_{θ^*} and its estimator mse_{θ^*} along the lines of the method used for the FH linking model. Note that the transformed vector $\theta_i^* = A_i\theta_i$ has length $n_i - 1$ unlike the vector θ_i with elements $\theta_{ij}, j = 1, \ldots, n_i$, and as a result each area loses one degree of freedom after transformation. The variable selection criteria can then be computed using mse_{θ^*}, as in the case of the FH model. Ref. [21] report simulation results showing that the proposed transformation method performs well in variable selection, unlike the naive method treating the linking model as a FH-type model ignoring the correlations, especially as σ_v^2 increases.

Three-fold models linking sub-subarea means to related covariates and random effects at the area, subarea and sub-subarea levels are also used in practice to estimate sub-subarea means as well as subarea means. For example, the Program for the International Assessment of Adult Competencies (PIAAC) in the United States used a three-fold model with census divisions as areas, states within a census division as subareas and counties within a state as sub-subareas. Ref. [20] extended the two-fold model variable selection method of [21] to variable selection to variable selection under three-fold models.

6.7 Informal model checking

According to [173], model diagnostics include informal checking and diagnostic tests. Informal checking uses graphical tools to inspect the data, and apparent violation of the model assumptions. Diagnostic tests carry out hypothesis testing, such as goodness-of-fit tests, on specific aspects of the model assumption. In this section, we discuss some existing methods of informal model checking applicable to SAE models.

Generally speaking, the literature on informal model checking for mixed effects models is not extensive. Several authors have used the EBLUP or empirical Bayes estimators (EB) for diagnosing distributional assumptions regarding the random effects [49], [22]. The approach is reasonable because the EBLUP or EB are natural predictors, or estimators, of the random effects. Below we describe some methods developed based on this idea.

A standard assumption regarding the random effects and errors is that they are normally distributed. In fact, the BP under the normality assumption (see Section 1.7) is a basis for the EBLUP, although the latter can be derived without the normality assumption (e.g., [123], p. 90). Ref. [152] considered the so-called longitudinal model, that is, (6.9), assuming that $v_i \sim N(0, G)$, where G is a covariance matrix, and developed a weighted normal for assessing normality of the random effects. First note that, under the assumed model and normality, one can derive the BP, or Bayes estimators, of the random effects α_i $i = 1, \dots, m$ (see Section 4.1), assuming that β and θ, the vector of variance components, are known. This is given by

$$\tilde{v}_i = \mathrm{E}(\alpha_i|y_i) = GZ_i'V_i^{-1}(y_i - X_i\beta),$$

where $V_i = \mathrm{Var}(y_i) = \tau^2 I_{n_i} + Z_i G Z_i'$. Next, the covariance matrix of $\tilde{\alpha}_i$ is given by

$$\mathrm{Var}(\tilde{v}_i) = GZ_i'V_i^{-1}Z_iG.$$

Ref. [152] proposed to examine a Q–Q plot of standardized linear combinations in the form of

$$z_i = \frac{c'\tilde{\alpha}_i}{\{c'\mathrm{Var}(\tilde{\alpha}_i)c\}^{1/2}}, \qquad i = 1, \dots, m, \tag{6.17}$$

where c is a known vector. The authors argued that, through appropriate choices of c, the plot can be made sensitive to different types of model departures.

For example, for a model with two random effect factors, a random intercept and a random slope, one may choose $c_1 = (1, 0)'$ and $c_2 = (0, 1)'$, hence producing two Q–Q plots. On the other hand, such plots may not reveal possible nonzero correlations between the random intercept and slope. To solve this problem, [152] suggests producing a set of plots by letting $c = (1 - u, u)'$ for some values $0 \le u \le 1$.

Furthermore, [152] suggests that the normal plot should be weighted to reflect the differing sampling variances of $\tilde{\alpha}_i$. Following the idea, the authors proposed a generalized weighted normal plot. They suggested plotting z_i against $\Phi^{-1}\{F^*(z_i)\}$, where F^* is the weighted empirical cdf defined by

$$F^*(x) = \frac{\sum_{i=1}^m w_i 1_{(z_i \le x)}}{\sum_{i=1}^m w_i},$$

and $w_i = c'\mathrm{Var}(\tilde{\alpha}_i)c = c'GZ_i'V_i^{-1}Z_iGc$.

In practice, however, the fixed effects β and variance components θ are unknown. In [152], it was suggested using the ML or REML estimators in place of these parameters. The authors argued that, under suitable conditions, the limiting distribution of $\sqrt{n}\{\hat{F}^*(x) - \Phi(x)\}$ is normal with mean zero and variance equal to the variance of $\sqrt{n}\{F^*(x) - \Phi(x)\}$ minus an adjustment, where $\hat{F}^*(x)$ is $F^*(x)$ with the unknown parameters replaced by their ML (REML) estimators. This suggests that, in the case of unknown parameters, the Q–Q plot will be $\hat{z}_i$ against $\Phi^{-1}\{\hat{F}^*(\hat{z}_i)\}$, where $\hat{z}_i$ is z_i with the unknown parameters replaced by their ML (REML) estimates.

Note that, however, the asymptotic variance of $\hat{F}^*(x)$ is different from that of $F^*(x)$, as indicated above. Therefore, if one wishes to include, for example, a ± 1 SD bound in the plot, the adjustment for estimation of parameters must be taken into account. See [152]for details. We consider an example.

Example 6.2. Consider a one-way random effects model with normality assumption. The model can be expressed as

$$y_{ij} = \mu + v_i + e_{ij}, \tag{6.18}$$

$i = 1, \ldots, m$, $j = 1, \ldots, n_i$, where μ is an unknown mean, v_i is a random effect and e_{ij} is an error. The random effects and errors are assumed independent such that $v_i \sim N(0, \sigma^2)$ and $e_{ij} \sim N(0, \tau^2)$. Because v_i is real-valued, $c = 1$ in (6.17). If μ, σ^2, τ^2 are known, the EB estimator of v_i is given by

$$\hat{v}_i = \frac{n_i\sigma^2}{\tau^2 + n_i\sigma^2}(\bar{y}_{i\cdot} - \mu),$$

where $\bar{y}_{i\cdot} = k_i^{-1}\sum_{j=1}^{k_i} y_{ij}$, with

$$w_i = \mathrm{var}(\hat{\alpha}_i) = \frac{k_i\sigma^4}{\tau^2 + k_i\sigma^2}.$$

Thus, in this case, we have

$$z_i = \frac{\hat{v}_i}{\mathrm{sd}(\hat{v}_i)} = \frac{\bar{y}_{i\cdot} - \mu}{\sqrt{\sigma^2 + \tau^2/k_i}},$$

$i = 1, \ldots, m$ and

$$F^*(x) = \left(\sum_{i=1}^m \frac{n_i\sigma^4}{\tau^2 + n_i\sigma^2}\right)^{-1} \sum_{i=1}^m \frac{n_i\sigma^4}{\tau^2 + n_i\sigma^2} 1_{(z_i \le x)}.$$

In practice, μ, σ^2, and τ^2 are unknown and therefore replaced by their REML (ML) estimators when making the Q–Q plot.

Ref. [238] developed a diagnostic tool for Bayesian analysis called posterior predictive p value. The goal was for checking the hierarchical structure under a Bayesian model. The posterior predictive distribution (PPD) of the unobserved outcomes, $y^* = (y_i)_{1 \leq i \leq N}$, given the observed data, $y = (y_i)_{1 \leq i \leq N}$, is defined as

$$p(y^*|y) = \int p(y^*|\theta)p(\theta|y)d\theta, \tag{6.19}$$

where θ is an unobserved variable, $p(y^*|\theta)$ is the pdf of y^* given θ, and $p(\theta|y)$ is the posterior of θ given y. For any index $1 \leq i \leq N$, the corresponding posterior predictive p value (PPP) is defined as

$$p_i = \mathrm{P}(y_i^* \leq y_i|y), \tag{6.20}$$

where $P(\cdot|y)$ is the posterior predictive distribution defined via (6.19).

It was suggested (e.g., [238]) that, under certain conditions of large sample size, the distribution of the PPP is approximately Uniform[0, 1]. While this may be justified theoretically, it also seems intuitive. Recall the well-known result that, for a continuous random variable, X with cdf F_X, the distribution of $U = F_X(X)$ is Uniform[0, 1]. Suppose that the large-sample conditions ensure that the PPD of y^* is approximately the same as the true distribution of y^*, which is assumed to be an i.i.d. copy of y. It follows that the right side of (6.20) is approximately the same as $F_{y_i^*}(y_i) = F_{y_i}(y_i) \sim$ Uniform[0, 1].

The above result is based on assuming that the underlying distribution, or hierarchical structure in the case of [238], is correctly specified. This leads to a graphical tool for checking the distributional assumption by comparing the PPP with the Uniform[0, 1] reference distribution via a Q-Q plot. [238] showed with empirical results that the tool works well for checking a certain hierarchical structure of a Bayesian model. Specifically, the null hypothesis assumes a simple hierarchical model,

$$\begin{aligned} y_{ij}|\mu, \phi &\overset{\text{i.i.d.}}{\sim} N(\mu, \phi), \quad i = 1, \dots, m, j = 1, \dots, n; \\ \pi(\mu, \phi) &\sim \text{constant} \end{aligned} \tag{6.21}$$

(that is, a non-informative prior). Three alternative models were considered: (i) a single-level model, which is the same as the first level of the null but with $\mu = \mu_0$ and $\phi = \phi_0$ for some fixed μ_0, ϕ_0; (ii) a two-level hierarchical model:

$$\begin{aligned} y_{ij}|\theta_i, \phi_0 &\overset{\text{i.i.d.}}{\sim} N(\theta_i, \phi_0), \\ \theta_i|\mu_0, \tau_0 &\overset{\text{i.i.d.}}{\sim} N(\mu_0, \tau_0), \quad i = 1, \dots, m; \end{aligned}$$

and (iii) a two-level hierarchical model with change point:

$$\begin{aligned} y_{ij}|\theta_i, \phi_0 &\overset{\text{i.i.d.}}{\sim} N(\theta_i, \phi_0), \\ \theta_i|\mu_{01}, \tau_0 &\overset{\text{i.i.d.}}{\sim} N(\mu_{01}, \tau_0), \quad i = 1, \dots, m_1, \\ \theta_i|\mu_{02}, \tau_0 &\overset{\text{i.i.d.}}{\sim} N(\mu_{02}, \tau_0), \quad i = m_1 + 1, \dots, m \end{aligned}$$

for some $1 < m_1 < m$.

Although the PPP was proposed for diagnostics of Bayesian models, a similar idea may be used for frequentist diagnostics as well. In fact, in such a case, the method may be closely related to a standard frequentist diagnostic tool. We consider an example.

Example 6.3 (Fay-Herriot model). Consider the Fay-Herriot (F-H) model for SAE (see Section 1.5.1). Let $\theta = (\theta_i)_{1\leq i\leq m}$, $y = (y_i)_{1\leq i\leq m}$ and $y^* = (y_i^*)_{1\leq i\leq m}$. First assume that β, A are known. The $p(y^*|\theta)$ in (6.19) is now understood as the conditional pdf of y^* given θ, denoted by $f(y^*|\theta)$; similarly, the posterior $p(\theta|y)$ in (6.19) is understood as the conditional pdf of θ given y, denoted by $f(\theta|y)$. Likewise, the left side of (6.19) is now denoted by $f(y^*|y)$. It is easy to show that

$$f(y^*|y) = \prod_{i=1}^{m} f(y_i^*|y_i), \tag{6.22}$$

where $f(y_i^*|y_i)$ is the pdf of $y_i^* \sim N(\mu_i, \sigma_i^2)$ with

$$\mu_i = B_i y_i + (1 - B_i)x_i'\beta, \quad \sigma_i^2 = (1 + B_i)D_i,$$

and $B_i = A/(A + D_i)$ (see [134]). At first sight, this may not seem encouraging, as the distribution of $y_i^*|y_i$ is clearly different from that of y_i^*, which is $N(x_i'\beta, A + D_i)$, and it does not matter even if the sample size, m, is large.

The problem is that the PPP idea works in an i.i.d. situation. We do not have i.i.d. data here, because $y_i \sim N(x_i'\beta, A + D_i)$ and x_i, D_i are different. In order to implement the diagnostic tool, we need to first "convert" the data to i.i.d. Again, let us first assume that β, A are known. Then, $z_i = (y_i - x_i'\beta)/\sqrt{A + D_i}, 1 \leq i \leq m$ are independent $N(0, 1)$. This may be regarded as a special case of the F-H model with $\beta = 0$, $A = 0$ and $D_i = 1$. It follows, from the above result, that $\mu_i = 0$ and $\sigma_i^2 = 1$; hence, $z_i^*|z_i$ has the same distribution as z_i or z_i^*. Furthermore, by (6.20) and (6.22), it can be shown that $p_i, 1 \leq i \leq m$ are independent. It follows that p_i, $1 \leq i \leq m$ are independent Uniform$[0, 1]$.

Of course, in practice, β and A are unknown. It is customary to replace β and A by their consistent estimators, $\hat{\beta}$ and $\hat{A}$, under the hypothesis that the assumed F-H model holds. If m is large, the PPD, $p_i, 1 \leq i \leq m$ are expected to be approximately independent Uniform$[0, 1]$. Thus, a Q-Q plot of the PPP vs the Uniform$[0, 1]$ reference is appropriate. Of course, in this case, it is simpler to simply plot the standard residuals, $\hat{z}_i = (y_i - x_i'\hat{\beta})/\sqrt{\hat{A} + D_i}$, vs the $N(0, 1)$ reference, which is a standard frequentist diagnostic tool.

Some variations of the PPP, also proposed by [238], replace (6.20) by $\mathrm{P}[c(y^* \geq c(y)|y]$, or $\mathrm{P}[d(y^*, \theta) \geq d(y, \theta)|y]$, where $c(\cdot)$ is a checking function and $d(\cdot, \cdot)$ is a discrepancy measure. Ref. [238] suggested that the F-statistic works well as a checking function. Similar diagnostic tools were proposed with the checking function of discrepancy measure, and the Uniform$[0, 1]$ reference distribution for the Q-Q plots.

Following the same idea, [239] developed a residual diagnostic tool similar to the z_i in Example 6.3, called Bayesian standardized residual (BSR). This is defined as

$$r_{ij} = \frac{y_{ij} - \mathrm{E}(y_{ij}^*|y)}{\sqrt{\mathrm{var}(y_{ij}^*|y)}}, \tag{6.23}$$

where the $\mathrm{E}(\cdot|y)$ and $\mathrm{var}(\cdot|y)$ are under the PPD (6.19). The authors provided theoretical justification of the BSR. For example, it was shown that, under the null hypothesis distribution, (6.21), (6.19 can be specified as

$$y^*|y \sim t_N\left(\bar{y}1_N, \frac{N-1}{N-3}s_y^2(I_N + \bar{J}_N), N-3\right),$$

where $N = mn$; $\bar{y}$, s_y^2 are the sample mean, sample variance of the y observations, respectively; 1_N and I_N denotes the $N \times 1$ vector of 1's and N-dimensional identity matrix, respectively, $\bar{J}_N = N^{-1}1_N1_N'$; and $t_k(\mu, V, d)$ denotes the k-dimensional multivariate t-distribution with mean vector μ, covariance matrix V, and d degrees of freedom, which has the joint pdf

$$f(u) = \frac{\Gamma\{(k+d)/2\}}{(d\pi)^{k/2}\Gamma(d/2)|V|^{1/2}}\left\{1 + \frac{1}{d}(u-\mu)'V^{-1}(u-\mu)\right\}^{-(k+d)/2}.$$

It then follows that, for any fixed set of BSRs, $r_1, \dots, r_k$, their joint distribution converges, as $N \to \infty$, to $N(0, I_k)$. Therefore, one may use the Q-Q plot vs the $N(0,1)$ reference as a diagnostic tool for checking that the null hypothesis distribution, (6.21), holds.

It should be noted that the diagnostic tools of [238] and [239] were developed for checking violation of the distributional assumption. As is well known, the residuals can also be used for detecting potential outliers. If the latter is the main purpose, the authors of [238] suggested to replace the y in the condition of (6.20) by $y_{-(i,j)}$, that is, y without y_{ij}.

6.8 Goodness-of-fit tests for mixed models: Overview

Formal diagnostic methods in terms of the goodness-of-fit tests have been developed for LMM and some types of GLMM. Of particular interest is checking the distributional assumptions about the random effects and errors. Consider a LMM involving s random effect factors, each associated with a vector of random effects, say, $\alpha_r = (\alpha_{rj})_{1\le j\le m_r}$, such that the α_{rj}s are i.i.d. with mean 0, variance σ_r^2, which is unknown, and distribution $F_r = F_r(\cdot\,|\sigma_r)$. In addition, the LMM involves a vector of errors, $\epsilon = (\epsilon_j)_{1\le j\le N}$, where the ϵ_js are i.i.d. with mean 0, variance τ^2, which is unknown, and distribution distribution $G = G(\cdot\,|\tau)$; and $\alpha_1, \dots, \alpha_s, \epsilon$ are independent. One is interested in testing hypothesis,

$$\begin{aligned} H_0 : F_r(\cdot|\sigma_r) &= F_{0r}(\cdot|\sigma_r), \qquad 1 \le r \le s, \\ \text{and} \quad G(\cdot|\tau) &= G_0(\cdot|\tau); \end{aligned} \tag{6.24}$$

where $F_{0r}, 1 \le r \le s$ and G_0 are known distributions. This means that the distributions of the random effects and errors, up to a set of unknown variance components $\sigma_1^2, \dots, \sigma_s^2, \tau^2$, are as assumed.

Such distributional assumptions are vital in SAE. Consider, for example, the NER model of Section 1.5.2, which a special case of the LMM. A mixed effect may be in the form $\eta = x'\beta + v_i$, where x is known. If the sample size is large (i.e., m is large), one may consistently estimate β and even obtain an asymptotic confidence interval for it, and this does not rely on the distributional assumptions, such as normality. However, large sample results may not help, for example, in obtaining a prediction interval for η, because the effective sample size for estimating v_i is n_i, the sample size for the ith small area, which is often quite small. Therefore, unless one knows the form of the distribution of v_i (e.g., normal), an accurate prediction interval for η cannot be obtained no matter how large m is, if n_i is bounded. To see another example, consider the estimation of the MSPE of the EBLUP. Prasad and Rao [196] derived an approximation formula for the MSPE of EBLUP, which is second-order correct, that is, the approximation error is $o(m^{-1})$. Although the result is asymptotic, assuming that m is large, the normality distributional assumption remain critical for the formula to be second-order correct.

Ref. [119] developed an asymptotic theory of Pearson's χ^2-test with estimated cell frequencies. The authors applied the method to the case of NER model for checking the distributions of v and e. The procedure requires splitting the data into two parts, one used for estimation and the other for testing. Ref. [106] developed a method that applies to a general mixed ANOVA model (e.g., [123], sec. 1.2.2.1). In a way, the procedure is similar to Pearson's χ^2-test with estimated cell probabilities (e.g., [180]). Let $E_1, \dots, E_M$ be a partition of R, the real line. Let a_n be a sequence of normalizing constants that is determined later on. Define

$$\hat{\chi}^2 = \frac{1}{a_n} \sum_{j=1}^{M} \{N_j - \mathrm{E}_{\hat{\psi}}(N_j)\}^2, \tag{6.25}$$

where $N_j = \sum_{i=1}^n 1_{(y_i \in E_j)} = \#\{1 \le i \le n : y_i \in E_j\}$, and $\hat{\psi}$ is the REML estimator of the vector of parameters involved in the LMM. Although (6.25 has some similarity of to Pearson's χ^2-statistic, there are major differences. First and most importantly, the observed count N_k is not a sum of independent random variables. In Pearson's χ^2-test, one deals with i.i.d. observations, so that N_k is a sum of i.i.d. random variables. Under a LMM, however, the observations are correlated. Second, unlike Pearson's χ^2-statistic, the normalizing constant in (6.25) is the same for all the squares in the sum. The choice of the normalizing constants in Pearson's χ^2-test is such that the asymptotic distribution is χ^2. However, even in the i.i.d. case, the asymptotic distribution of Pearson's χ^2-statistic is not necessarily χ^2, if the parameters are to be estimated (see [180]). This issue will be further discussed later. Thus, for simplicity, a unified normalizing constant a_n is used.

Note that, because of the dependence among the observations, a_n may not increase at the same rate as n, the sample size. Third, in a LMM, the number of fixed effects may increase with n (e.g., [104]). As a consequence, the dimension of ψ may increase with n. This shows, from another angle, that one can no longer expect an asymptotic distribution such as χ^2_{M-q-1}, where q is the number of free parameters being estimated.

Ref. [106] showed that, under suitable conditions, the asymptotic distribution of $\hat{\chi}^2$ is a weighted χ^2, that is, the distribution of $\sum_{j=1}^{M} \lambda_j Z_j^2$, where $Z_1, \dots, Z_M$ are independent $N(0,1)$ random variables, and $\lambda_1 \geq \cdots \geq \lambda_M$ are eigenvalues of some non-negative definite matrix, which depends on θ. Because the latter is unknown in practice, the author developed a method of estimating the critical value of the asymptotic distribution, and showed that

$$\mathrm{P}(\hat{\chi}^2 > \hat{c}_\rho) \to \rho$$

as $n \to \infty$, where $\rho \in (0,1)$ is the level of significance of the test. The estimated critical value, $\hat{c}_\rho$ is determined as

$$c_\rho(\hat{\lambda}_1, \dots, \hat{\lambda}_M),$$

where for any given $\lambda_1 \geq \cdots \geq \lambda_M$ and $0 < \rho < 1$, $c_\alpha(\lambda_1, \dots, \lambda_M)$ is the ρ-critical value of the random variable $\xi = \sum_{j=1}^{M} \lambda_j Z_j^2$, and $\hat{\lambda}_1 \geq \cdots \geq \hat{\lambda}_M$ are the eigenvalues of a matrix $\hat{\Sigma}_n = \Sigma_n(\hat{\theta})$. See, for example, [123] (sec. 2.7) for the definition of $\Sigma_n(\theta)$.

It remains to specify the normalizing constant a_n. Ref. [106] noted that the choice of a_n is not unique. However, in some special cases there are natural choices. For example, in the case of linear regression, which may be regarded as a special case of the LMM, one has $a_n = n$. In the case of the one-way random effects model of Example 6.2, if the n_is are bounded, one has $a_n = m$. The choice is less obvious in the case of multiple random effects factors. Ref. [106] proposed the following principle that in many cases either uniquely determines a_n or at least narrows the choices. Note that there are a number of integers that contribute to the total sample size, n, for example, m, n_i in the NER model. Usually, a_n is a function of these integers. It is required that a_n depend on these integers in a way as simple as possible. In particular, no unnecessary constant is allowed in the expression of a_n. This is called a "natural choice". See [123] (p. 122) for further discussion.

Several other methods of mixed model diagnostics have been proposed since [106]. In particular, [36] proposed an alternative approach to the χ^2 test for checking the normality assumption in LMM. The authors showed that, in some cases, the above χ^2-test is not sensitive to certain departures from the null distribution; as a result, the test may have low power against those alternatives. As a new approach, the authors considered a class of distributions that include the normal distribution as a special case. The test is based on the likelihood-ratio (LRT) that compares the "estimated distribution" and the null distribution (i.e., normal). Here the estimated distribution is an identified member of the class of distributions, derived from the Hermite expansion in the form of

$$f_U(u) = \phi(u)\{1 + \kappa_3 H_3(u) + \kappa_4 H_4(u) + \cdots\}, \tag{6.26}$$

where $\phi(\cdot)$ is the pdf of $N(0,1)$; $\kappa_3, \kappa_4, \ldots$ are related to the cumulants of U and the Hermite polynomials, H_j, satisfy $H_j(u)\phi(u) = (-1)^j(d^j\phi/du^j)$. In particular, we have $H_3(u) = u^3 - 3u$, $H_4(u) = u^4 - 6u^2 + 3$.

In practice, a truncated version of (6.26) is used, which is a polynomial of order M. Ref. [36] suggests to determine M using a model selection approach via the information criteria, such as the AIC [4] or BIC [208]. The U in (6.26) may correspond to either a random effect or an error. Thus, the estimated distribution in the LRT is obtained under the order M chosen by the information criterion. Ref. [36] indicated that a suitably normalized LRT statistic has an asymptotic null distribution in the form of the distribution of $\sup_{r\geq 1}\{2Q_r/r(r+3)\}$, where $Q_r = \sum_{q=1}^{r} \chi^2_{q+1}$, and $\chi^2_2, \chi^2_3, \ldots$ are independent such that χ^2_j has a χ^2 distribution with j degrees of freedom, $j \geq 2$.

Note that the forms of the asymptotic null distribution in [106] and [36] are not simple. In the former case, the asymptotic null distribution is a weighted χ^2; in the latter case, the asymptotic null distribution is given above, which also is not simple. In both cases, a Monte-Carlo method is needed to obtain the critical value of the test. Aiming for a simple asymptotic null distribution, [38] proposed a modified Pearson's χ^2 test that is guaranteed to have an asymptotic χ^2 distribution. The test applies not only to LMM but also to GLMM diagnostics. We discuss some further extension of the latter method in the next section.

6.9 The tailoring method

In this section, we describe a general approach to obtaining a test statistic that has an asymptotic χ^2 distribution under the null hypothesis. The original idea can be traced back to R. A. Fisher [63], who used the method to obtain an asymptotic χ^2 distribution for Pearson's χ^2-test, when the so-called minimum chi-square estimator is used. However, Fisher did not put forward the method that he originated under a general framework, as we do here.

Suppose that there is a sequence of s-dimensional random vectors, $B(\vartheta)$, which depend on a vector ϑ of unknown parameters such that, when ϑ is the true parameter vector, one has $\mathrm{E}\{B(\vartheta)\} = 0$, $\mathrm{Var}\{B(\vartheta)\} = I_s$, and, as the sample size increases,

$$|B(\vartheta)|^2 \xrightarrow{\mathrm{d}} \chi^2_s, \tag{6.27}$$

where $|\cdot|$ denotes the Euclidean norm. However, because ϑ is unknown, one cannot use (6.27) directly for goodness-of-fit test. What is typically done, such as in Pearson's χ^2-test, is to replace ϑ by an estimator, $\hat{\vartheta}$. The question is: what is $\hat{\vartheta}$?

The ideal scenario would be that, after replacing ϑ by $\hat{\vartheta}$ in (6.27), one has a reduction of degrees of freedom (d.f.), which leads to

$$|B(\hat{\vartheta})|^2 \xrightarrow{\mathrm{d}} \chi^2_\nu, \tag{6.28}$$

where $\nu = s - r > 0$ and $r = \dim(\vartheta)$. This is the famous "subtract one degree of freedom for each parameter estimated" rule taught in many elementary statistics books (e.g., [205], p. 242). However, as is well known (e.g., [180]), depending on what $\hat{\vartheta}$ is used, (6.28) may or may not hold, regardless of what d.f. is actually involved. In fact, the only method that is known to achieve (6.28) without restriction on the distribution of the data is Fisher's minimum χ^2 method. In a way, the method allows one to "cut-down" the d.f. of (6.27) by r, and thus convert an asymptotic χ^2_s to an asymptotic χ^2_ν. For such a reason, we have coined the method, under the more general setting below, *tailoring*. Below develop the method with a heuristic derivation, and refer the rigorous justification to the Appendix of [134].

The "right" estimator of ϑ for tailoring is supposed to be the solution to an estimating equation of the following form:

$$C(\vartheta) \equiv A(\vartheta)B(\vartheta) = 0, \tag{6.29}$$

where $A(\vartheta)$ is an $r \times s$ non-random matrix that plays the role of tailoring the s-dimensional vector, $B(\vartheta)$, to the r-dimensional vector, $C(\vartheta)$. The specification of A will become clear at the end of the derivation. Throughout the derivation, ϑ denotes the true parameter vector. For notation simplicity, we use A for $A(\vartheta)$, $\hat{A}$ for $A(\hat{\vartheta})$, etc. Under regularity conditions, one has the following expansions, which can be derived from the Taylor series expansion and large-sample theory (e.g., [113]):

$$\hat{\vartheta} - \vartheta \approx -\left\{\mathrm{E}_\vartheta\left(\frac{\partial C}{\partial \vartheta'}\right)\right\}^{-1} C, \tag{6.30}$$

$$\hat{B} \approx B - \mathrm{E}_\vartheta\left(\frac{\partial B}{\partial \vartheta'}\right)\left\{\mathrm{E}_\vartheta\left(\frac{\partial C}{\partial \vartheta'}\right)\right\}^{-1} C. \tag{6.31}$$

Because $\mathrm{E}_\vartheta\{B(\vartheta)\} = 0$ [see above (6.27)], one has

$$\mathrm{E}_\vartheta\left(\frac{\partial C}{\partial \vartheta'}\right) = A\mathrm{E}_\vartheta\left(\frac{\partial B}{\partial \vartheta'}\right). \tag{6.32}$$

Combining (6.31) and (6.32), we get

$$\hat{B} \approx \{I_s - U(AU)^{-1}A\}B, \tag{6.33}$$

where $U = \mathrm{E}_\vartheta(\partial B/\partial \vartheta')$. We assume that A is chosen such that

$$U(AU)^{-1}A \text{ is symmetric.} \tag{6.34}$$

Then, it is easy to verify that $I_s - U(AU)^{-1}A$ is symmetric and idempotent. If we further assume that the following limit exists:

$$I_s - U(AU)^{-1}A \longrightarrow P, \tag{6.35}$$

then P is also symmetric and idempotent. Thus, assuming that $B \stackrel{\rm d}{\to} N(0, I_s)$, which is typically the argument leading to (6.27), one has, by (6.35), $\hat{B} \stackrel{\rm d}{\to} N(0, P)$, hence (e.g., [123], sec. B.2) $|\hat{B}|^2 \stackrel{\rm d}{\to} \chi^2_\nu$, where $\nu = \mathrm{tr}(P) = s - r$. This is exactly (6.28).

It remains to answer one last question: Is there such a non-random matrix $A = A(\vartheta)$ that satisfies (6.34) and (6.35)? We show that, not only the answer is yes, there is an optimal one. Let $A = N^{-1}U'W$, where W is a symmetric, non-random matrix to be determined, and N is a normalizing constant that depends on the sample size. By (6.30) and the fact that $\mathrm{Var}_\vartheta(B) = I_s$ [see above (6.27)], we have

$$\mathrm{var}_\vartheta(\hat{\vartheta}) \approx (U'WU)^{-1}U'W^2U(U'WU)^{-1} \geq (U'U)^{-1}, \tag{6.36}$$

by, for example, Lemma 5.1 of [113]. The equality on the right side of (6.36) holds when $W = I_s$, giving the optimal A:

$$A = A(\vartheta) = \frac{U'}{N} = \frac{1}{N}\mathrm{E}_\vartheta\left(\frac{\partial B'}{\partial \vartheta}\right). \tag{6.37}$$

The A given by (6.37) clearly satisfies (6.34) [which is $U(U'U)^{-1}U'$]. In the next section, it is shown that, with $N = m$, (6.35) is expected to be satisfied. It should be noted that the solution to (6.29), $\hat{\vartheta}$, does not depend on the choice of N.

We remark that a basic assumption for the tailoring method to work is that $\mathrm{E}\{B(\vartheta)\} = 0$ when ϑ is the true parameter vector. However, from the proof of the result (see the Appendix of [134]) it is seen that the condition "ϑ is the true parameter vector" is not critical. For example, in case there is a model misspecification, a "true parameter vector" may not exist. Nevertheless, what is important is that there is some parameter vector, ϑ, which is not necessarily the true parameter vector, such that the equation

$$A(\vartheta)\mathrm{E}\{B(\vartheta)\} = 0 \tag{6.38}$$

holds. This equation holds, of course, when ϑ is the true parameter vector, but it can also hold when the true parameter vector does not exist, as in the case of model misspecification. In fact, in the latter case, one may define the "true parameter vector" as the unique ϑ, assumed exist, that satisfies (6.38). Note that the number of equations in (6.38) is the same as the dimension of ϑ; thus, one expects that a solution exists and is unique, under some regularity conditions.

To see that (6.38) is the key, note that under (6.38), (6.29) is equivalent to $A(\vartheta)[B(\vartheta) - \mathrm{E}\{B(\vartheta)\}] = 0$, where the expectation is with respect to the true underlying distribution. It follows that one can replace $B(\vartheta)$ by $B(\vartheta) - \mathrm{E}\{B(\vartheta)\}$, which has mean zero, and all of the arguments in the proof go through. This property has given tailoring some unexpected robustness feature, that is, it can work correctly in spite of some model misspecification. We illustrate this more specifically in the next section.

As another remark, there is a connection between the tailoring method and the specification test (ST) based on GMM (e.g., [86]). However, there is also a difference. The difference is that ST is equivalent to (6.29) with $A(\vartheta)$ given by (6.37) without the expectation, while for tailoring the expectation is taken first before using it in (6.29). One may compare this difference to that between the observed Fisher information and expected one in maximum likelihood (ML) estimation [53]. Although it may be argued that, asymptotically, this difference may be of lower order—in fact, ST may also be viewed as an extension of the original idea of Fisher [63], and it has the same asymptotic null distribution as tailoring, finite-sample performance may differ. We demonstrate this difference in the next section.

Furthermore, because, after taking the expectation, some terms in $\partial B'/\partial\vartheta$ in (6.37) may vanish, the form of $A(\vartheta)$ in (6.29) may be substantially simplified. One may, again, compare this to Fisher scoring in ML. For example, [173] (p. 42) noted that Fisher scoring, which is implemented in the well-known GLM algorithm, simplifies the numerical computation of the MLE. Ref. [134] also observed that tailoring is computationally more efficient than ST, apparently also due to the simplification of taking the expectation.

6.10 A robust goodness-of-fit test

We follow up the tailoring method introduced in the previous section by applying it to checking the appropriateness of an area-level model in SAE. In the case of linear models, this is known as the Fay-Herriot (F-H) model (see Section 1.5.1). The model can be written as a hierarchical model (although there is no prior as in a Bayesian model): (i) $y_i|\theta_i \sim N(\theta_i, D_i)$; (ii) $\theta_i \sim N(x_i'\beta, A)$; plus the assumption that $(y_i, \theta_i), i = 1, \dots, m$ are independent. We focus on checking stage (ii) of the model via a goodness-of-fit test.

A standard approach is to embed the F-H model into a larger class of models. Ref. [134] considered a skew normal distribution (SN), denoted by $N(\mu, \sigma^2, \alpha)$, where α is the skew parameter. The pdf of $N(\mu, \sigma, \alpha)$ is given by

$$f_{\mu,\sigma,\alpha}(x) = \frac{1}{\sigma}\phi_\alpha\left(\frac{x-\mu}{\sigma}\right), \quad -\infty < x < \infty, \tag{6.39}$$

where $\phi_\alpha(x) = 2\phi(x)\Phi(\alpha x)$ with $\phi(\cdot)$ and $\Phi(\cdot)$ being the pdf and cdf of $N(0,1)$, respectively. The fact that (6.39) is a pdf can be verified by utilizing the following result (e.g., [5]):

Lemma 6.1. For any constants $a, b \in R$, we have

$$\int_{-\infty}^{+\infty} \phi(x)\Phi(a+bx)dx = \Phi\left(\frac{a}{\sqrt{1+b^2}}\right).$$

The tailoring is based on the B function, and there is some flexibility in choosing an appropriate B function depending on the problem of interest. Because, in SAE, one is generally interest in predicting the mixed effects associated with the small area means, it is desirable to choose something from a predictive perspective. The PPD, introduced in Section 6.7, provides a reasonable path. This is given by $p(y^*|y)$ of (6.19. If one let $y^* = y$, one has what is called the predictive likelihood (PL; [134]), $p(y|y)$. The latter authors gave some explanations on why this is called predictive likelihood. If the primary interest were to estimate a parameter, θ, one would use a prior distribution, $p(\theta)$, to obtain the (marginal) likelihood,

$$p(y) = \int p(y|\theta)p(\theta)d\theta. \tag{6.40}$$

Now, because

$$p(y|y) = \int p(y|\theta)p(\theta|y)d\theta, \tag{6.41}$$

the difference is replacing the $p(\theta)$ in (6.40) by $p(\theta|y)$, the predictive distribution of θ. (6.41) is called predictive likelihood because it is the outcome of replacing the prior distribution of θ in the likelihood expression, (6.40), by the predictive distribution of θ.

It should be noted that the PL is not necessarily a likelihood, as it does not always possess some of the well-known properties of the likelihood. However, one can adjust the score equation of the PL to make it unbiased. To make the notation consistent, we use f, instead of p, for the (conditional) pdfs. Let ψ denote the vector of parameters involved in either $f(y|\theta)$ or $f(\theta)$, where $\theta = (\theta_i)_{1\leq i\leq m}$ denotes the vector of small area means. Accordingly, write $f(y|y) = f(y|y,\psi)$. The adjusted PL score is given by

$$s_{\text{a}}(\psi) = \frac{\partial}{\partial\psi}\log f(y|y,\psi) - \text{E}_{\psi}\left\{\frac{\partial}{\partial\psi}\log f(y|y,\psi)\right\}. \tag{6.42}$$

The estimator of ψ obtained by solving the adjusted PL score equation, $s_{\text{a}}(\psi) = 0$, or, equivalently, the following equation:

$$\frac{\partial}{\partial\psi}\log f(y|y,\psi) = \text{E}_{\psi}\left\{\frac{\partial}{\partial\psi}\log f(y|y,\psi)\right\} \tag{6.43}$$

is called the maximum adjusted PL estimator, or Maple, in view of its analogy to the MLE.

Recall our focus is on checking the null hypothesis H_0, which states part (ii) of the F-H model, assuming that the other parts of the model hold. It is easy to show that $f(\theta|y) = \prod_{i=1}^m f(\theta_i|y_i)$. Thus, one has

$$\begin{aligned} f(y|y) &= \int \prod_{i=1}^m f(y_i|\theta_i) f(\theta_i|y_i) d\theta \\ &= \prod_{i=1}^m \int f(y_i|\theta_i) f(\theta_i|y_i) d\theta_i \\ &= \prod_{i=1}^m f(y_i|y_i), \end{aligned} \tag{6.44}$$

where $f(y_i|y_i) = \int f(y_i|\theta_i)f(\theta_i|y_i)d\theta_i$. Here, $f(\theta_i)$ is the pdf of $N(x_i'\beta, A, \alpha)$ and the null hypothesis corresponds to $\alpha = 0$. It can be shown [134] that

$$\begin{aligned} f(y_i|y_i) &= \frac{1}{\sqrt{D_i(1+B_i)}}\phi\left(\frac{y_i - x_i'\beta}{\sqrt{D_i(1+B_i)/(1-B_i)}}\right) \\ &\quad \times \frac{\Phi(\alpha_{2i}(y_i - x_i'\beta))}{\Phi(\alpha_{3i}(y_i - x_i'\beta))}, \end{aligned} \tag{6.45}$$

where

$$\alpha_{si} = \frac{(4-s)\sqrt{A}\alpha}{\sqrt{\{(4-s)A + D_i\}\{(4-s)A + (1+\alpha^2)D_i\}}}, \quad s = 2, 3$$

with $B_i = A/(A + D_i)$. Note that, when $\alpha = 0$, (6.45) reduces to that under normality, and $f(y_i|y_i) \neq f(y_i)$ even in this case.

To test the null hypothesis, $\mathrm{H}_0 : \alpha = 0$, let

$$b_i(y_i, \vartheta) = \frac{\partial}{\partial \psi} \log f(y_i|y_i, \psi)\Big|_{\alpha=0} - \mathrm{E}\left\{\frac{\partial}{\partial \psi} \log f(y_i|y_i, \psi)\Big|_{\alpha=0}\right\}.$$

The PL equation, (6.43), can be specified as (see [134])

$$\begin{aligned} &\frac{\partial \log f(y_i|y_i, \psi)}{\partial \beta}\Big|_{\alpha=0} - \mathrm{E}\left\{\frac{\partial \log f(y_i|y_i, \psi)}{\partial \beta}\Big|_{\alpha=0}\right\} \\ &= a_i(A)x_i(y_i - x_i'\beta), \end{aligned}$$

$$\begin{aligned} &\frac{\partial \log f(y_i|y_i, \psi)}{\partial \sigma^2}\Big|_{\alpha=0} - \mathrm{E}\left\{\frac{\partial \log f(y_i|y_i, \psi)}{\partial \sigma^2}\Big|_{\alpha=0}\right\} \\ &= b_i(A)(y_i - x_i'\beta)^2 - c_i(A), \end{aligned}$$

$$\begin{aligned} &\frac{\partial \log f(y_i|y_i, \psi)}{\partial \alpha}\Big|_{\alpha=0} - \mathrm{E}\left\{\frac{\partial \log f(y_i|y_i, \psi)}{\partial \alpha}\Big|_{\alpha=0}\right\} \\ &= d_i(A)(y_i - x_i'\beta), \end{aligned}$$

where $a_i(A) = (1 - B_i)^2/D_i(1 + B_i)$,

$$b_i(A) = \frac{(1 - B_i)^3(3 + B_i)}{2D_i^2(1 + B_i)^2}, \quad c_i(A) = \frac{(1 - B_i)^2(3 + B_i)}{2D_i(1 + B_i)^2},$$

and $d_i(A) = \sqrt{2/\pi}\sigma(1 - B_i)^2/D_i(1 + B_i)$.

Here, we want to point out some robustness feature of the tailoring test. Let ϑ denote the true ϑ. If the model is correctly specified under the null hypothesis, then, under the null hypothesis, $\sum_{i=1}^m b_i(y_i, \vartheta)$ is a sum of independent random vectors with mean zero. On the other hand, if there is some misspecification in the mean function that the true β, hence the true ϑ, does not exist (under the null hypothesis), one can define the "true ϑ" as the unique solution to (6.38). All of the arguments in the derivation of Section 6.9 go through by replacing $\sum_{i=1}^m b_i(y_i, \vartheta)$ with $\sum_{i=1}^m [b_i(y_i, \vartheta) - \mathrm{E}\{b_i(y_i, \vartheta)\}]$. Furthermore, one has

$$V_b(\vartheta) = \mathrm{Var}_\vartheta\{\sum_{i=1}^m b_i(y_i, \vartheta)\} = \sum_{i=1}^m \mathrm{Var}_\vartheta\{b_i(y_i, \vartheta)\},$$

where (see [134])

$$\mathrm{Var}_\vartheta\{b_i(y_i, \vartheta)\} = \begin{bmatrix} g_i(A)x_i x_i' & 0_p & g_i(A)(x_i\sqrt{2A/\pi}) \\ 0_p' & h_i(A) & 0 \\ g_i(A)(x_i'\sqrt{2A/\pi}) & 0 & g_i(A)(2A/\pi) \end{bmatrix},$$

0_p denotes the $p \times 1$ vector of zeros, $g_i(A) = (1 - B_i)^3/D_i(1 + B_i)^2$ and

$$h_i(A) = \frac{(1 - B_i)^4(3 + B_i)^2}{2D_i^2(1 + B_i)^4}.$$

Thus, if one define $B(\vartheta) = V_b^{-1/2}(\vartheta)\sum_{i=1}^m b_i(y_i, \vartheta)$, one has

$$B(\vartheta) \stackrel{\mathrm{d}}{\longrightarrow} N(0, I_s),$$

where $s = \dim(\psi)$. It follows that (6.27) holds. Because $r = \dim(\vartheta) < s$, the tailoring applies to yield (6.28) with $\nu = s - r = 1$. In particular, one has

$$A(\vartheta) = \frac{1}{m}\left\{\sum_{i=1}^m \mathrm{E}_\vartheta\left(\frac{\partial b_i'}{\partial \vartheta}\right)\right\} V_b^{-1/2}(\vartheta),$$

where b_i is define above and

$$\mathrm{E}_\vartheta\left(\frac{\partial b_i'}{\partial \vartheta}\right) = -\begin{bmatrix} a_i(A)x_i x_i' & 0_p & d_i(A)x_i \\ 0_p' & b_i(A) & 0 \end{bmatrix}.$$

This gives $A(\vartheta)$ for solving the tailoring equation (6.29).

Note. It is said that one begins to truly understand a concept when she/he is not confused by the notation. The variance of v_i, A, under the F-H model is not to be confused with the $A(\cdot)$ function in the tailoring method, even though the same letter A is used (to keep consistent in both contexts).

Ref. [134] carried out a series of simulation studies intended to demonstrate the robustness feature of their proposed tailoring test [see the paragraph including (6.38) and the one following it]. Partial results are presented here. The goodness-of-fit test aims to test the hypothesis $\mathrm{H}_0 : \alpha = 0$, assuming that other parts of the model assumptions hold. This is case (I). Another scenario, case (II), was considered, in which case there is a model misspecification under the null hypothesis. Specifically, the assumed model is

$$y_i = \beta_1 x_i + v_i + e_i, \quad i = 1, \dots, m;$$

however, the data are generated under the following model:

$$\begin{aligned} y_i &= \beta_1 x_i + v_i + e_i, \quad 1 \le i \le n, \\ y_i &= \beta_2 x_i + v_i + e_i, \quad n+1 \le i \le m, \end{aligned}$$

where $m = 2n, D_i = D_{i1}$ for $1 \le i \le n$ and $D_i = D_{i2}$ for $n+1 \le i \le m$. The variance of v_i, $A = 10$, noting that $v_i \sim SN(0, A, \alpha)$ and $e_i \sim N(0, D_i)$. The D_{i1}s were generated from the Uniform$(3.5, 4.5)$ distribution. There were two scenarios for D_{i2}, one (S1) generated from Uniform$(3.5, 4.5)$ and the other (S2) from Uniform$(0.5, 1.5)$. Let $\beta_1 = 1$, and $\beta_2 = 1$ (Case I) or 3 (Case II). The true value of α is 0 under the null hypothesis, and 0.5 under the alternative. Finally, the x_i's were generated from the Uniform$(0, 1)$ distribution. The x_i's and D_i's were fixed throughout the simulation study.

The performance of the tailoring test based on PL, denoted by $\hat{B}^2_{\mathrm{PL}}$, was compared with four other tests of goodness-of-fit, including the χ^2 test of [106], denoted by $\hat{\chi}^2_{\mathrm{J}}$, the LRT test of [36], denoted by $\hat{\chi}^2_{\mathrm{CH}}$, and the ST test, denoted by $\hat{B}^2_{\mathrm{ST}}$. The last three tests were reviewed in Section 6.8. One additional test under consideration was an F-test, denoted by $\hat{F}$, which was proposed by [193]. The test statistic is given by $\hat{F} \equiv m\hat{T}^2_m / V$, where

$$\begin{aligned} \hat{T}_m &= \frac{1}{m} \sum_{i=1}^m \frac{\sqrt{D_i}(y_i - x_i'\hat{\beta})}{\hat{A} + D_i}, \\ V &= \frac{1}{m} \sum_{i=1}^m \frac{D_i}{\hat{A} + D_i} - m\hat{P}'\{\mathrm{var}(\hat{\psi} - \psi)\}\hat{P}, \end{aligned}$$

where $\hat{A}$ is the Prasad-Rao estimator [196] and $\hat{\beta}$ is given by the BLUE (1.37) with A replaced by $\hat{A}$. Furthermore, $P = \lim \mathrm{E}(\partial T_m / \partial \psi)$, which, in the current case, is given by

$$P' = -\frac{1}{m} \left[\sum_{i=1}^m \frac{\sqrt{D_i} x_i'}{A + D_i}, 0 \right],$$

TABLE 6.9
Size (power) of different tests (level of significance 0.05).

D_{i2}	β_2	m	$\hat{B}^2_{\rm PL}$	$\hat{F}$	$\hat{\chi}^2_{\rm J}$	$\hat{\chi}^2_{\rm CH}$	$\hat{B}^2_{\rm ST}$
S1	1	50	0.050 (0.855)	0.048 (0.896)	0.049 (0.023)	0.054 (0.380)	0.014 (0.634)
		100	0.048 (0.966)	0.047 (0.979)	0.051 (0.017)	0.044 (0.398)	0.000 (0.969)
		200	0.053 (0.998)	0.053 (0.999)	0.051 (0.004)	0.007 (0.108)	0.001 (0.991)
		500	0.050 (1.000)	0.050 (1.000)	0.048 (0.0002)	0.007 (0.096)	0.054 (1.000)
	3	50	0.048 (0.849)	0.045 (0.889)	0.053 (0.021)	0.044 (0.377)	0.016 (0.629)
		100	0.046 (0.963)	0.045 (0.979)	0.060 (0.013)	0.046 (0.332)	0.000 (0.967)
		200	0.051 (0.998)	0.051 (0.999)	0.053 (0.003)	0.003 (0.149)	0.000 (0.990)
		500	0.048 (1.000)	0.046 (1.000)	0.049 (0.000)	0.000 (0.155)	0.046 (1.000)
S2	1	50	0.051 (0.799)	0.052 (0.849)	0.051 (0.025)	0.060 (0.461)	0.000 (0.813)
		100	0.046 (0.936)	0.049 (0.967)	0.052 (0.017)	0.034 (0.372)	0.000 (0.932)
		200	0.053 (0.994)	0.049 (0.999)	0.048 (0.006)	0.008 (0.112)	0.000 (0.993)
		500	0.049 (1.000)	0.049 (1.000)	0.048 (0.000)	0.004 (0.159)	0.033 (1.000)
	3	50	0.048 (0.792)	0.092 (0.772)	0.061 (0.021)	0.062 (0.467)	0.000 (0.802)
		100	0.046 (0.928)	0.122 (0.935)	0.074 (0.011)	0.047 (0.329)	0.000 (0.924)
		200	0.051 (0.993)	0.209 (0.993)	0.070 (0.003)	0.009 (0.114)	0.000 (0.992)
		500	0.047 (1.000)	0.415 (1.000)	0.081 (0.000)	0.004 (0.370)	0.031 (1.000)

and $\hat{P}$ is P with A replaced by $\hat{A}$. The asymptotic null distribution of the test statistic is χ^2_1. The same estimators, $\hat{\beta}$ and $\hat{A}$, were also used in $\hat{\chi}^2_{\rm J}$.

Different sample sizes, $m = 50, 100, 200, 500$, were considered. The results for the level of significance 0.05 are presented in Table 6.9 (from Table 2 of [134]). It is seen that $\hat{B}^2_{\rm PL}$ has approximately the right size under different scenarios, while all of the other tests, especially $\hat{F}$, $\hat{\chi}^2_{\rm CH}$, and $\hat{B}^2_{\rm ST}$, do not have the right size if there are misspecification in the mean and significant change in the range of the sampling variances for the small areas. The size also does not seem to improve for $\hat{\chi}^2_{\rm CH}$ with increasing sample size. As for $\hat{B}^2_{\rm ST}$, it seems that the size does not seem even close to the nominal level (0.05) until $m = 500$.

Regarding the power, $\hat{B}^2_{\rm PL}$ seems to perform very well under all scenarios. The power performance of $\hat{\chi}^2_{\rm J}$ and $\hat{\chi}^2_{\rm CH}$ seems to be poor compared to $\hat{B}^2_{\rm PL}$, while the power performance of $\hat{F}$ and $\hat{B}^2_{\rm ST}$ is similar to that of $\hat{B}^2_{\rm PL}$.

To show the relevance of the practice of SAE, [134] considered the median income data of four-person families at the state level of the U.S., previously discussed in [76]. The first application has to do for choosing an appropriate model. The outcome variable, y_i, is sample survey of four-person median income at state i in year 1989. A covariate, x_i, is the census four-person median income at state i in year 1979 ($i = 1, ..., m = 51$). A plot of y_i vs x_i is presented in Figure 6.6

The first question is whether a linear model is appropriate. The plot seems to suggest that a linear mean function would be appropriate without the apparent "outlier" to the right. Does the outlier put an end to the linear model? Alternatively, one may accommodate the outlier by considering a quadratic mean function. To answer these questions, the authors considered the

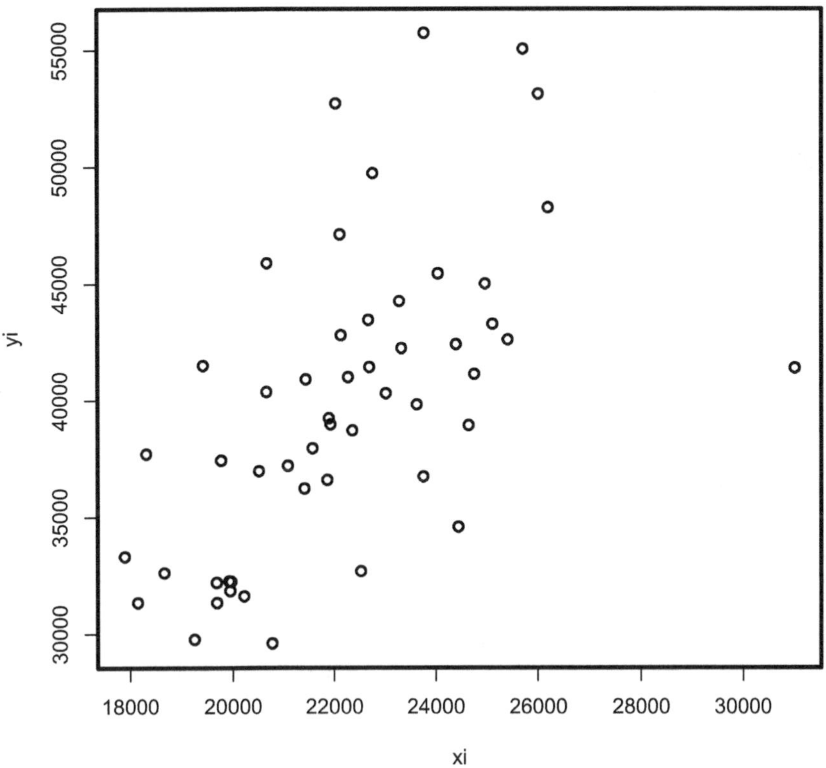

FIGURE 6.6
Four-person family median income survey data at the state level plotted against the 1979 census.

following F-H model,

$$y_i = \beta_0 + \beta_1 x_i + \beta_2 x_i^2 + v_i + e_i, \quad i = 1, \ldots, 51, \tag{6.46}$$

under the standard assumptions of the F-H model, and tested the hypothesis H_0: $\beta_2 = 0$. The tailoring/PL test rejected H_0 with $\hat{B}_{PL}^2 = 5.36 > 3.84$, which is the critical value of χ_1^2 with the level of significance equal to 0.05. Thus, the test suggests that the linear model is inappropriate.

The next question is whether the quadratic model is appropriate. To answer the question, the authors extended model (6.46) to

$$y_i = \beta_0 + \beta_1 x_i + \beta_2 x_i^2 + \beta_3 x_i^3 + v_i + e_i, \quad i = 1, \ldots, 51, \tag{6.47}$$

and tested the hypothesis H_0: $\beta_3 = 0$. This time, the tailoring/PL test yielded a test statistic $\hat{B}_{PL}^2 = 1.19 < 3.84$, the same critical value at the 0.05 level of significance. Thus, the test failed to reject H_0. The authors of [134] also considered another way of extending (6.46) via the quadratic-outlying (Q-O) model [126], that is,

$$y_i = \beta_0 + \beta_1 x_i + \beta_2 x_i^2 + \beta_3 1_{(x_i > 30000)} + v_i + e_i, \quad i = 1, \ldots, 51. \tag{6.48}$$

The tailoring/PL test, once again, failed to reject the null hypothesis.

Therefore, the authors concluded that the quadratic model is appropriate.

It is worth noting that the authors also applied the other four tests considered in the simulation study above to this data. None of these tests were able to reject the linear model null hypothesis. This seems to be consistent with the pattern observed in the simulation study that these tests appear to have lower power than the tailoring/PL test.

It is known that income data are typically non-normal. As a way of validating the proposed tailoring/PL test, [134] also considered checking the normality assumption under the quadratic model,(6.46). This is equivalent to testing H_0: $\alpha = 0$ vs H_1: $\alpha \neq 0$, where α is the skew parameter, assuming that other parts of the F-H model assumptions hold.

The tailoring/PL test resulted in a test statistic $\hat{B}^2_{\mathrm{PL}} = 4.67 > 2.70 = \chi^2_1(0.90)$; rejecting H_0 at the 10% significance level. This seems to validate the tailoring/PL test as being consistent with the "common sense" knowledge.

As for the other four tests considered in the simulation study reported in Table 6.9, the ST test statistic $\hat{B}^2_{\mathrm{ST}} = 4.93 > 2.70$, also rejecting H_0 at the 10% significance level. On the other hand, none of these were able to reject the null hypothesis as the same level of significance.

7

Other Topics

This final chapter covers a few other topics also related to robust SAE. These include bench-marking, Bayesian methods, machine-learning methods, and missing data.

7.1 Bench-marking

In his 2013 review on SAE topics, Pfeffermann wrote, "Benchmarking robustifies inference by forcing the model-based predictors to agree with a design-based estimator for an aggregate of the areas for which the design-based estimator is reliable". A bench-marking equation typically has the following form:

$$\sum_{i=1}^{m} w_i \hat{\theta}_i = \sum_{i=1}^{m} w_i y_i, \tag{7.1}$$

where $\hat{\theta}_i$ is a model-based estimator of the small-area mean for the ith small area, y_i is a design-based estimator for the same small-area mean, and w_is is a known weight. The right side of (7.1) is typically a reliable estimator for the aggregated population mean due to the (much) larger sample size. For example, if the small areas are states, the aggregated population may be the entire U.S.; if the small areas are counties, the aggregated population may be the states.

Furthermore, because the design-based estimators have nothing to do with the model, the right side of (7.1) is unaffected by model failures. If equation (7.1) is forced to satisfy when developing the model-based estimator, or predictors, $\hat{\theta}_i$, there is, at least, protection from model failure at the aggregated level of interest.

Sometimes, there are multiple bench-marking requirements, so one may have multiple equations like (7.1) that need to be incorporated.

The traditional EBLUP (see Section 1.7) does not, in general, satisfy the bench-marking requirement(s). For the most part, there have been two approaches to benchmark the EBLUP, which we briefly discuss below.

The first method is to make adjustments to the EBLUP to enforce the bench-marking requirement. For example, [233] proposed to minimizing the

DOI: 10.1201/9781003395171-7

overall loss function, defined as

$$Q(\hat{\theta}^a) = \sum_{i=1}^{m} \psi_i E(\hat{\theta}_i^a - \theta_i)^2, \tag{7.2}$$

where the ψ_i's are a chosen set of positive weights. Under the Fay-Herriot model with A known (see Section 1.5.1), this leads to the adjusted BLUP of the small-area mean, θ_i:

$$\tilde{\theta}_i^{\breve{a}} = \tilde{\theta}_i^H + \breve{a}_i \left(\sum_{i=1}^{m} w_i y_i - \sum_{i=1}^{m} w_i \tilde{\theta}_i^H \right),$$

where $\breve{a} = (\sum \psi_i^{-1} w_i^2)^{-1} \psi_i^{-1} w_i$. The bench-marked BLUP, $\hat{\theta}_i^{\breve{a}}$, is defined as the unique predictor among all linear unbiased predictors that satisfy the bench-marking requirements and minimizes the loss function defined in (7.2). When A is unknown, it is replaced by a suitable estimator, and the adjusted EBLUP is defined similarly as

$$\hat{\theta}_i^{\breve{a}} = \hat{\theta}_i^H + \breve{a}_i \left(\sum_{i=1}^{m} w_i y_i - \sum_{i=1}^{m} w_i \hat{\theta}_i^H \right).$$

The second approach is to modify the way that EBLUP is derived so that the resulting small area predictors are self-bench-marking, that is, they automatically satisfy the bench-marking requirement(s). For example, [242] proposed a pseudo-EBLUP using survey weights, which has the self-bench-marking property. In an alternative approach, [233] introduced an augmented model by adding the weights, w_i, as an additional covariate, to the Fay-Herriot model. The augmented model can be expressed as

$$y = X_2 \beta + v + e, \tag{7.3}$$

where $X_2 = (X \; w_e)$ and $w_e = (w_1 D_1^2, \cdots w_m D_m^2)'$, and X is the covariate matrix whose ith row is the x_i' in (1.16). Note that, under this notation, (1.16) can be expressed as $y = X\beta + v + e$ in its matrix form. The EBLUP derived under the augmented model is self-bench-marking.

As we know, the OBP is more robust against model misspecification than the EBLUP (see Chapter 4). However, like the EBLUP, the OBP does not necessarily satisfy the bench-marking requirement(s). Following [233], [7] proposed two methods to benchmark the OBP under the Fay-Herriot model. The first method adjusts the OBP to make it bench-marking; the second method develops an augmented model, leading to a self-bench-marking OBP. The two methods are described below. *1. Adjusting the OBP.* First assume that β and A are known. Similar to (7.2), consider

$$Q(\hat{\theta}^a) = \sum_{i=1}^{m} E(\hat{\theta}_i^a - \theta_i)^2, \tag{7.4}$$

which defines a loss function with equal weights. Assume that the true underlying true model is $y_i = \mu_i + v_i + e_i$, $i = 1 \cdots, m$. Then, the small area means under the true model are $\theta_i = E(y_i|v_i) = \mu_i + v_i$, $1 \leq i \leq m$. Here, the $v_i e_i$, $i = 1, \ldots, m$ are as under the Fay-Herriot model. Let

$$\tilde{\theta} = (\tilde{\theta}_1, \cdots, \tilde{\theta}_m)$$

be BP of θ, and

$$\tilde{\theta}_i^{\breve{a}} = \tilde{\theta}_i + \breve{a}_i \left(\sum_{i=1}^{m} w_i y_i - \sum_{i=1}^{m} w_i \tilde{\theta}_i \right), \tag{7.5}$$

where $\breve{a}_i = (\sum_{i=1}^{m} w_i^2)^{-1} w_i$, w_i's being sampling weights. Ref. [7] showed that $\tilde{\theta}^{\breve{a}} = (\tilde{\theta}_1^{\breve{a}}, \cdots, \tilde{\theta}_m^{\breve{a}})$ is the unique adjusted BP that satisfies bench-marking equation (7.1) and minimizes the loss (7.4).

The rationale of the adjusted OBP is the following: Suppose the observed MSPE approximates the true MSPE. Then, the Observed BP, or OBP, $\hat{\theta}$ would approximate the BP, $\tilde{\theta}$. In practice, one replaces the unknown β and A with their BPE's (see Section 4.1.2), leading to the adjusted OBP:

$$\hat{\theta}_i^{\breve{a}} = \hat{\theta} + \breve{a}_i \left(\sum_{i=1}^{m} w_i y_i - \sum_{i=1}^{m} w_i \hat{\theta}_i \right). \tag{7.6}$$

2. Self-bench-marking via augmented model. As said, the idea is to include the sampling weight as a covariate. Similar approaches have been used in literature, such as in the modeling approach to design-consistency in regression [62], and in a robust approach that includes a penalized spline of the weight as a covariate [244].

Note that the bench-marking equation, (7.1), can be re-written as follows:

$$\sum_{i=1}^{m} w_i y_i - \sum_{i=1}^{m} w_i \hat{\theta}_i = \sum_{i=1}^{m} w_i (y_i - \hat{\theta}_i) = 0. \tag{7.7}$$

This suggests that the bias is non-zero if there is non-zero correlation between w_i and $(y_i - \hat{\theta}_i)$. The bias can be reduced by including w_i in the model. Accordingly, one may consider an augmented Fay-Herriot model as follows:

$$y_i = x_i' \beta_1 + \beta_2 w_i (1 - B_i)^{-1} + v_i + e_i; \quad i = 1, \cdots, m, \tag{7.8}$$

where $B_i = A/(A + D_i)$. If A is known, define

$$X_1 = (x_i')_{1 \leq i \leq m}, \quad X_2 = X_2(A) = \left[w_i (1 - B_i)^{-1} \right]_{1 \leq i \leq m},$$

$X(A) = [X_1 \ X_2(A)]$ and $\beta = (\beta_1' \ \beta_2)'$. Then, (7.8) can be expressed as

$$y = X(A)\beta + v + e.$$

The BPE of β is given by

$$\tilde{\beta}(A) = \left\{X(A)'\Gamma^2(A)X(A)\right\}^{-1} X(A)'\Gamma^2(A)y, \tag{7.9}$$

where $\Gamma(A) = \text{diag}(1 - B_i; 1 \leq i \leq m)$. When A is unknown, as in Section 4.1.2, the BPE of A, $\hat{A}$, is the minimizer of

$$\tilde{Q}(A) = \{y - X(A)\tilde{\beta}(A)\}'\Gamma^2(A)\{y - X(A)\tilde{\beta}(A)\} + 2A\text{tr}\{\Gamma(A)\}. \tag{7.10}$$

The BPE of β is then given by $\hat{\beta} = \tilde{\beta}(\hat{A})$; the OBP of $\theta = (\theta_i)_{1 \leq i \leq m}$ is

$$\hat{\theta} = \hat{A}\hat{V}^{-1}y + D\hat{V}^{-1}X(\hat{A})\hat{\beta}, \tag{7.11}$$

where $\hat{V} = \hat{A}I_m + D$ and $D = \text{diag}(D_i, 1 \leq i \leq m)$.

With a few lines of arguments, [7] showed that (7.11) satisfies the bench-marking equation, (7.1). Write $\hat{X} = X(\hat{A})$ and $\hat{\Gamma} = \Gamma(\hat{A})$. Then we have

$$\begin{aligned} \hat{\theta} &= (\hat{V} - D)\hat{V}^{-1}y + D\hat{V}^{-1}\hat{X}\hat{\beta} \\ &= y + D\hat{V}^{-1}(\hat{X}\hat{\beta} - y) \\ &= y + D\hat{V}^{-1}\left\{\hat{X}(\hat{X}'\hat{\Gamma}^2\hat{X})^{-1}\hat{X}'\hat{\Gamma}^2 - I_m\right\}y, \end{aligned}$$

using 7.9. Let $w = (w_i)_{1 \leq i \leq m}$. Then, we have

$$w'\hat{\theta} = w'y + w'D\hat{V}^{-1}\left\{\hat{X}(\hat{X}'\hat{\Gamma}^2\hat{X})^{-1}\hat{X}'\hat{\Gamma}^2 - I_m\right\}y.$$

On the other hand, let $\hat{B}_i$ be B_i with A replaced by $\hat{A}$. We have

$$D\hat{V}^{-1} = \text{diag}\left(\frac{D_i}{\hat{A} + D_i}, 1 \leq i \leq m) = \text{diag}(1 - \hat{\gamma}_i, 1 \leq i \leq m\right),$$

hence

$$\begin{aligned} w'D\hat{V}^{-1} &= [w_i(1 - \hat{\gamma}_i)]'_{1 \leq i \leq m} \\ &= \left[w_i(1 - \hat{\gamma}_i)^{-1}\right]'_{1 \leq i \leq m} \text{diag}\left\{(1 - \hat{\gamma}_i)^2, 1 \leq i \leq m\right\} \\ &= \hat{X}_2'\hat{\Gamma}^2 \\ &= [0\ 1]\hat{X}'\hat{\Gamma}^2, \end{aligned}$$

where $\hat{X}_2 = X_2(\hat{A})$. Combining the above results, we have

$$w'\hat{\theta} = w'y + [0\ 1]\hat{X}'\hat{\Gamma}^2\left\{\hat{X}(\hat{X}'\hat{\Gamma}^2\hat{X})^{-1}\hat{X}'\hat{\Gamma}^2 - I_m\right\}y = w'y,$$

which is equivalent to equation (7.1).

Ref. [7] carried out a simulation study to compare the performance of several predictors of the small area means, including the EBLUP, OBP, and their bench-marked versions. The simulation design followed that of [233]. It

TABLE 7.1
Population and sample sizes of 50 states of the US in 2000.

State	Population (in 1000)	Sample size(m_i)	State	Population (in 1000)	Sample size(m_i)
1	33640	58	26	4000	20
2	21160	46	27	3610	19
3	19360	44	28	3240	18
4	16000	40	29	3240	18
5	12250	35	30	2890	17
6	12250	35	31	2890	17
7	11560	34	32	2560	16
8	10240	32	33	2560	16
9	8410	29	34	2250	15
10	8410	29	35	1960	14
11	7840	28	36	1690	13
12	7290	27	37	1690	13
13	6250	25	38	1690	13
14	6250	25	39	1210	11
15	5760	24	40	1210	11
16	5760	24	41	1210	11
17	5760	24	42	1210	11
18	5290	23	43	1000	10
19	5290	23	44	810	9
20	5290	23	45	810	9
21	4840	22	46	810	9
22	4410	21	47	640	8
23	4410	21	48	640	8
24	4410	21	49	640	8
25	4000	20	50	490	7

was designed to simulate a large national survey in the U.S., in which state estimates are of interest. From each of the 50 states in U.S., a sample was generated under the model:

$$y_{ij} = x_i'\beta + v_i + e_{ij} \quad i = 1\cdots 50, \quad j = 1\cdots n_i,$$

where, $x_i' = (1, z_i)$, $\beta' = (6.0, 3.0)$ and $z_i = pop_i^{0.2} - \overline{pop_i^{0.2}}$. pop_i is the approximate population of the state i in millions, $\overline{pop_i^{0.2}}$ is the mean of $pop_i^{0.2}$. See Table 7.2 (copied from [7]) for details. The sample sizes n_i's are approximately proportional to $\sqrt{pop_i}$. The random effects and errors were simulated such that $v_i \overset{iid}{\sim} N(0,1)$, $e_{ij} \overset{ind}{\sim} N(0,16)$, and v_i's and e_{ij}'s are independent.

The assumed model is a Fay-Herriot model with $x_i{}' = 1$. The data-generating model, however, has $x_i' = (1, z_i)$. Thus, the assumed model is misspecified. The goal was to investigate the performance of the different predictors in a case of model misspecification. The weights considered were $w_i = (\sum_{i=1}^{50} pop_i)^{-1} pop_i$. A total of 100 simulation runs were carried out,

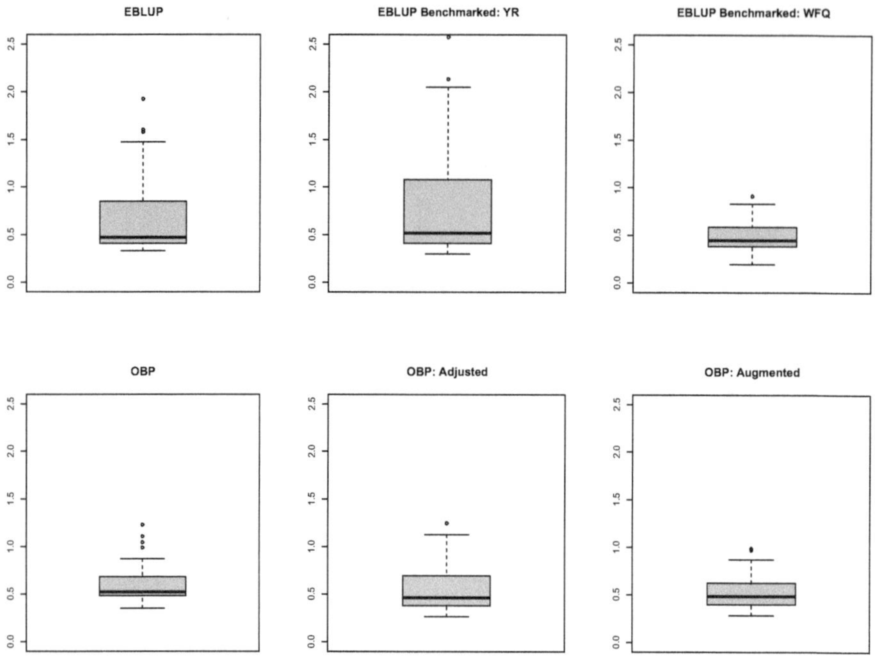

FIGURE 7.1
Boxplots of MSPE of different predictors: empirical MSPEs of different predictors. Top row: EBLUP and its bench-marked versions; bottom Row: OBP and its bench-marked versions.

and the simulated area-specific MSPEs of different small area predictors were reported.

Figure 7.1 compares the performances of the bench-marked OBPs with some existing predictors, including the EBLUP and its bench-marked versions, and OBP without bench-marking through boxplots of their simulated MSPEs for all of the small areas. Table 7.2 shows some summary statistics for the simulated MSPEs. As is evident from the figure, the OBP without bench-marking performs better than the EBLUP without bench-marking. As for the bench-marked versions, the bench-marked OBP via the augmented model seems to perform a little better than the adjusted one. The bench-marked OBP via the adjustment seems to perform better than its counterpart in EBLUP; however, the bench-marked OBP via the augmented model does not seem to perform better than its counterpart in EBLUP.

See [7] for a comparison of the different predictors under a different simulation setting.

TABLE 7.2
Summary statistics comparing the MSPE of different methods: study 1. Min. = Minimum; 1st Qu. = 1st Quartile; 3rd Qu. = 3rd Quartile; Max. = Maximum.

	Min.	1st Qu.	Median	Mean	3rd Qu.	Max.
EBLUP	0.3295	0.4079	0.4697	0.6666	0.8223	1.9260
EBULP: YR	0.2994	0.4107	0.5164	0.7983	1.0460	2.5760
EBLUP: WFQ	0.1957	0.3847	0.4470	0.4909	0.5849	0.9105
OBP	0.3489	0.4820	0.5196	0.5959	0.6646	1.2290
OBP: Adjusted	0.2643	0.3812	0.4617	0.5455	0.6684	1.2480
OBP: Augmented	0.2827	0.4023	0.4848	0.5295	0.6221	0.9881

In terms of measures of uncertainty, [7] proposed two approaches for estimating the MSPE of the bench-marked OBPs. Those approaches were developed with the potential model misspecification in mind.

The first approach is called Monte-Carlo Spline (McSpline). Following the non-parametric model approach discussed in Section 5.2, the McSpline was developed under the non-parametric model (5.3). This model is broader than the (linear) Fay-Herriot model; therefore, the risk, and extent, of model misspecification is lower. After the spline approximation (5.1), one estimates the coefficients of the spline model, and bootstrap parametrically under the estimated spline model.

The second approach follows the McJack, proposed by [117]. The jackknife is another well-known resampling method (e.g., [211]). Ref. [118] (JLW) extended the jackknife to estimate the MSPE of an EBP (see Section 1.8.1). A basic element of the JLW jackknife is the MSPE of the BP, which is assumed to have an analytic form. Often time, this is not the case. [117] proposed to approximate the $b(\psi)$ in (1.41) via a Monte-Carlo method; the authors also simplified the JLW jackknife to just one term; in other words, (1.42) is no longer needed. The name McJack refers to a combination of the Monte-Carlo and jackknife.

A key element of McJack is drawing bootstrap samples for the evaluation of $b(\psi)$. Similar to McSpline, [7] proposed to do this, under the spline-approximated non-parametric model. It is seen that the spline approximation plays a key role in both McSpline and McJack. Thus, a condition for both (or any) of the methods to work is that the spline model offers a good approximation to the true underlying model.

To investigate how the proposed methods work in a situation where the true underlying model is not a spline (hence the spline approximation cannot be exactly accurate), [7] carried out another simulation study, this time on the performance of the two proposed MSPE estimators for both the bench-marked OBP via the adjustment method, called adjusted OBP, and the bench-marked OBP via the augmented model, called augmented OBP. The data were generated under Case 3 of [125] (see Section 6.3).

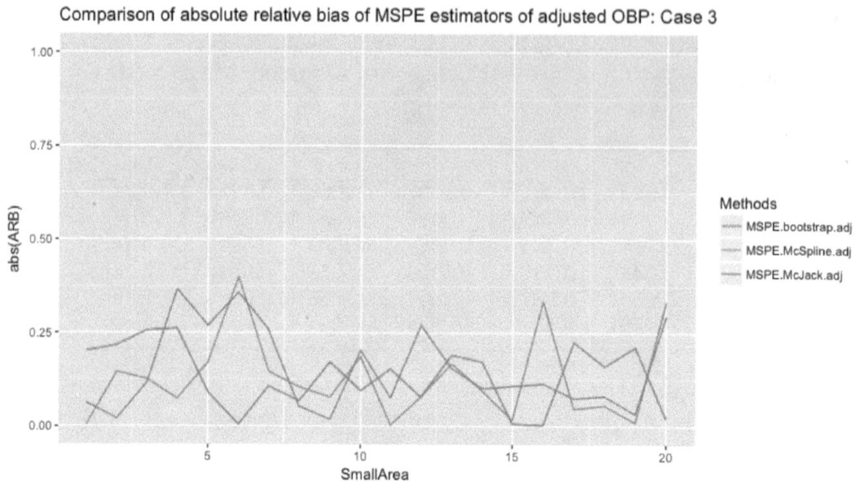

FIGURE 7.2
Comparison of ARB of different MSPE estimators: adjusted OBP.

A performance measure for the MSPE estimation is the absolute relative bias (ARB), defined as the absolute value of the relative bias, that is,

$$\text{ARB} = \left| \frac{\text{Mean of MSPE Estimator} - \text{True MSPE}}{\text{True MSPE}} \right|, \tag{7.12}$$

where the Mean of MSPE Estimator was obtained by the average of the MSPE estimators over $R = 200$ simulation runs; the True MSPE was approximated by the average of True MSPE over the same simulation runs. Three MSPE estimators were compared in the simulation study, including McJack, McSpline, and a bootstrap MSPE estimator proposed by [126] (see the third paragraph of Section 4.1.5). The latter MSPE estimator (here called bootstrap) was for OBP without bench-marking. The ARB is plotted again the small area index in Figure 7.2 for the adjusted OBP, and Figure 7.3 for the augmented OBP.

It is seen that, in terms of the ARB, the two MSPE estimators performed well for some areas but not for the other (a rule-of-thumb cut-off for good performance is ARB less than 0.1). The three MSPE estimators seem to perform similarly. Although, in theory, McJack is second-order unbiased [117] while the other two MSPE estimators do not have such a property, the conditions of the theory may not be met here due to multiple reasons. Firstly, the spline approximation to the true underlying model may not be very accurate. Secondly, the Monte Carlo sample size may not be large enough to obtain accurate results. According to [117], in order to achieve second-order unbiasedness, the Monte Carlo sample size, K, needs to satisfy $m^2/K \to 0$; in our case, $m^2/K = 400/200 = 2$, which may not be small enough. On the other

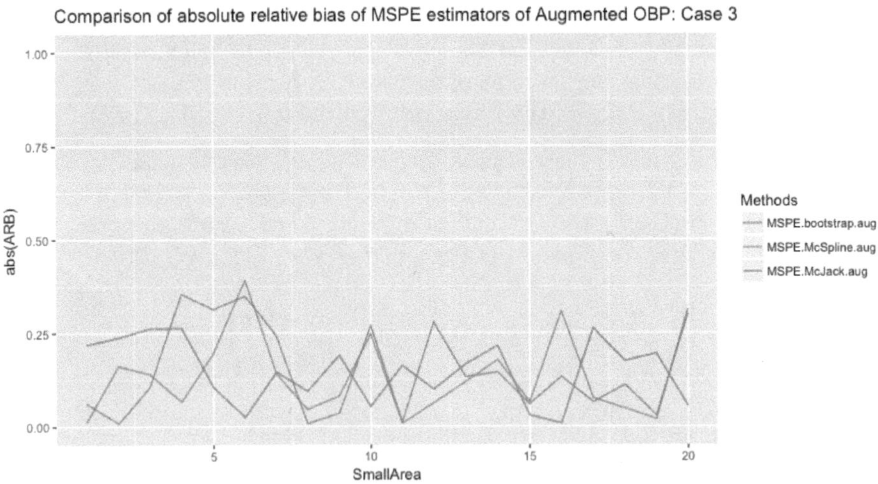

FIGURE 7.3
Comparison of ARB of different MSPE estimators: augmented OBP.

hand, McJack becomes computationally intensive when K is large. Thus, the McSpline method was preferred over the other two methods according to [7].

7.2 Bayesian methods

[44] proposed a robust hierarchical Bayes (HB) approach for SAE in the presence of covariates and outliers. The authors considered an extended version of the Fay-Herriot model, expressed as a hierarchical model. Namely, it is assumed that $y|\theta \sim N(\theta, \Sigma)$, where $y = (y_i)_{1\le i\le m}$ and $\theta = (\theta_i)_{1\le i\le m}$ as in the Fay-Herriot model, but Σ is a known matrix, which is not necessarily diagonal.

The authors suggested a way to achieve robustness to second-level model failure in the multivariate Fay-Herriot model by replacing normality with a scale mixture of normal distributions. Specifically, it is assumed that $\theta_1, \dots, \theta_m$ are independent such that

$$\theta_i \sim \frac{1}{\sigma} p_i \left(\frac{\theta_i - x_i'\beta}{\sigma} \right), \tag{7.13}$$

where $p_i(\cdot)$ is a scale mixture of normal distributions with a mixing pdf $g_i(\cdot)$ in the form of

$$p_i(x) = \int_0^\infty \sqrt{u}\phi(\sqrt{u}x) g_i(u) du,$$

where $\phi(\cdot)$ is the pdf of $N(0,1)$. The authors noted that, by choosing different mixing distributions, the scale normal mixture can yield many different distributions, such as the normal, t, Laplace, and Logistic distributions. The mixture of normal distribution is similar to some approaches to relaxing the normality assumption for the random effects discussed in Section 2.4.

Ref. [44] derived an expression of the Bayes estimator of θ under the square error loss, which is

$$\mathrm{E}(\theta|y) = Wy + (I - W)X\beta, \tag{7.14}$$

where $W = \mathrm{E}(H)$, $H = (\Sigma^{-1} + \Gamma)^{-1}\Sigma^{-1}$, $\Gamma = \sigma^{-2}\mathrm{diag}(u_1, \dots, u_m)$, and the latter expectation is with respect to the pdf $g(u|y, \beta, \sigma^2) \propto$

$$\frac{1}{|\Sigma + \Gamma^{-1}|^{1/2}} \exp\left\{-\frac{1}{2}(y - X\beta)'(\Sigma + \Gamma^{-1})^{-1}(y - X\beta)\right\}\prod_{i=1}^{m} g_i(u_i).$$

Furthermore, as a measure of uncertainty, the authors derived an expression of the posterior covariance matrix of θ as

$$\begin{aligned}\mathrm{Var}(\theta|y) &= \mathrm{E}\{(\Sigma^{-1} + \Gamma)^{-1} \\ &\quad +(H - W)(y - X\beta)(y - X\beta)'(H - W)'\}\end{aligned} \tag{7.15}$$

with the expectation with respect to the same pdf as the above.

In (7.14) and (7.15), it is assumed that all of the unknown parameters involved are known. The next step, in the standard Bayesian procedure, is to assign priors to these unknown parameters. The robustness of the Bayes method is achieved through the assignment of those priors. The authors propose the following (improper) prior:

$$\psi(\beta, \sigma) \propto \sigma^{b-1} \exp\left(-\frac{a}{2}\sigma^2\right), \quad \beta \in R^p, \sigma > 0,$$

where $p = \dim(\beta)$ and showed that the resulting posterior is proper.

The authors derived interesting theoretical properties. For example, it was shown that if the model fails for a given area due to an outlier, one still retains the benefit of shrinking for all areas except for the outlying area, which does not happen if one had the normal prior (in which case the HB estimators for all areas converge to the direct estimators). For the outlying area, the HB estimator converges to the direct estimator.

In the case of unit-level models, [24] showed that an HB small area predictor proposed by [41] is not robust to outliers in a way similar to the EBLUP. The authors proposed a robust Bayesian method under the NER model (see Section 1.5.2). The standard Bayesian NER model is the same as the NER model until the prior stage. As for the prior, the standard assumption is that

$$\pi(\beta, \sigma_v^2, \sigma_e^2) \propto \frac{1}{\sigma_e^2}. \tag{7.16}$$

This prior is non-informative, hence improper.

Ref. [24] noted that the NER model is unable to explain the outlier behavior of unit-level error terms. Recall that [70] (see Section 2.4) intends to

address a similar problem using a mixture of normal distribution approach. Similarly, [24] proposed a Bayesian NER model involving a two-component normal mixture distribution. Specifically, it is assumed that
(i) conditional on β, $v = (v_i)_{1\le i\le m}$, z_{ij}, $i = 1, \dots, m$, $j = 1, \dots, n_i$, and σ_1^2, σ_2^2, σ_v^2, the responses y_{ij}, $i = 1, \dots, m$, $j = 1, \dots, n_i$ are independent such that $y_{ij} \sim z_{ij}N(x'_{ij}\beta + v_i, \sigma_1^2) + (1 - z_{ij})N(x'_{ij}\beta + v_i^2, \sigma_2^2)$;
(ii) the z_{ij}'s are i.i.d. such that $z_{ij}|p_e \sim$ Bernoulli(p_e), and are independent with β, v, $\sigma_r^2, r = 1, 2$ and σ_v^2;
(iii) conditional on β, the z_{ij}'s, p_e, $\sigma_r^2, r = 1, 2$ and σ_v^2, $v_i, i = 1, \dots, m$ are independent $N(0, \sigma_v^2)$.
It is seen that, up to this stage, a difference from the standard Bayesian NER model is the distributional assumption about the e_{ij}'s, which are now assumed to follow a normal mixture distribution. Next, for the prior distribution, the authors assumed

$$\pi(\beta, \sigma_1^2, \sigma_2^2, \sigma_v^2, p_e) \propto \frac{1_{(0<\sigma_1^2<\sigma_2^2<\infty)}}{\sigma_2^4}. \tag{7.17}$$

This prior is improper for β and the σ^2's but proper for p_e (Uniform[0, 1]). However, as indicated by the authors, the conditional distributions, $\sigma_r^2|\sigma_{3-r}^2$, $r = 1, 2$ are proper, which is necessary for the parameters involved in the mixture distribution to ensure the propriety of the posterior resulting from the HB model. The prior was also chosen carefully so that the conditional distributions, used in the Markov-Chain Monte-Carlo (MCMC) computation are simple.

Ref. [24] studied the frequentist properties of their proposed Bayesian predictors via empirical studies, and showed that they performed similarly to the robust EBLUP of [215] when outliers were present, and both methods outperformed the HB method of [41]and the M-quantile method of [29]. In the absence of outliers, the proposed method performed similarly to that of [41]. As an application, the authors of [24] reanalyzed the Iowa crops data of [8] (see Section 6.2) and found/confirmed a suspected outlier.

Under the notion of Bayesian semi-parametric models, [168] considered SAE problems associated with the National Health Interview Survey. The NHIS is a multistage interview survey conducted annually for the National Center for Health Statistics to provide health and health care information for the civilian and non-institutionalized population in the United States. For example, the 1985–1994 NHIS sample involved about 200 primary sampling units (PSUs), selected from a stratified population of 1983 PSUs. Each PSU consists essentially of a single county or a group of contiguous counties. Within each sampled PSU, groups of households are aggregated into areal segments and sampled. Each year there is a new sample of approximately 50,000 households, or about 120,000 individuals. Although the NHIS emphasizes national estimates, there is also a need for estimates for small geographical areas or sub-populations. The usual design-based estimator, which uses only the sample survey data for the particular small area of interest, is often unreliable due to the relatively small samples that are available from the area. See [170] for more information.

A feature of NHIS is that most of the variables are binary. A mixed logistic model for SAE was introduced in Section 1.5.3. Similar models under the Bayesian framework have also been used in SAE. See, for example, [169]. However, the authors of [168] were concerned that these models oversimplified the geographic diversity of the U.S. population. The latter authors focused on an application to estimation of the proportions of women in sub-populations of the U.S. whose mammograms had been obtained within the past two years.

A binomial model was proposed for estimating the proportion of having the mammogram, namely, $y_{sid} \sim \text{Binomial}(n_{sid}, p_{sid})$, where $i = 1, \dots, I$ corresponds to county, $d = 1, \dots, D$ to a domain, and $s = 1, \dots, S$ to strata. The authors selected $D = 6$ demographic groups defined by the three age groups (30–39, 40–49, 50+) crossed with race (black, non-black) as the domains. As for the strata, they were defined by grouping the county-based PSUs into $S = 198$ strata. The counties are nested within the strata, and the domains are crossed with the counties. Furthermore, n_{sid} is the number of individuals interviewed in stratum s, county i, and domain d, and y_{sid} is the total number of positive responses (yes for the mammogram) among the n_{sid} individuals.

Next, the p_{sid} is modeled as

$$\text{logit}(p_{sid}) = x'_{sid}\beta + v_{s(i)} + e_{sid} \tag{7.18}$$

with $v_s \sim N(0, \sigma^2_{m(s)})$. Here $s(i)$ denotes the stratum containing county i, and $m(s)$ denotes the mega-stratum containing stratum s. There were 2 mega strata formed by grouping the 198 strata into those representing the very large metropolitan areas (e.g., the New York City area) versus all others. The x_{sid} represents a vector of county-level covariates, whose components had been selected via the BIC. Two components, namely, % of the workforce in white collar jobs ($x_{sid,1}$) and % of persons aged 25+ with no more than a ninth-grade education ($x_{sid,2}$) were selected out of a total of 22 covariates. Furthermore, priors were assigned to β and σ^2_m as

$$\beta \sim N(m_\beta, V_\beta), \quad \sigma_m^{-2} \sim \text{Gamma}(a_\sigma, b_\sigma),$$

where Gamma(a, b) denotes the Gamma distribution with mean a/b.

The distribution of e_{sid} is where the robustness concern has focused on. Let $e_i = (e_{sid})_{s=s(i), 1 \le d \le D}$. Let $\phi_\theta(\cdot)$ denote the D-dimensional multivariate normal pdf with mean vector μ and covariance matrix Σ, with $\theta = (\mu, \Sigma)$. It is assumed that

$$e_i \sim f(u) = \int \phi_\theta(u) dG(\theta), \tag{7.19}$$

where $G(\cdot)$ is a prior distribution. Ref. [168] propose use a Dirichlet process (DP) prior, that is, $G \sim \text{DP}(\alpha, G_\nu)$, where $\alpha > 0$ is the total mass parameter and G_ν is the standardized base measure, which depends on a vector ν of hyper-parameters. The DP is a distribution of a random distribution, P, over a finite partition of the sample space, say, $A_1, \ldots, A_K$ such that $P \sim \text{DP}(\alpha, G_\nu)$ iff

$$(P(A_1), \ldots, P(A_k)) \sim \text{Dir}(\alpha G_\nu(A_1), \ldots, \alpha G_\nu(A_K)).$$

Here, $\text{Dir}(\alpha_1, \ldots, \alpha_K)$ denotes the Dirichlet distribution with joint pdf

$$\begin{aligned} f(u_1, \ldots, u_K | \alpha_1, \ldots, \alpha_K) &= \frac{1}{B(\alpha_1, \ldots, \alpha_K)} \prod_{k=1}^{K} u_k^{\alpha_k - 1}, \\ & \quad u_k \geq 0, 1 \leq k \leq K, \sum_{k=1}^{K} u_k = 1, \end{aligned}$$

where $B(\alpha_1, \ldots, \alpha_K)$ is the K-variate Beta function defined as

$$B(\alpha_1, \ldots, \alpha_K) = \frac{\prod_{k=1}^{K} \Gamma(\alpha_k)}{\Gamma(\sum_{k=1}^{K} \alpha_k)}$$

with $\Gamma(\cdot)$ being the Gamma function.

Note that, here, the distribution, P, under consideration is a multivariate normal distribution corresponding to θ. Thus, a distribution of P is the same as a distribution of θ.

A challenge, from a computational perspective, is how to implement the Bayesian procedure with the DP prior. Ref. [168] propose to use posterior simulation. First, the DP can be expressed as

$$e_i \sim \phi_{\theta_i}(\cdot), \quad \theta_i = (\mu_i, \Sigma_i) \sim \text{DP}(\alpha, G_\nu).$$

which is more convenient to sample from. The posterior simulation is done with an MCMC procedure, namely, the Gibbs sampler (e.g., [160]), with the following component-wide sampling steps:

$$\begin{aligned} &(a) \quad \beta | e, v, y; \\ &(b) \quad e_i | \mu_i, \Sigma_i, \beta, v, y_i; \\ &(c) \quad v | \beta, e, \sigma, y; \\ &(d) \quad \sigma | v; \\ &(e) \quad \mu, \Sigma, v, \alpha | e. \end{aligned}$$

A point that the authors of [168] tried to make is that the semi-parametric model approach, which refers to the DP prior (rather than a normal prior), is needed. The authors used the NHIS data to illustrate. First, graphically presented empirical results (boxplots) showed that assuming the random effects for different domains to be i.i.d. generated from some un-ivariate distribution would be inappropriate. Also, the "heat-map" of the empirically estimated G function suggested some features of the G function. A fully parametric model would correspond to either a point mass or a conjugate multivariate normal (in μ) distribution, corresponding to a hierarchical model. "Neither seems to be a good approximation to G", as the authors stated.

Finally, [56] proposed a robust Bayesian method based on quantile regression. Recall the M-quantile regression method discussed in Section 3.3. Here, [56] adopted a model from [202] for the quantile function, expressed as

$$q(\tau|x_i) = x_i'\beta_0 + \sum_{l=1}^{L+1} x_i'\beta_l B_l(\tau), \tag{7.20}$$

where $B_l(\cdot)$ are specified functions. Here, L is the number of piece-wise intervals that divide $[0, 1]$, and is up to one's choice. Model (7.20) is a population model. A sub-population, or small-area version of the model is derived by allowing β_0 and some of the β_l, $0 \leq l \leq L+1$, say, $\beta_l, 1 \leq l \leq p$ to be area-specific. Here, the unit-level model is considered, so the index i is replaced by ij with i corresponding to the small area, and the index l replaced by il.

The subsequent developments followed the standard Bayesian procedure. First, the random coefficients were assigned certain distributions. The distributions under consideration included, for $0 \leq l \leq p$, $1 \leq i \leq m$, (i) $\beta_{il} \sim N(\beta_l, \sigma_l^2)$; or (ii) $\beta_{il} \sim t(\beta_l, d_l, \sigma_l^2)$, where d_l is the degrees of freedom (d.f.); or (iii) $\beta_{il} \sim \text{TDP}(\alpha_l, G_l, T_l)$, where TDP denotes the truncated DP distribution (see above) with truncation bound T.

Prior distributions were then assigned to the hyper-parameters involved. Specifically, the variance σ_l^2 was assumed to have the inverse Gamma distribution (IG), $\sigma_l^2 \sim \text{IG}(0.01, 0.01)$ distribution; the d.f., d_l, was assumed to have the Exp(0.1) distribution; the DP parameter α_l was assumed to have the Uniform(0.3, 10) distribution; the distribution G_l was assumed to be $N(\mu_l, \tau_l^2)$, where μ_l was assumed to have the $N(0, 100)$ distribution and $\tau_l \sim \text{Uniform}(0, 50)$; finally, the truncation bound T_l was chosen to be 50.

The authors of [56] carried out a simulation study under the same simulation setting of [28]. The authors compared their proposed predictors of the small-area mean with several existing predictors, including the EBLUP, the HB-NER method of [41], the robust EBLUP of [215], the M-quantile of [29] and [28]. The simulation had four different scenarios similar to those of [215] (see Section 3.2). Specifically, in the (0,0) setting, there was no outlier in v or e; in the (0,1) setting, the errors e are outlying with a $\delta N(0,6)+(1-\delta)N(20,150)$ distribution, where $\delta \sim \text{Bernoulli}(0.97)$; in the (1,0) setting, the v is outlying such that, for the first 36 of the total of $m = 40$ small areas, $v_i \sim N(0, 6)$, and for the last 4 small areas, $v_i \sim N(9, 20)$; finally, in the (1,1) setting, the v and e were having the worse of both. For the proposed method, which required posterior MCMC simulation, the authors of [56] used $K = 10,000$ simulations with 5,000 burn-ins. The simulation results showed that the proposed method has advantage in terms of the root MSE under the "contaminated" scenarios [all except (0,0)], especially for the "contaminated" areas (last four small areas).

Ref. [56] discussed an application to Italian income data. Income data are known to be skewed, for which quantile-based methods often work better than mean-based. The specific data set was related to people 60 years or older who received financial benefits, called pension earners (PE). The outcome variable was equivalized income (EI), defined as the total disposable household income divided by an equivalence factor. The goal was to estimate the average EI of PE at a local level called travel to work areas. There were $m = 113$ such local areas, which were treated as small areas, and there were good reasons for such a consideration. The total sample size was 2,779, but the sample sizes for the local areas ranged from 0 to 496, and there were 39 zeros, that is, 39 of the 113 areas had no samples. The average sample size was 19. Plots of EI vs PE showed the presence of outliers, a situation where the proposed robust Bayesian method may be needed.

As noted, the number of piece-wise intervals dividing $[0, 1]$, L in (7.20), is up to one's choice. Here, the authors considered $L = 3$, which corresponded to the standard quantiles of Q1 (0.25), Q2 (0.5), and Q3 (0.75), as well as $L = 5$, which added 0.1 and 0.9 to the $L = 3$ case. The authors compared the TDP distribution with the other two distributions considered (i.e., normal and t; see above) and found that TDP worked better in terms of the deviance information criterion (DIC; e.g., [216]); they also found that $L = 5$ worked more favorably than $L = 3$, also in terms of the DIC. The authors also discussed the feasibility of their specifications of the priors (see above).

A question that one always wants answered is whether the proposed method helps. In this case, the authors of [56] used areas, both with fairly large sample sizes ($n_i = 127$ and $n_i = 79$), to show the impact of the outliers. It was reported that the posterior standard deviation (s.d.) of the proposed Bayesian predictor was much smaller than the s.d. of the direct estimator (i.e., the sample mean), and considerably smaller than that of the HB-NER predictor. It should be noted that these s.d. were estimated s.d.'s, which were not necessarily the same as the true s.d.'s. In terms of the bias of the predictors, the authors of [56] compared the mean of their proposed predictor and that of the direct estimator via hypothesis testing, and showed the difference is statistically insignificant.

7.3 Machine learning methods

Traditional regression-based approaches to SAE require the selection of an appropriate model (discussed in detail in Ch 6) and do not handle non-linear relationships between the auxiliary data and the outcome. In addition, complex interactions are difficult to capture as they are rarely known apriori. To overcome these limitations usually requires a statistician well versed in the subject matter or who works closely with a subject matter expert who has a good feel for these complexities.

Machine learning alternatives have been gaining more attention because of their potential benefits: more accurate predictions, faster training times for large datasets, and their data-adaptive nature which can require less subject matter knowledge from the statistician. The class of machine learning methods is quite wide but specific types have found use in SAE research. These include gradient boosting, random forests, and neural networks. The rationale to focus mostly on these has been a result of extensive empirical comparisons between methods where these three have come out as the winners.

7.3.1 Mixed effects random forests (MERFs)

[142] introduced a general mixed model framework allowing for the estimation of random forests (RFs) while simultaneously allowing for structural dependencies of survey data. Here, the LMM-based models in SAE are treated as a special case.

Consider a generalization of the NER model of [8]. As there, we assume a finite population U of size N consisting of m domains $U_1, U_2, \ldots, U_m$ with $N_1, N_2, \ldots, N_m$ units where $i = 1, \ldots, m$ indicates respective areas. Outcome variable y_{ij} for individual observation j in area i is available for every unit within the sample.

Sample s is drawn from U and consists of n units partitioned into sample sizes $n_1, n_2, \ldots, n_m$. Thus s_i is the sub-sample from area i.

A vector of auxiliary variables $x = (x_1, x_2, \ldots, x_p)$ is available for every unit j within sample s. Then we can generalize the usual NER model which captures the relationship between the x_{ij} and y_{ij} as follows:

$$y_{ij} = f(x_{ij}) + u_i + e_{ij}, \tag{7.21}$$

with $u_i \sim N(0, \sigma_u^2)$ and $e_{ij} \sim N(0, \sigma_e^2)$. If the assumptions of the LMM are met, then optimal estimates of the fixed effects $\hat{\beta}$ and the variance components $\hat{\sigma}_u^2$ and $\hat{\sigma}_e^2$ [199].

If we are suspicious of the LMM assumptions holding, then we could instead fit model (7.21) using a semi-parametric approach where f is estimated as a random forest - in theory, combining the predictive advantage of RFs within the mixed model framework.

Ref. [85] proposed an EM-type algorithm for estimating the model components $\hat{f}, \hat{u}, \hat{\sigma}_u^2, \hat{\sigma}_e^2$ which iterates between estimating the a) the forest function assuming the random effects to be correct and b) the random effects part assuming predictions from the forest part of the model to be correct/accurate. Here, out-of-bag (OOB) predictions [17] are used to assess the forest whereby unused observations from the construction of each forest's sub-tree are utilized. Estimation of the variance components is done by taking the expectation of the ML estimators given the data. To monitor convergence, [142] use the generalized log-likelihood criterion of the composite model.

The resulting estimator for model-based predictions from the MERF is as follows,

$$\begin{aligned}\hat{\mu}_{ij}^{MERF} &= \hat{f}(x_{ij}) + \hat{u}_i \\ &= \hat{f}(x_{ij}) + \frac{\hat{\sigma}_u^2}{\hat{\sigma}_u^2 + \hat{\sigma}_e^2/n_i}(1/n_i \sum_{j \in s_i}(y_{ij} - \hat{f}^{OOB}(x_{ij})).\end{aligned} \tag{7.22}$$

Assuming unit-level population data (from census or administrative data) is available, the estimate of the area level means for each area i is,

$$\begin{aligned}\hat{\mu}_{i}^{MERF} &= \bar{\hat{f}}(X_{ij}) + \hat{u}_i \\ &= \bar{\hat{f}}(X_{ij}) + \frac{\hat{\sigma}_u^2}{\hat{\sigma}_u^2 + \hat{\sigma}_e^2/n_i}(1/n_i \sum_{j \in s_i}(y_{ij} - \hat{f}^{OOB}(x_{ij})).\end{aligned} \tag{7.23}$$

From an interpretation perspective, the RF part expresses the conditional mean of the fixed effects while the random effect is the BLUP for the linear part of the model (7.21) [142]. For non-sampled areas, rather than use the fixed part of the model, one could use a classified mixed model prediction (CMMP) approach (Section 7.5) to get more accurate estimates.

As noted in [130], not many methods in SAE can cope with the dual problem of providing robustness against model-failure while simultaneously working under limited auxiliary data. Ref. [143] addressed this by introducing aggregate census-level covariate information through calibration weights w_{ij}, which are meant to balance the unit-level predictions from MERFs trying to achieve coherence with census data covariate means. This estimator can be written as,

$$\hat{\mu}_i^{MERFagg} = \sum_{j=1}^{n_i} \hat{w}_{ij}[\hat{f}(X_{ij}) + \hat{u}_i]. \tag{7.24}$$

Here, auxiliary population information is incorporated through the weights $\hat{w}_{ij}$ maximizing the profile empirical likelihood,

$$\Pi_{j=1}^{n_i} w_{ij},$$

under three constraints.

To obtain optimal weights, a Lagrange multiplier method (Owen 2001; Emerson and Owen, 2009) is used. Ref. [143] discuss technical conditions for the solutions and give best practice guidelines.

7.3.1.1 Estimate of uncertainty

Estimated MSE for each area is the typical estimate of uncertainty but these can only be approximated (*e.g.* [196]), even in the case of LMMs with block diagonal covariance matrices and estimated variances. The introduction of RFs to model (7.21) adds further complexity. Typically, bootstrap algorithms of one kind or another are used to estimate MSEs. Specifically, [142] proposed a non-parametric block bootstrap for estimating MSE of area-level mean estimates from sampled and unsampled areas. This builds on work by [25] who developed a non-parametric bootstrap for the random components. As [174] noted, one usually centers and scales the empirical error components by a bias-adjusted residual variance. This is because the estimator of error variance under the MERF is positively biased. When working with aggregated census-level data (7.24), [143] construct pseudo-true values by generating only error components instead of simulating full bootstrap populations.

Indeed, given the data adaptiveness/model selection aspects inherent in building the MERF, one could turn to the SUMCA approach [133] for estimating uncertainty for area-level mean estimates.

7.3.1.2 Application

[143] analyzed synthetic Austrian European Union Statistics on Income and Living Conditions (EU-SILC) from 2006 on the household level. This data is freely available using the R package *emdi* [144]. Household income is a highly skewed variable which can lead to model violations for the classical NER model, thus providing a platform to showcase the benefits of using MERFs instead. Their target variable was equivalized house hold income which is defined as the ratio of total household disposable income and the equivalized house hold size. This target was available in the survey but not in the census. The dataset has 25000 households distributed across 94 districts and 1945 households. 70 districts are included in the sample with sample sizes varying between 14 and 200 households (the median was 22.5 households). Thus, direct area-level estimates are not available for 24 out-of-sample districts, making the use of SAE methods necessary.

Methods compared include the direct estimation, the two variations of the MERF above, and the EBP method of [179] using a data-driven Box-Cox transformation (EBP-BC).

Variable selection was performed for the two EBP methods resulting in 13 auxiliary variables remaining based upon selection being done with the Bayesian Information Criterion (BIC). For the MERFs, variable importance measures were estimated and the variables for cash assets, the receiving of age benefits, self-employment status, and income from a rental property were found to be the most important.

Figure 7.4 shows that the two MERF estimators perform very similarly to the EBP-BC estimator. Figure 7.5 shows that the estimates of uncertainty have slightly lower median MSE than the EBP-BC and much lower than the direct estimation or the MERF-agg.

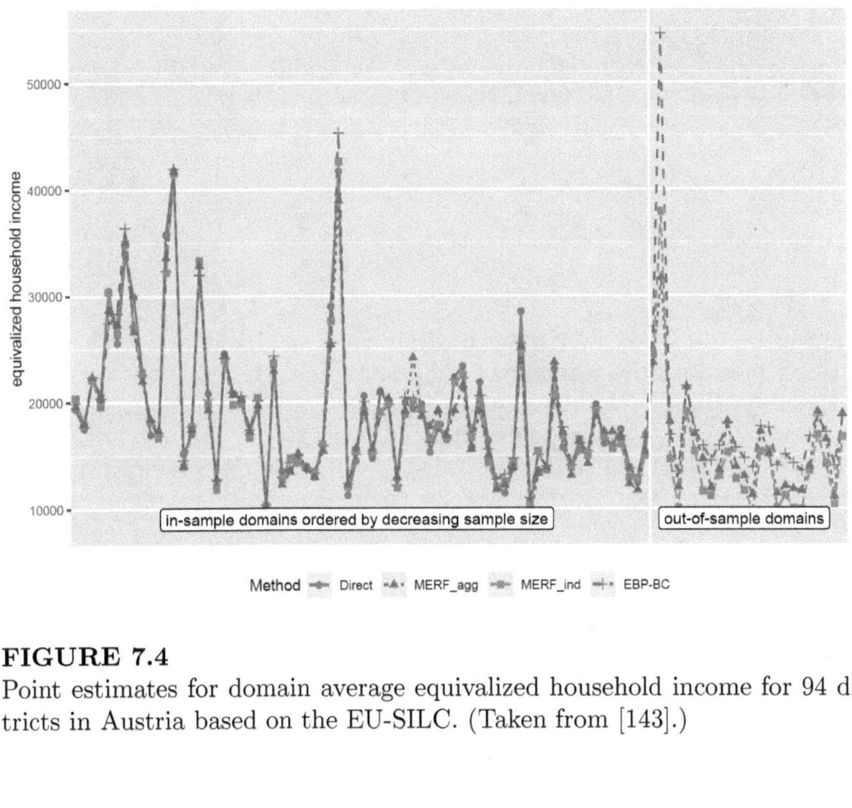

FIGURE 7.4
Point estimates for domain average equivalized household income for 94 districts in Austria based on the EU-SILC. (Taken from [143].)

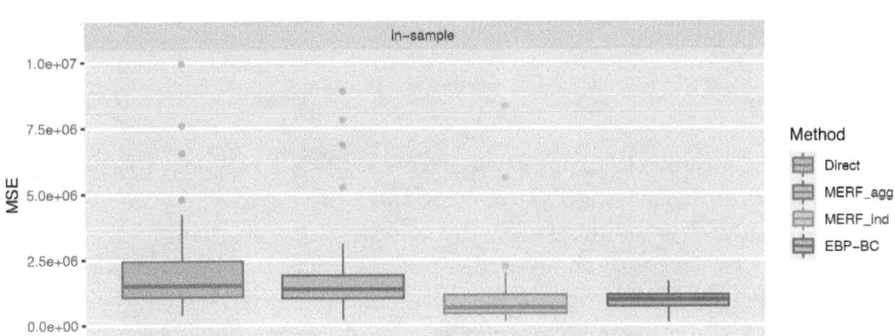

FIGURE 7.5
MSE estimates for the point estimates of average equivalized household income. (Taken from [143].)

7.3.2 Neural networks

[190] developed an extension of the Fay-Herriot (FH) model using random weight neural networks using a Bayesian hierarchical approach that allows for nonlinear relationships between the outcome and the auxiliary variables.

Recall, the FH model written hierarchically as,

$$\begin{aligned} y_i|\theta_i,\sigma_i^2 &\overset{ind}{\sim} N(\theta_i,\sigma_i^2), i=1,\ldots,m \\ \theta_i &= x_i'\beta + v_i \\ v_i &\overset{iid}{\sim} N(0,\tau^2). \end{aligned} \tag{7.25}$$

This model can be estimated in a Bayesian fashion as described in Section 7.2 by putting a prior over the regression coefficients β and the random effect variances σ_i^2.

Ref. [190] proposed to assume a more flexible structure for the unconditional mean: $E(\theta_i) = f(x_i)$ and estimate this using a single layer feed-forward neural network described as,

$$f(x_i) = \sum_{j=1}^{N} g(a_j' x_i + b_j)\beta_j, \tag{7.26}$$

where N is the number of hidden nodes (number of different affine transformations of the input data), and a_j and b_j are layer-specific parameters termed hidden layer weights. Each affine transformation is followed by a nonlinear transformation according to the function $g(\cdot)$ known as the activation function. It's very common to invoke the sigmoid function,

$$g(x) = \frac{1}{1+e^{-x}}, \tag{7.27}$$

or the so-called ReLU function,

$$g(x) = max(0,x). \tag{7.28}$$

The β_j parameters are termed output layer weights. It's also typical to scale the input variables (auxiliary variables) to have mean zero and variance one.

Fitting of this model is done usually by a stochastic gradient descent method that is used to estimate the entire set of weights [79] but new techniques using random generation of the fixed and hidden layer weights thus focusing attention on the output layer weights have been use [98]. Since the model is conditional on the hidden layer weights, estimation of the output layer weights is easy.

Ref. [190] developed a fully Bayesian implementation of the random weight feed-forward neural network for the FH model (termed NFH) as follows,

$$\begin{aligned} y_i|\theta_i, \sigma_i^2 &\overset{ind}{\sim} N(\theta_i, \sigma_i^2), i = 1, \ldots, m \\ g_i' &= \frac{1}{1 + e^{-Ax_i}} \\ v_i &\overset{iid}{\sim} N(0, \tau^2) \\ \beta_j &\overset{iid}{\sim} N(0, \kappa^2), j = 1, \ldots, N. \end{aligned} \tag{7.29}$$

The hyperparameters τ and κ require prior distributions (like vague inverse gamma priors) whose details can be found in [190]. The $N \times p$ matrix A comprises the hidden layer weights which are generated random and independently from a standard normal distribution before model fitting and these values are fixed. Therefore the hidden layer term g_i is fully determined before model fitting. The hierarchical model induces shrinkage/regularization for β which is a key component of fitting neural networks.

Ref. [190] showed that this hierarchical NFH model is conditionally conjugate allowing for the posterior distribution to be sampled easily using Gibbs sampling.

7.3.2.1 Application

Ref. [190] re-analyzed data from the Small Income and Poverty Estimates program (SAIPE) [12] where the goal was to estimate household income from various geographic areas focusing on the state of California. The direct estimates of census tract median household income for the 2021 five year period from the American Community Survey were used as the response. Auxiliary variables included census tract level poverty rate and the proportion of the population associated with four different race groups (white, black, Hispanic and Asian). California has 9129 census tracts but only 8927 were available with direct estimates (others could be unavailable due to privacy concerns due to very small sample sizes or unpopulated areas). Ref. [190] used a value of $N = 100$ for fitting the NFH model. Posterior means and posterior standard deviations were extracted for both the FH and NFH models using the hierarchical representations above.

Figure 7.6 shows a comparison of the direct estimates against the FH and NFH estimates with the line of identity drawn in red. In census tracts with higher, more extreme incomes (part of the long tail where the usual FH model assumptions may fail), we see that the NFH estimates are slightly more shrunken (relative to the direct estimates) than the FH model estimates are. Figure 7.7 shows the comparison of the standard error estimates. It reveals that the direct estimates have the largest standard errors (as expected) but the NFH model results in a greater reduction in standard errors as compared to the FH standard errors (points falling above the red line).

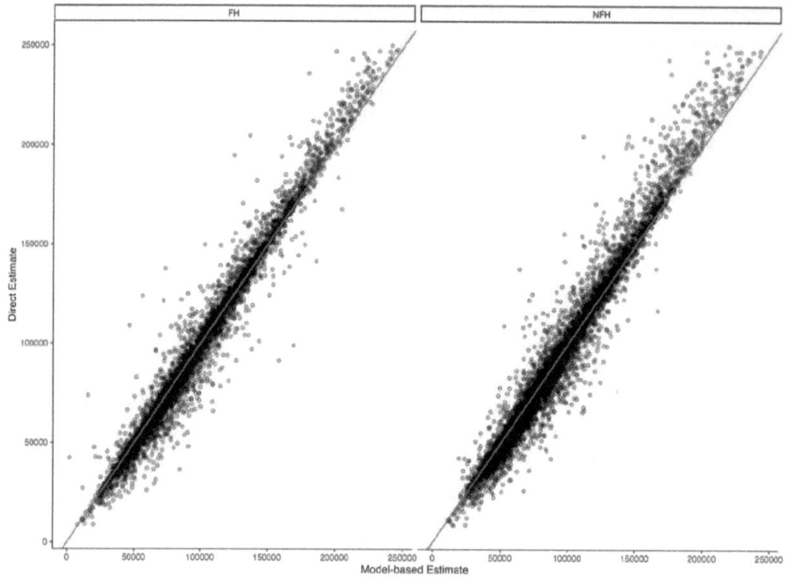

FIGURE 7.6
A comparison of direct estimates and FH and NFH estimates of census tract median incomes from California using 2021 last five year data. (Taken from [190].)

Finally, [190] noted that the posterior mean of τ^2 was nearly twice as large under the FH model as compared to the NFH model. This is an empirical indication that the nonlinearity being modeled by the NFH approach is decreasing the importance of the random effects.

7.3.3 Gradient boosting

Gradient boosting is a machine learning technique applicable to both regression and classification tasks. Boosting is a process of creating a learning ensemble consisting of so-called weak prediction models where at each iteration of building the ensemble, a new weak prediction model is added to address the errors made the current set of weak prediction models.

Gradient boosting generalizes this to allow for the optimization of arbitrary loss functions. Given a loss function $L(y_i, f(x_i))$, the algorithm can be described as follows (here we focus on gradient boosted tree-based models [65]:

1. Initialization: $f_0(x) = argmin_{\gamma_0} L(y_i, \gamma_0)$. This is typically a constant function.
2. For iteration $t; t = 1, \ldots, T$, construct a new tree.

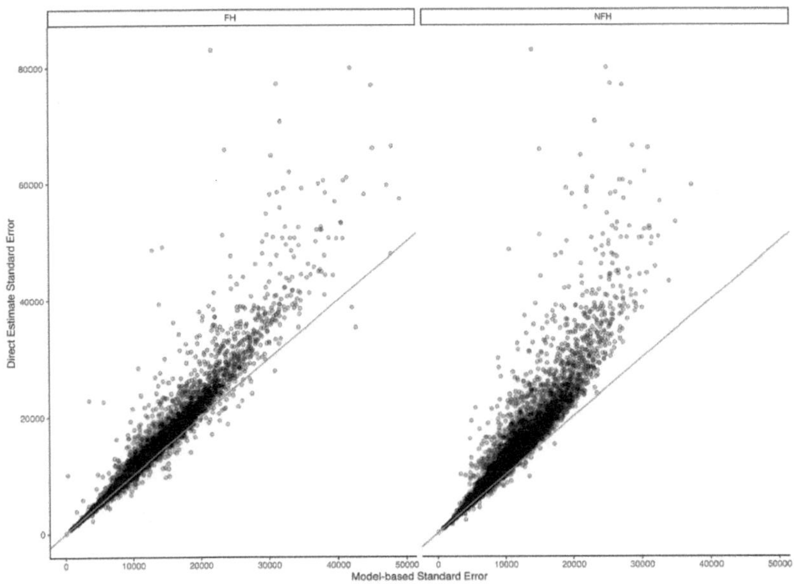

FIGURE 7.7
A comparison of direct estimates and FH and NFH estimates of census tract median incomes from California using 2021 last five year data. (Taken from [190].)

(a) Compute the negative gradient,

$$r_{it} = -\left[\frac{\delta L(y_i, f(x_i))}{\delta f(x_i)}\right]_{f=f_{t-1}}.$$

(b) Fir a tree model to the r_{it} resulting in terminal nodes $j = 1, \ldots, J_t$ and terminal regions $R_{j,t}$.

(c) For $j = 1, \ldots, J_t$, computer terminal node estimates

$$\gamma_{jt} = argmin_\gamma \sum_{x_i \in R_{j,t}} L(y_i, f_{t-1}(x_i + \gamma).$$

(d) Update a new function $f_t(x)$ using learning rate *alpha* as

$$f_t(x) = f_{t-1}(x) + \alpha \sum_{j=1}^{h} \gamma_{jt} I(x \in R_{j,t}).$$

The learning rate α and number of trees T are tuning parameters. The optimal T depends on both the predicted outcome and the learning rate. It's typical to set the learning rate at a low value like $\alpha = 0.01$ and use a stopping rule based on (multi-fold) cross-validation [230]. Fitting can be done using the XGBoost R package (Chen *et al.* 2021).

7.3.3.1 Monitoring the health and housing of the Dutch population

[230] applied tree-based gradient boosting for the analysis of the Adult and Elderly Health Monitor (HeMo) survey in the Netherlands. This is an extensive national survey about self-reported health and well being. It's administered every four years by the MHS region in collaboration with the National Institute for Public Health and Environment (RIVM) and Statistics Netherlands (CBS). The authors considered data from the 2020 survey. The authors also constructed a corresponding population dataset with features based on administrative data provided by Statistics Netherlands. This population dataset includes the entire Dutch population aged 18 or older for the year of the survey (excluding institutionalized people). A secured ID number to allow linking between the survey and the administrative data was assigned to each respondent.

For the HeMo survey, they have 539,895 respondents using an online questionnaire. The researchers focused on 34 binary health indicators including alcohol use, smoking behaviors, body weight, physical and mental health, and many other indicators.

To validate their gradient tree boosting models and others they compared it against including the raw estimates and structured additive regression model (STAR). The STAR model explicitly models spatial correlations due to geography via random effects. For the gradient boosted trees, the authors used instead either simple xy coordinates or oblique geographic coordinates (ogc) [230]. They used the dataset of survey respondents with known binary health indicators. This was split using five-fold cross-validation. Then out-of-sample predictions were made across each fold and averaged. Given the binary outcome, receiver operating characteristic (ROC) curves were used focusing on the area under the curve (AUC) estimates of model discrimination prowess. Accuracy was also considered and was defined as the proportion of correctly classified individuals for a given threshold value using calibration curves.

The authors also sought to try and use measures that would combine discrimination and accuracy including MSE which in the binary outcome case is the Brier score, and negative log likelihood (NLL) (normalized by sample size). Figure 7.8 shows a visual comparison of the prevalence estimates of the "drinker" indicator. Some mild differences between STAR and XGBoost are evident but they are relatively mild.

Table 7.3 shows the results of their analysis focusing on the HeMo indicator "drinker". As the table shows, the gradient boosting models do quite well as compared to the standard approaches with improvements along all measures (no statistical comparisons were formally done however). The authors actually found that XGBoost was the best model for all 34 health indicators in the HeMo survey.

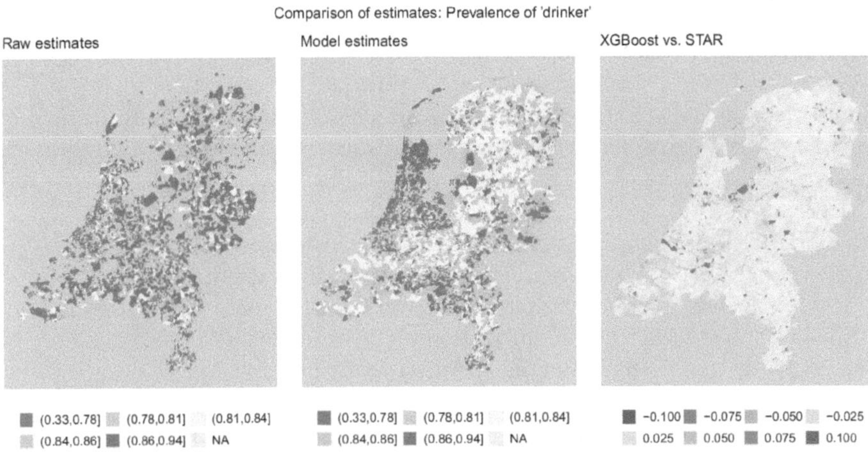

FIGURE 7.8
A comparison of estimates of drinking prevalence from 2020 HeMo survey. (Taken from [230].)

TABLE 7.3
Validation measures for the analysis of drinking health indicators from 2020 HeMo survey.

Model	Accuracy	AUC	MSE	NLL
Null	0.801	0.500	0.160	0.500
STAR	0.814	0.737	0.138	0.437
$XGBoost_{xy}$	0.816	0.744	0.136	0.433
$XGBoost_{ogc}$	0.816	0.744	0.136	0.433

Note: Adapted from [230].

7.4 Missing data

Survey data are often incomplete in the sense that some of the responses, or covariates, are missing. The standard approaches to handling missing data are based on the E-M algorithm and multiple imputation (e.g., [23]). However, such procedures often require strong, unverifiable assumptions, called missing-data mechanisms (MDM).

The framework of MDM was set up by [207] and followed extensively in the literature (e.g., [159]). Let Y denote the complete data. The data are missing completely at random (MCAR) if the probability distribution of the missing does not depend on either the observed or missing values in Y. Let Y_{obs} and Y_{mis} denote the observed and missing parts of Y. The data are missing at random (MAR) if the probability distribution of the missing depends only on Y_{obs}; otherwise, the data are not missing at random (NMAR). Of course, MCAR is a stronger assumption than MAR, but even the latter is often unverifiable, in practice.

A subject area that has been developed to address the robustness of the results of statistical analysis to various MDM is called sensitivity analysis. While such a procedure is fairly standard in medical studies (e.g., [177]), applications to SAE problems so far have been limited.

Ref. [195] adapted a cautious likelihood approach (CLA; [194]) to non-responses in SAE problems. The authors considered the case of binary responses, of which some are missing. As an example, the authors discussed a study on area-specific ratio of people at risk of poverty from the German General Social Survey. Here the binary response corresponds to the status of "poor" or "rich" according to a certain definition. Out of a total of 3,466 intended responses, 454 were missing. Instead of assuming the standard MAR, or NMAR, mechanisms, CLA makes either no assumption or a weaker assumption regarding the MDM. In the case of binary responses with no MDM assumption, the responses are assumed to be either 0, or 1, or "NA" (no answer), hence in three categories. A multinomial likelihood approach is then used for inference. In the case of weaker MDM assumptions, the assumption is in terms of setting up constraints on the multinomial likelihood.

The CLA method is, for the most part, applied to the design-based logistic generalized regression (LGREG)-synthetic estimator [154]. Application to model-based methods is limited, according to [195], due to some "technical" and computational difficulties.

In a way, NMAR is the most difficult situation to handle because any method that works under NMAR would also work in case "life is good", that is when MAR or MCAR actually hold. In this sense, the method developed under NMAR would be robust against violation of the MAR or MCAR assumptions. There has been work in SAE developed under the NMAR scenario. For example, [223] considers SAE in the case of informative sampling and NMAR non-responses. As indicated by the latter authors, in general, an NMAR model "cannot be tested by use of the observed data since the outcomes are missing for the non-respondents". Therefore, some assumptions have to be made, and the developed method depends on these assumptions.

Basically, these assumptions state that there is a connection between the observed and missing, and a model for the connection is proposed. This allows one to "impute" the missing values based on the observed data. Specifically, [223] considered an extended version of the NER model, expressed as

$$
\begin{aligned}
y_{ij}|x_{ij}, v_i &\sim f(y_{ij}|x_{ij}, v_i), \\
v_i &\sim f(v_i) \text{ with } \mathrm{E}(v_i) = 0, \mathrm{var}(v_i) = \sigma_v^2,
\end{aligned}
$$

$i = 1, \dots, M$, $j = 1, \dots, N_i$. The model is assumed to hold for all of the M sub-populations (small areas) and all of the N_i units in the i sub-population. Let s denote the subset of sampled index i, and s_i for the subject of sampled index j for $i \in s$. Let $I_i = 1$ if $i \in s$, and $I_{ij} = 1$ if $i \in s$ and $j \in s_i$. For $i \in s$, $j \in s_i$, let $R_{ij} = 1$ if unit j responded to the survey, and $R_{ij} = 0$ otherwise.

It is assumed that the probability of response depends on an unknown vector of parameters, γ. Also let $R = \{(i,j) : I_i = 1, I_{ij} = 1, R_{ij} = 1\}$, that is, the subset of indexes of the respondents among the samples. The key assumptions are

(I) For $j \neq k$, one has

$$\begin{aligned} & \mathrm{P}(R_{ij} = 1, R_{ik} = 1 | x_{ij}, y_{ij}, x_{ik}, y_{ik}, \gamma) \\ = \; & \mathrm{P}(R_{ij} = 1 | x_{ij}, y_{ij}, \gamma) \mathrm{P}(R_{ik} = 1 | x_{ik}, y_{ik}, \gamma); \end{aligned}$$

(II) For any $(i,k) \in R^c$, one has

$$f(y_{ik}|O, u_i) = f(y_{ik}|x_{ik}, u_i),$$

where $u_i = v_i - \mathrm{E}(v_i|i \in s)$ and

$$O = \{y_{ij}, \pi_i, \pi_{j|i}, n_i, (i,j) \in R; x_{ij}, i = 1, \ldots, m, j = 1, \ldots, N_i\}.$$

Here π_i is the selection probability for i, $\pi_{j|i}$ is the selection probability of j within area i, and n_i is the sample size for area i. Assumption (I) essentially states that R_{ij}, R_{ik} are independent given the relevant data, and depend on data with the respective indexes only. Assumption (II) is the key assumption that connects the observed and missing. Basically, it states that the unobserved outcomes in a sampled area are independent of the observed outcomes, given the area random effect and the covariates.

It is natural to ask if these assumptions, especially (II), are verifiable. Ref. [223] referred to some sufficient conditions for (II) given in another reference, and claimed that those conditions are mild. Another underlying assumption is that the parametric form of the response probability, $p(y_{ij}, x_{ij}, \gamma) = \mathrm{P}(R_{ij} = 1|y_{ij}, x_{ij}, \gamma)$ is known. The authors considered logistic response probability in their simulation study and real-data application.

If the missing value were actually observed, one may estimate γ based on the response probability distribution by solving the associated likelihood equation. In reality, of course, the missing values are unobserved. Using the E-M idea [48], one thus replaces the likelihood equation with its conditional expectation given the observed in order to estimate γ. The prediction of the small-area means followed essentially the EB (or BP) procedure for a finite population (e.g., [123], sec. 4.4.4; also see Section 4.1.4), taking into consideration of the joint distribution of the data and the response probability derived under the assumptions.

As for measures of uncertainty, [223] adopted a semi-parametric bootstrap methods proposed earlier by the same authors [222] for estimating the MSPE, where the bootstrap model is based on the predicted random effects (also see [126]).

To demonstrate their method, [223] carried out a simulation study, in which the outcome variable is binary following a mixed logistic model (see Section 1.5.3). The response probability associated with the missing data also follows a logistic model, as follows.

$$p(y_{ij}, x_{ij}, \gamma) = \frac{e^{\gamma_0+\gamma_1 x_{ij}+\gamma_2 y_{ij}}}{1+^{\gamma_0+\gamma_1 x_{ij}+\gamma_2 y_{ij}}}, \tag{7.30}$$

with $\gamma = (\gamma_0, \gamma_1, \gamma_2)'$. The authors compared their proposed predictor of the small-area means with three alternative methods. The first alternative is the EBP (or EB) based on the mixed logistic model, which ignores the sampling and response process, assuming that the population distribution also holds for the respondents. The second alternative is the Hajek-Brewer (H-B) estimator [84], which assumes that the non-response is completely random. The third alternative derives the estimator by assuming that the non-response is MAR, in which case the response probability is often referred to as the propensity score. The authors considered different scenarios, in which the "size" of the random effect was either large or medium, and the response probability was either (7.30) or

$$p(y_{ij}, x_{ij}, \gamma) = \frac{e^{-x_{ij}y_{ij}/2}}{1 + e^{-x_{ij}y_{ij}/2}}. \tag{7.31}$$

The authors concluded that their proposed method has good performance and showed robustness to misspecification of the response model [regarding (7.30) and (7.31)].

It is often said that a simulation study "always works"; however, relevance to the real-life practice is more important. On this occasion, [223] illustrated the relevance with an application to the prediction of the number of married people in small statistical areas in Israel. It might be worth noting that one of the coauthors had been the National Statistician of Israel (equivalent to the Director of the Census in the U.S.). It was indicated that Israel, in general, has a very good register system; however, the system encountered difficulties with the emigrants, who were often slow to report their whereabouts when they moved. To deal with the problem, an integrated system was conducted in 2008 by the Israel Central Bureau of Statistics, which combined the census data with surveys. Letters were sent to the sampled individuals asking them to respond to phone interviews, and non-responses were among the sampled individuals. The authors reported the response rates from different statistical areas, that is, area samples of addresses, as about 0.75 with a standard deviation of about 0.14.

In this case, the non-responses seemed to be NMAR because they were more likely to be those who were not registered correctly (e.g., living in another area), and hence did not receive the notice letter in the first place. As in their simulation study, [223] proposed the logistic models for both the outcome variable, y_{ij}, which is the indicator that the corresponding person was married, and the response probability. Both models involved the same covariates, which are indicator variables associated with the person's age. Specifically, there are two indicators, one for age greater than 25 and the other for age greater than 40, but the response probability also involved y_{ij} as in (7.30).

The target of practical interest was the total number of people who were registered as living in the area and actually living there. However, the authors had no information on the true target numbers. Thus, as a way to demonstrate the performance of different predictors, the authors considered predicting the number of married people who were registered as living in the area, for which the true counts were known. The authors compared four different predictors, including their proposed predictor, the H-B predictor, and two other predictors derived under the MCAR and MAR mechanisms, respectively, in terms of the prediction bias and root MSPE (RMSPE). The proposed predictor was shown to perform much better both in terms of the bias and in terms of the RMSPE.

7.4.1 Model selection with missing data

Model selection was considered as a strategy to achieve robustness by reducing the chance or impact of model misspecification (see Chapter 6). While there is an extensive literature on statistical analysis with missing or incomplete data, including that in the SAE context (see above), the literature on model selection in the presence of missing data is relatively sparse. Ref. [102] proposed a class of model selection criteria based on the output of the E-M algorithm. Also see [68]. A potential drawback with the above E-M approach is that the conditional expectation in the E-step is taken under the assumed (candidate) model, rather than an objective (true) model. Note that the complete-data log-likelihood is also based on the assumed model. Thus, by taking the conditional expectation, again, under the assumed model, it may bring false supporting evidence for an incorrect model. This type of problem is sometimes referred to as "double-dipping". We illustrate this with an example.

Example 7.1. Suppose that one attempts to select a logistic model, $\text{logit}(p_i) = x_i'\beta$, where $p_i = \text{P}(y_i = 1)$, $y_1, \dots, y_n$ being independent, binary, observations, and x_i is a vector of covariates to be selected. Suppose that $y_1, \dots, y_5$ are observed, and the rest of the y_i's are missing. Also, for simplicity, assume that all the x_i's are observed. Suppose that the true model is not the intercept model (i.e., $x_i = 1$), denoted by M_0; let β_0 be the intercept. The complete-data log-likelihood under M_0 has the expression

$$l = \sum_{i=1}^{n} \{y_i \log(p_0) + (1 - y_i)\log(1 - p_0)\},$$

where $p_0 = e^{\beta_0}/(1 + e^{\beta_0})$. Note that, under M_0, one has

$$\begin{aligned} \text{E}(l|y_1, \dots, y_5, \text{ all } x_i's) &= \sum_{i=1}^{5} \{y_i \log(p_0) + (1 - y_i)\log(1 - p_0)\} \\ &\quad + (n-5)\{p_0 \log(p_0) + (1 - p_0)\log(1 - p_0)\}. \end{aligned}$$

If $y_i = 1, 1 \le i \le 5$, then, as $p_0 \to 1$, one has

$$\text{E}(l|y_1, \dots, y_5, \text{ all } x_i's) \to 0.$$

On the other hand, under any other model, M, the corresponding log-likelihood satisfies

$$l = \sum_{i=1}^{n} \{y_i \log(p_i) + (1 - y_i) \log(1 - p_i)\} \leq 0,$$

hence $\mathrm{E}(l|y_1, \dots, y_5, \text{ all } x_i's) \leq 0$, under M.

This means that the maximized conditional expectation of l under M_0. which is 0, is $\geq$ the maximized conditional expectation of l under M, which is ≤ 0. Thus, the first term of any information criterion [see (6.1)] under M_0 is $\leq$ that under M. Furthermore, M_0 certainly has the smallest dimension. Therefore, M_0 will be selected as the optimal model by the criteria of [102], which, of course, is an incorrect model.

Ref. [129] developed two model selection strategies for model selection with incomplete (or missing) data. One is called EMAF; the other is called E-MS.

The EMAF method is closely related to the adaptive fence, or AF method, discussed in Section 6.1. In fact, the AF idea works naturally with the incomplete data. For the simplicity of illustration, let us assume, for now, that the candidate models include a correct model as well as a full model. It then follows that the full model is, at least, a correct model, even though it may not be the most efficient one. In the presence of missing data, one can run the E-M to obtain the MLE of the parameters, under the full model. Note that, here, we do not have the double-dipping problem as illustrated by Example 7.1, because the conditional expectation under the full model is "objective".

Once the parameter estimates are obtained, we can use the model-based (or parametric) bootstrap to draw samples, under the full model, as in the AF (see Section 6.1). The best part of this strategy is that, when one draws the model-based bootstrap samples, one draws samples of the complete data, rather than data with the missing values. Therefore, one can apply any existing model selection procedure that is built for the complete-data situation (see Section 6.1) to the bootstrap sample.

Suppose that B bootstrap samples are drawn. The model with the highest frequency of being selected as the optimal model, among the bootstrap samples, is the (final) optimal model. The procedure is called EMAF algorithm due to its connection to the EM algorithm, and similarity to the AF idea. The idea can be extended to situations where a correct model may not exist, or exists but not among the candidates. In such a case, the bootstrap samples may be drawn under the model M_*, where M_* is the model with the minimum Q as in (6.2); in other words, M_* is the model among the candidate models with the best approximation in terms of fitting the current data.

The E-MS algorithm may be viewed as an extension of the E-M algorithm [48]. The basic idea is to extend the concept of the parameter to the model plus the parameters under the model. The E-M algorithm is well known for parameter estimation in the presence of incomplete or missing data. On the other hand, model selection, as another component of model identification, may also be viewed as parameter estimation, with the parameter being [the identification (ID) number of] the model and the parameter space being the (ID numbers of the) model space. Unlike the EMAF, E-MS can be incorporated with any existing model selection procedure, not just the fence methods.

Specifically, one combines the parameters with the model under which the parameters are defined. So, at the current stage of the iteration, one has the current model, $M_{\rm c}$, as well as the current estimates of the parameters, $\hat{\theta}_{\rm c}$, under $M_{\rm c}$. Let $Q(M) = Q(Y, M, \theta_M)$ be a measure of lack-of-fit, where Y represents the data, M a candidate model, and θ_M the vector of parameters under M. One takes the conditional expectation of $Q(M)$ under $M_{\rm c}$, with the parameters under $M_{\rm c}$, $\theta_{M_{\rm c}} \equiv \theta_{\rm c}$, being $\hat{\theta}_{\rm c}$, given the observed data, $Y_{\rm o}$, denoted by $\mathrm{E_c}\{Q(M)|Y_{\rm o}\}$. This is the E-step.

In the next step, one carries out model selection using $\mathrm{E_c}\{Q(M)|Y_{\rm o}\}$ as the measure of lack-of-fit. This is the MS-step, where MS stands for model selection. The procedure iterates between the E-step and MS-step until converges. The convergence property has been established in [129].

7.4.2 Simulation study: Regression variable selection with incomplete data

The classical linear regression is a conditional model, in which the distribution of the covariates, or predictors, is not specified. Such a model may not directly work with the E-M algorithm, if some of the covariates are also missing. Ref. [159] proposed the following model for the joint distribution of the response and predictors in a linear regression. Suppose that the candidate predictors can be listed as $x_1, \dots, x_p, x_{p+1}, \dots, x_{p+q}$ such that $x_1, \dots, x_p$ are continuous and $x_{p+1}, \dots, x_{p+q}$ are discrete, or categorical, predictors (in case there is an intercept, the corresponding constant, 1, is considered as the first discrete/categorical predictor). Furthermore, let $v_1, \dots, v_s$ be all the possible (vector-valued) values for $x_{\rm d} = (x_{p+1}, \dots, x_{p+q})'$. Let $x_{i,\rm d}$ be the $x_{\rm d}$ corresponding to the ith observation, and $x_{i,\rm c}$ be the vector $(x_1, \dots, x_p)'$ corresponding to the ith observation, and $x_i = (x'_{i,\rm c}, x'_{i,\rm d})'$. The assumptions are:

(i) $y_i, x_i, i = 1, \dots, n$ are independent.

(ii) For each i, $x_{i,\rm d}$ has the probability distribution $\mathrm{P}(x_{i,\rm d} = v_r) = \pi_r, 1 \leq r \leq s$, where the π_r's are unknown probabilities such that $\sum_{r=1}^s \pi_r = 1$.

(iii) Given $x_{i,\rm d} = v_r$, $x_{i,\rm c}$ has a multivariate normal distribution with mean μ_r and covariance matrix Ω, where $\mu_r, 1 \leq r \leq s$ are unknown vectors, and Ω is an unknown covariance matrix that does not depend on r.

(iv) Given x_i, y_i is normal with mean $x_i'\beta$ and variance σ^2, where β is an unknown $(p + q)$-dimensional vector of regression coefficients, and σ^2 is an unknown variance.

The assumptions (i)–(iii) are for the full model. More generally, we are interested in a model, M, for the conditional distribution (iv). Write $x_{i,M} = (x'_{i,M,\rm c}, x'_{i,M,\rm d})'$, and $\beta_M = (\beta'_{M,\rm c}, \beta'_{M,\rm d})'$. Then, under M, (iv) is replaced by

(iv-M) Given x_i, $y_i \sim N(x'_{i,M}\beta_M, \sigma^2)$,

while parts (i)–(iii) of the model are unchanged.

Suppose that the candidate predictors consist of two continuous variables and two indicator variables. So, x_1, x_2 are continuous (with $p = 2$) and x_3, x_4 are indicators (0 or 1). Thus, $x_i = (x_{i1}, x_{i2}, x_{i3}, x_{i4})'$ with $x_{i,\mathrm{c}} = (x_{i1}, x_{i2})'$ and $x_{i,\mathrm{d}} = (x_{i3}, x_{i4})'$. The distinct possible values for $x_{i,\mathrm{d}}$ are $v_1 = (0,0)'$, $v_2 = (0,1)'$, $v_3 = (1,0)'$, and $v_4 = (1,1)'$. Here, assumption (iii) means that there are 2×1 vectors $\mu_1, \mu_2, \mu_3, \mu_4$ and 2×2 covariance matrix Ω such that, given $x_{i,\mathrm{d}} = v_r$, $x_{i,\mathrm{c}} \sim N(\mu_r, \Omega)$, $r = 1, 2, 3, 4$.

A simulation study was carried out with the sample size $n = 100$ and the true model being x_1 and x_3. The true parameters are $\beta_1 = \beta_3 = \sigma^2 = 1$. The true μ_r is the same as v_r, $r = 1, 2, 3, 4$; and the true Ω is the 2×2 identity matrix. After the complete data is generated, we randomly select a subset of indexes from $\{1, \dots, n\}$ for the response as well as for each of the candidate predictors, which correspond to the missing data, so that $100 p_{\mathrm{m}}$ % of the data are missing for the response and each of the candidate predictors. Here we consider two cases: $p_{\mathrm{m}} = 0.1$ and $p_{\mathrm{m}} = 0.2$.

Here, we study the performance of E-MS incorporated with a variation of the fence method, called invisible fence (IF; [127]). The basic idea of IF is that, if parsimony is used as the criterion of selecting the model within the fence, only a (much smaller) subset of the candidate models have nonzero chance of being selected; those are the models that minimize Q at each dimension. Thus, one only needs to select the best dimension, which is a uni-variate selection problem, via the AF.

It is known that IF may suffer from a dominant-factor effect Namely, although the true coefficients for the true predictors (x_1 and x_3) are both equal to 1, it turns out that the continuous predictor is a dominant factor. Thus, with a moderate sample size, such as in the current case, the IF tends to overwhelmingly select x_1 at dimension 1, leading to a potentially under-fitting model. To overcome such a problem, a modification of the IF is considered to make it more "aggressive". Let α be a chosen number between 0 and 1. Let d^* be the selected dimension by IF and p^* be the corresponding maximum empirical probability (that a given model is selected at the dimension). Let p_1^* be the largest maximum empirical probability corresponding to a dimension greater than d^* (thus, $p_1^* < p^*$). Let d_1^* be the corresponding dimension to p_1^*. If $p_1^* \geq (1 - \alpha) p^*$, then d_1^* is selected instead of d^*. It is clear that the modified IF is more in favor of a "larger" model, and in this sense it is more aggressive.

A same-data comparison of E-MS with a number of different procedures was carried out in the simulation study. The first is an IF version of EMAF, following a very similar procedure as the EMAF (see the previous section), with the modified IF, as described above. This is called EMIF. Three different values of α were considered in the simulation study: $\alpha = 0$ (corresponding to IF without the modification), $\alpha = 0.1$, and $\alpha = 0.5$, for EMIF as well as each of the comparing procedures described below, except the E-MS. Of course, this raises a question of what α is the best, which one may not know in practice. On the other hand, the E-MS seems to have some advantage in this regard. The idea is to start with a relatively large α (say, $\alpha = 0.5$), in order

TABLE 7.4
Summary of performance ($p_{\mathrm{m}} = 0.1$).

Method	α	TP	OF	UF	MC	s.d.	MIC	s.d.
CRNIF	0	0.67	0.00	0.33	1.67	0.47	0.00	0.00
	0.1	0.85	0.00	0.15	1.85	0.36	0.00	0.00
	0.5	0.73	0.27	0.00	2.00	0.00	0.27	0.45
IMIF	0	0.61	0.00	0.39	1.61	0.49	0.00	0.00
	0.1	0.88	0.00	0.12	1.88	0.33	0.00	0.00
	0.5	0.72	0.28	0.00	2.00	0.00	0.28	0.45
EMIF	0	0.66	0.00	0.34	1.66	0.48	0.00	0.00
	0.1	0.88	0.00	0.12	1.88	0.33	0.00	0.00
	0.5	0.75	0.25	0.00	2.00	0.00	0.25	0.44
CDIF	0	0.71	0.00	0.29	1.71	0.46	0.00	0.00
	0.1	0.92	0.00	0.08	1.92	0.27	0.00	0.00
	0.5	0.69	0.31	0.00	2.00	0.00	0.31	0.46
E-MS		0.98	0.01	0.01	1.99	0.10	0.01	0.10

to be more conservative in dropping the predictors, and gradually reduce α as the iteration progresses. More specifically, we begin with $\alpha = 0.5$; with each iteration, we reduce α by half, until convergence.

Two other methods are also included in the simulation comparison, in addition to E-MS and EMIF. One is IF based on the complete-record-only analysis (CRNIF). Another is IF with the missing data replaced by the imputed data (IMIF). The latter is based on a method of multivariate imputation developed by [229], implemented in the R package, aregImpute(). The last comparing procedure is IF based on the complete data that were generated before the missing values were taken out (CDIF). The latter is, of course, not possible in a practical situation, but it would be interesting to see how much loss of efficiency there is for a method, if any, compared to the "gold standard". Note that all of the comparing methods are in conjunction with the IF, the only difference being how the missing data are handled.

The results based on 100 simulation runs are presented in Tables 7.4 and 7.5, where OF standards for over-fitting, that is, the empirical probability that the selected model includes all of the true predictors, plus at least one extraneous predictor; UF standards for under-fitting, that is, the empirical probability that the selected model misses at least one true predictor [but may include extraneous predictor(s) as well]. Other performance measures are the empirical mean number of correctly identified predictors (MC); and the empirical mean number of incorrectly identified predictors (MIC), along with the corresponding empirical standard deviation (s.d.)'s.

It is seen in this comparison that, overall, E-MS outperformed not only all of the methods that are practically feasible (CRNIF, IMIF, EMIF) but also the "gold standard" (CDIF) that is practically infeasible. In fact, in terms of the overall performance, the order seems to be (from best to worst) E-MS, CDIF, EMIF, IMIF, CRNIF. Of course, it is not surprising that CRNIF took

TABLE 7.5
Summary of performance ($p_{\mathrm{m}} = 0.2$).

Method	α	TP	OF	UF	MC	s.d.	MIC	s.d.
CRNIF	0	0.56	0.00	0.44	1.56	0.50	0.01	0.10
	0.1	0.73	0.07	0.20	1.80	0.40	0.09	0.29
	0.5	0.68	0.31	0.01	1.99	0.10	0.32	0.47
IMIF	0	0.59	0.00	0.41	1.59	0.49	0.01	0.10
	0.1	0.85	0.00	0.15	1.85	0.36	0.01	0.10
	0.5	0.81	0.19	0.00	2.00	0.00	0.19	0.39
EMIF	0	0.65	0.00	0.35	1.65	0.48	0.00	0.00
	0.1	0.88	0.00	0.12	1.88	0.33	0.00	0.00
	0.5	0.82	0.17	0.01	1.99	0.10	0.18	0.39
CDIF	0	0.75	0.00	0.25	1.75	0.44	0.00	0.00
	0.1	0.94	0.00	0.06	1.94	0.24	0.00	0.00
	0.5	0.74	0.26	0.00	2.00	0.00	0.26	0.44
E-MS		0.95	0.01	0.04	1.96	0.20	0.03	0.17

the last place, but what seems a little unexpected is that E-MS even (slightly) outperformed CDIF.

An explanation for this seemingly surprising result is that the performance of CDIF still suffers, to some extent, the dominant-factor effect (see above), but E-MS is able to overcome this (see below). A key to this "super-performance" is α, which may be viewed as a tuning parameter. It appeared that, in this case, the best α for CRNIF, IMIF, EMIF, and CDIF was somewhere between 0.1 and 0.5. Of course, in the simulation study, one could explore this best value, but it would not be possible in practice. On the other hand, the E-MS seemed to be able to get the best out of this "α-business" during its iterations.

It may also be worthy of reporting that, in all of the simulation runs, the E-MS converged in 2-3 iterations. Also, it seems that the performance of IMIF, EMIF, and E-MS are not affected much by the increase of p_{m}. This, too, is a bit surprising, as a larger p_{m} means less observed data. In fact, with $p_{\mathrm{m}} = 0.1$, one expects about 59% complete data records; with $p_{\mathrm{m}} = 0.2$, the % of the complete data records drops to less than 33%. On the other hand, it took about twice the computing time to run the E-MS for $p_{\mathrm{m}} = 0.2$, compared to $p_{\mathrm{m}} = 0.1$. This is reasonable, as more data are missing under $p_{\mathrm{m}} = 0.2$; therefore, the conditional expectations, which have to be dealt with via the Monte-Carlo method, need to be evaluated more often than under $p_{\mathrm{m}} = 0.1$.

7.5 Classified mixed model prediction (CMMP)

Prediction of mixed effects for out-of-sample areas is a challenging problem. Historically, one typically assigns the prediction as the predicted fixed effect part of the small area model (FH or NER model for instance) using the set of auxiliary covariates associated with the out-of-sample area. Clearly, there is information lost in doing so especially if the random effects are important (i.e. have large enough variance component(s)).

Recently [131] developed a technique to *classify* a random effect associated with in-sample areas to the out-of-sample data. This method is called *classified mixed model prediction.*

Suppose that we have a set of training data, $y_{ij}, i = 1, \ldots, m, j = 1, \ldots, n_i$ in the sense that their classifications are known, that is, one knows which group, i, that y_{ij} belongs to. The assumed model is a linear mixed model (LMM; e.g., [109]):

$$y_i = \mathrm{E}(y_i|\alpha) + \epsilon_i = X_i\beta + Z_i\alpha_i + \epsilon_i, \tag{7.32}$$

where $y_i = (y_{ij})_{1\leq j\leq n_i}$, $X_i = (x'_{ij})_{1\leq j\leq n_i}$ is a matrix of known covariates, β is a vector of unknown regression coefficients (the fixed effects), Z_i is a known $n_i \times q$ matrix, α_i is a $q \times 1$ vector of group-specific random effects, ϵ_i is a $n_i \times 1$ vector of errors, and $\alpha = (\alpha_i)_{1\leq i\leq m}$. It is assumed that the α_i's and ϵ_i's are independent, with $\alpha_i \sim N(0, G)$ and $\epsilon_i \sim N(0, R_i)$, where the covariance matrices G and R_i depend on a vector ψ of dispersion parameters, or variance components. Our goal is to make a classified prediction for a mixed effect associated with a new observation, y_{n}. Suppose that

$$y_{\mathrm{n}} = \mathrm{E}(y_{\mathrm{n}}|\alpha) + \epsilon_{\mathrm{n}} = x'_{\mathrm{n}}\beta + z'_{\mathrm{n}}\alpha_{\mathrm{I}} + \epsilon_{\mathrm{n}}, \tag{7.33}$$

where $x_{\mathrm{n}}, z_{\mathrm{n}}$ are known vectors, $I \in \{1, \ldots, m\}$ but one does not know which element i, $1 \leq i \leq m$, is equal to I. Furthermore, ϵ_{n} is a new error that is independent of $y_i, 1 \leq i \leq m$, and has mean zero. The mixed effect of interest is $\theta = \mathrm{E}(y_{\mathrm{n}}|\alpha) = y_{\mathrm{n}} - \epsilon_{\mathrm{n}}$. From the training data, one can estimate the parameters, β and ψ. Thus, we can assume that estimators $\hat{\beta}, \hat{\psi}$ are available for β, ψ, respectively. Suppose that $I = i$. Then, it can be shown that $\mathrm{E}(\theta|y_1, \ldots, y_m) = \mathrm{E}(\theta|y_i)$ and, according to the normal theory,

$$\mathrm{E}(\theta|y_i) = x'_{\mathrm{n}}\beta + z'_{\mathrm{n}}GZ'_i(R_i + Z_iGZ'_i)^{-1}(y_i - X_i\beta). \tag{7.34}$$

The right side of (7.34) is the BP under the assumed LMM, if the true parameters, β and ψ, are known. Because the latter are unknown, we replace them by $\hat{\beta}$ and $\hat{\psi}$, respectively. The result is called an empirical best predictor (EBP), noted by $\tilde{\theta}_{(i)}$. In practice, however, I is unknown. In order to identify I, we consider the MSPE of predicting θ by the BP when I is classified as i, that is

$$\mathrm{MSPE}_i = \mathrm{E}\{\tilde{\theta}_{(i)} - \theta\}^2 = \mathrm{E}\{\tilde{\theta}^2_{(i)}\} - 2\mathrm{E}\{\tilde{\theta}_{(i)}\theta\} + \mathrm{E}(\theta^2). \tag{7.35}$$

Using the expression $\theta = y_{\mathrm{n}} - \epsilon_{\mathrm{n}}$, we have

$$\mathrm{E}\{\tilde{\theta}_{(i)}\theta\} = \mathrm{E}\{\tilde{\theta}_{(i)}y_{\mathrm{n}}\} - \mathrm{E}\{\tilde{\theta}_{(i)}\epsilon_{\mathrm{n}}\} = \mathrm{E}\{\tilde{\theta}_{(i)}y_{\mathrm{n}}\}. \tag{7.36}$$

Thus, we have

$$\text{MSPE}_i = \text{E}\{\tilde{\theta}^2_{(i)} - 2\tilde{\theta}_{(i)} y_{\text{n}} + \theta^2\}. \tag{7.37}$$

Note that the E in (7.37) denotes the true expectation, which may be unknown; nevertheless, the observed MSPE corresponding to (7.37) is the expression inside the expectation. Therefore, a natural idea is to identify I as the index i that minimizes the observed MSPE. Because θ^2 does not depend on i, this is equivalent to

$$I = \text{argmin}_i\{\tilde{\theta}^2_{(i)} - 2\tilde{\theta}_{(i)} y_{\text{n}}\}. \tag{7.38}$$

Denote the I identified by (7.38) by $\hat{I}$. Then, the classified mixed-effect predictor (CMMP) of θ is given by $\hat{\theta} = \tilde{\theta}_{(\hat{I})}$.

An important concept being exploited by CMMP is the idea that it *captures what is not captured by the fixed effect (the uncaptured)* through the classified random effect. This uncaptured information could also include missing covariates.

7.5.1 Robustness of CMMP

It is important to note that the primary interest is not to identify the correct "match", I. In fact, in many applications such a match may not exist, that is, there is no group among the training that matches exactly the group corresponding to the new observations. Even if the exact match does exist, as the number of training data groups, m, increases, the probability of identifying the correct group, that is, $\text{P}(\hat{I} = I)$, goes to zero (as opposed to going to one, as one might expect). But, regardless, the CMMP of the mixed effect, θ, is consistent (in fact, converges in L^2 to the true mixed effect). For example, it was demonstrated by [131] that CMMP significantly outperforms the traditional regression prediction (RP) whether or not the true match exists. The rationale behind the mismatch-led-correct-prediction is because, as m increases, the difference between different groups becomes smaller and smaller; thus, even though there is no exact match, there is a "close match" between one of the training data groups and the new group, of which CMMP is able to take advantage. In mathematical notation, this is $|\tilde{\theta}_{(i)} - \bar{y}_{\text{n}}|^2$, where $\tilde{\theta}_{(i)}$ is the estimated θ assuming $I = i$ and $\bar{y}_{\text{n}}$ is the observed θ. This important, and interesting, feature makes the CMMP idea practically more attractive because, in practice, an exact match may not exist but a close resemblance may well be expected.

7.5.2 Simulation study: Prediction of mixed effects associated with new data - matched and unmatched scenarios.

Consider a simulation study carried out under the following model:

$$y_{ij} = 1 + 2x_{1,ij} + 3x_{2,ij} + \alpha_i + \epsilon_{ij}, \tag{7.39}$$

TABLE 7.6
Average MSPE for prediction of mixed effects. %MATCH = % of times that the new observations were matched to some of the groups in the training data.

	Scenario	σ_α^2	0.1	1	2	3
$m = 10$	I	RP	0.157	1.002	1.940	2.878
	I	CMMP	0.206	0.653	0.774	0.836
	I	%MATCH	91.5	94.6	93.6	93.2
	II	RP	0.176	1.189	2.314	3.439
	II	CMMP	0.225	0.765	0.992	1.147
	II	%MATCH	91.2	94.1	92.6	92.5
$m = 50$	I	RP	0.112	1.013	2.014	3.016
	I	CMMP	0.193	0.799	0.897	0.930
	I	%MATCH	98.7	98.5	98.6	98.2
	II	RP	0.113	1.025	2.038	3.050
	II	CMMP	0.195	0.800	0.909	0.954
	II	%MATCH	98.8	98.7	98.4	98.4

$i = 1, \dots, m$, $j = 1, \dots, n$, with $n = 5$, $\alpha_i \sim N(0, G)$, $\epsilon_{ij} \sim N(0, 1)$, and α_i's, ϵ_{ij}'s are independent. The $x_{k,ij}, k = 1, 2$ were generated from the $N(0, 1)$ distribution, and then fixed throughout the simulation. There are $K = 10$ new observations, generated under two scenarios. Scenario I: The new observations have the same α_i as the first K groups in the training data ($K \leq m$), but independent ϵ's; that is, they have "matches". Scenario II: The new observations have independent α's and ϵ's; that is, they are "unmatched". Note that there are K different mixed effects. We use CMMP (and RP) to predict each of the K mixed effects, and report the average of the simulated MSPEs, obtained based on $T = 1000$ simulation runs. We considered $m = 10$ and $m = 50$.

The results are shown in Table 7.6 which it clearly demonstrates that regardless of whether the new observations actually have matches or not, CMMP matches them anyway. More importantly, the results show that even a "fake" match still helps which is empirical evidence of the point brought up earlier.

7.5.3 Estimating uncertainty for CMMP

A standard uncertainty measure for a predictor is the MSPE. A "gold standard" for the MSPE estimation is to produce a second-order unbiased MSPE estimator, that is, the order of bias of the MSPE estimator is $o(m^{-1})$, where m is the total number of clusters in the training data. Typically, the $o(m^{-1})$ term is, in fact, $O(m^{-2})$, but this difference is usually ignored. For the most part, there have been two approaches for producing a second-order unbiased MSPE estimator. The first is the Prasad-Rao linearization method [196]. The approach uses Taylor series expansion to obtain a second-order approximation to the MSPE, then corrects the bias, again to the second-order, to produce an MSPE estimator whose bias is $o(m^{-1})$. Various extensions of the Prasad-Rao method have been developed; see, for example, [45], [115], [39], and [42]. Although the method often leads to an analytic expression of the MSPE estimator, the derivation is tedious, and the final expression is likely to be complicated. More importantly, errors often occur in the process of analytic derivations as well as computer programming based on lengthy expressions. Furthermore, the linearization method does not apply to situations where a non-differentiable operation is involved in obtaining the predictor, such as shrinkage estimation (e.g., [224]), CMMP (Jiang [131]), as well as the CMMP.

The second approach to second-order unbiased MSPE estimation is resampling methods. Ref. [118]; hereafter JLW proposed a jackknife method to estimate the MSPE of an empirical best predictor (EBP). The method avoids tedious derivations of the Prasad-Rao method, and is "one formula for all". On the other hand, there are restrictions on the class of predictors to which JLW applies. Namely, JLW only applies to empirical best predictor (EBP), that is, the predictor obtained by replacing the parameters involved in the best predictor (BP), which is the conditional expectation, by their (consistent) estimators. The CMMP predictor, however, is not an EBP, because it involves a matching process. Ref. [117] proposed a Monte-Carlo jackknife method, called McJack, which potentially applies to CMMP; however, the method is computationally very expensive. Another resampling-based approach is double bootstrapping (DB; [88]). Although DB is capable of producing a second-order unbiased MSPE estimator, it is, perhaps, computationally even more intensive than the McJack. It is also unclear whether DB can be extended to CMMP.

Ref. [185]Nguyen *et al.* proposed a method that may be viewed as a hybrid of the linearization method and resampling method, by combining the best part of each method. In short, this uses a simple, analytic approach to obtain the leading term of our MSPE estimator, and a Monte-Carlo method to take care of a remaining, lower-order term.

Specifically, let θ be the mixed effect corresponding to the new observations, and $\hat{\theta}$ the CMMP predictor of θ. The MSPE of $\hat{\theta}$ can be expressed as $\text{MSPE} = \text{E}(\hat{\theta} - \theta)^2 = \text{E}\left[\text{E}\{(\hat{\theta} - \theta)^2|y\}\right]$, where y represents the available data. Suppose that the underlying distribution of y depends on a vector of unknown parameters, ϕ. Then, the conditional expectation inside the expectation is a function of y and ϕ, which can be written as $a(y, \phi) = \text{E}\{(\hat{\theta} - \theta)^2|y\} = \hat{\theta}^2 - 2\hat{\theta}\text{E}(\theta|y) + \text{E}(\theta^2|y) = \hat{\theta}^2 - 2\hat{\theta}a_1(y, \phi) + a_2(y, \phi)$, where $a_j(y, \phi) = \text{E}(\theta^j|y), j = 1, 2$. If we replace the ϕ in $a(y, \psi)$ by $\hat{\phi}$, a consistent estimator of ϕ, the result is a first-order unbiased estimator, that is, we have $\text{E}\{a(y, \hat{\phi}) - a(y, \phi)\} = O(m^{-1})$. On the other hand, both $\text{MSPE} = \text{E}\{a(y, \phi)\}$ and $\text{E}\{a(y, \hat{\phi})\}$ are functions of ϕ, denoted by $b(\phi)$ and $c(\phi)$, respectively. It follows that $d(\phi) = b(\phi) - c(\phi) = O(m^{-1})$; thus, if we replace, again, to replace ϕ by $\hat{\phi}$ in $d(\phi)$, the difference is a lower-order term, that is, $d(\hat{\phi}) - d(\phi) = o_{\text{P}}(m^{-1})$ [see, e.g., [125], sec. 3.4 for notation like o_{P}

and $O_{\rm P}$]. Now consider the estimator

$$\widehat{\rm MSPE} = a(y, \hat{\phi}) + d(\hat{\phi}) = a(y, \hat{\phi}) + b(\hat{\phi}) - c(\hat{\phi}). \tag{7.40}$$

We have ${\rm E}(\widehat{\rm MSPE}) = {\rm E}\{a(y, \phi)\} + {\rm E}\{a(y, \hat{\phi}) - a(y, \phi)\} + {\rm E}\{d(\hat{\phi})\} = {\rm MSPE} + {\rm E}\{d(\hat{\phi}) - d(\phi)\} = {\rm MSPE} + o(m^{-1})$. Essentially, this one-line, heuristic derivation shows the second-order unbiasedness of the proposed MSPE estimator, (7.40), provided that the terms involved can be evaluated.

Note that the leading term, $a(y, \hat{\phi})$, in (7.40) is guaranteed positive, a desirable property for an MSPE estimator. The lower-order term, $b(\hat{\phi}) - c(\hat{\phi})$, corresponds to a bias correction to the leading term. This term is typically much more difficult to evaluate than the leading term. We propose to approximate this term using a Monte-Carlo method. Let P_ϕ denote the distribution of y with ϕ being the true parameter vector. Given ϕ, one can generate y under P_ϕ. Let $y_{[k]}$ denote y generated under the kth Monte-Carlo sample, $k = 1, \dots, K$. Then, by the law of large numbers, we have $b(\phi) - c(\phi) \approx K^{-1} \sum_{k=1}^K \left\{a(y_{[k]}, \phi) - a(y_{[k]}, \hat{\phi}_{[k]})\right\} \equiv d_K(\phi)$, where $\hat{\phi}_{[k]}$ denotes $\hat{\phi}$ based on $y_{[k]}$. If K is sufficiently large, which one has control over during the Monte-Carlo simulation, the difference between the two sides of the approximation is $o(m^{-1})$. Note that $y_{[k]}, k = 1, \dots, K$ also depend on ϕ. Then, a Monte-Carlo assisted MSPE estimator is given by

$$\begin{aligned} \widehat{\rm MSPE}_K &= a(y, \hat{\phi}) + d_K(\hat{\phi}) \\ &= a(y, \hat{\phi}) + \frac{1}{K} \sum_{k=1}^K \left\{a(y_{[k]}, \hat{\phi}) - a(y_{[k]}, \hat{\phi}_{[k]})\right\} \end{aligned} \tag{7.41}$$

where $y_{[k]}, k = 1, \dots, K$ are generated as above with $\phi = \hat{\phi}$, and $\hat{\phi}_{[k]}$ is, again, the estimator of ϕ based on $y_{[k]}$. Equation (7.41) is called the Sumca estimator of the MSPE of $\hat{\theta}$ (Sumca is an abbreviation of "simple, unified, Monte-Caro assisted").

7.5.3.1 A small simulation illustrating Sumca

A simulation study was carried out, in which the Sumca method was applied to the same simulation model considered in the previous subsection with $\beta_0 = 5$ and $\beta_1 = 1$; $w_i = -2.5 + 0.1i$; $\alpha_i = w_i^3 + v_i$ with $v_i \sim N(0, 10^{-4})$. The new observation satisfies the same model as in the previous case. Furthermore, we consider two cases of sample size configurations for n_i, namely, Case 1: $n_i = 5$, $1 \le i \le m/2$ and $n_i = 25$, $m/2 + 1 \le i \le m$; Case 2: $n_i = 50$, $1 \le i \le m/2$ and $n_i = 250$, $m/2 + 1 \le i \le m$. The results, based on 1000 simulation runs, are summarized below in terms of the percentage relative bias defined as $\%{\rm RB} = 100 \times \{(\widehat{\rm MSPE} - {\rm MSPE})/{\rm MSPE}\}$, where MSPE is the MSPE of the CMMP, $\hat{\theta}$, of $\theta = \beta_0 + \beta_1 w_1 + \alpha_I$ based on the simulation runs, and $\widehat{\rm MSPE}$ is the mean of the Sumca estimator over the simulation runs. The number in the parentheses is the value of R.

Case 1: -0.66 (0.25); -7.17 (0.5); -10.14 (1); -17.28 (2); -27.83 (3).

Case 2: -8.45 (0.25); -4.02 (0.5); -4.38 (1); 0.56 (2); -7.92 (3).

The performance of Sumca is satisfactory with the %RB decreasing to single digits as n_i increases (indicating that the size of training data increases).

7.5.4 CMMP for binary outcomes

[221] extended the CMMP to the case of binary responses. Suppose that, given the subject-specific (e.g., patient, or group of patients) random effects, $\alpha_1, \dots, \alpha_m$, binary responses $y_{ij}, i = 1, \dots, m, j = 1, \dots, n_i$ are conditionally independent with the conditional probability satisfying $\mathrm{P}(y_{ij} = 1|\alpha) = p_{ij}$ with $\mathrm{logit}(p_{ij}) = x'_{ij}\beta + \alpha_i$,where $\mathrm{logit}(p) = \log\{p/(1-p)\}$. Here x_{ij} is a vector of observed covariates (i.e. defined by the bump signature), and β is a vector of unknown parameters (the fixed effects). Furthermore, $\alpha_i, 1 \le i \le m$ are independent and distributed as $N(0, \sigma^2)$, where σ^2 is an unknown variance. For any known vector x and function $g(\cdot)$, the best predictor (BP) of the mixed effect, $\theta = g(x'\beta + \alpha_i)$, in the sense of minimum mean squared prediction error(MSPE), is

$$\tilde{\theta} = \mathrm{E}(\theta|y) = \frac{\mathrm{E}[g(x'\beta + \sigma\xi)\exp\{y_{i\cdot}\sigma\xi - \sum_{j=1}^{n_i}\log(1 + e^{x'_{ij}\beta + \sigma\xi})\}]}{\mathrm{E}[\exp\{y_{i\cdot}\sigma\xi - \sum_{j=1}^{n_i}\log(1 + e^{x'_{ij}\beta + \sigma\xi})\}]}, \quad (7.42)$$

where $y_{i\cdot} = \sum_{j=1}^{n_i} y_{ij}$, and the expectations are taken with respect to $\xi \sim N(0, 1)$. The parameters β and σ are understood as the true parameters, which are typically unknown in practice. It is then customary to replace β, σ by their consistent estimators.

Let the new, binary observations be $y_{\mathrm{n},k}, k = 1, \dots, n_{\mathrm{new}}$, and the corresponding covariates be $x_{\mathrm{n},k}, k = 1, \dots, n_{\mathrm{new}}$ such that, conditional on a random effect α_I that has the same $N(0, \sigma^2)$ distribution, $y_{\mathrm{n},k}, 1 \le k \le n_{\mathrm{new}}$ are independent with $\mathrm{P}(y_{\mathrm{n},k} = 1|\alpha_I) = p_{\mathrm{n},k}$ and $\mathrm{logit}(p_{\mathrm{n},k}) = x'_{\mathrm{n},k}\beta + \alpha_I$.

We consider a general case in which x_{ij} may depend on both i and j (the cluster level covariate case can easily be handled as well). We call this case unit-level covariates. Ref. [221] proposed the following strategy for matching the index I to one of the indexes $1 \le i \le m$, regardless of whether or not there is an actual match. Write $\bar{p}_{\mathrm{n}} = n_{\mathrm{new}}^{-1}\sum_{k=1}^{n_{\mathrm{new}}} p_{\mathrm{n},k}$, and $X_i = (x'_{ij})_{1 \le j \le n_i}$, $y_i = (y_{ij})_{1 \le j \le n_i}$. Define

$$\begin{aligned}
\psi(X_i, y_i, \beta, \alpha_i) &= \sum_{j=1}^{n_i}\{\alpha_i y_{ij} - \log(1 + e^{x'_{ij}\beta + \alpha_i})\} \\
&= \alpha_i y_{i\cdot} - \sum_{j=1}^{n_i}\log(1 + e^{x'_{ij}\beta + \alpha_i}).
\end{aligned}$$

If $I = i$, then, it can be shown that the BP of $\bar{p}_{\rm n}$ is given by

$$\tilde{\bar{p}}_{{\rm n},i} = \frac{1}{n_{\rm new}} \sum_{k=1}^{n_{\rm new}} \frac{\int \text{logit}^{-1}(x'_{{\rm n},k}\beta + \sigma\xi)\exp\{\psi(X_i, y_i, \beta, \sigma\xi)\}\phi(\xi)d\xi}{\int \exp\{\psi(X_i, y_i, \beta, \sigma\xi)\}\phi(\xi)d\xi}, \quad (7.43)$$

where, again, $\phi(\cdot)$ denotes the pdf of $N(0,1)$. The EBP of $\bar{p}_{\rm n}$, denoted by $\hat{\bar{p}}_{{\rm n},i}$, is obtained by (1) with β, σ replaced by their consistent estimators $\hat{\beta}, \hat{\sigma}$, respectively. We then identify I by $\hat{I} = \text{argmin}_{1 \le i \le m} |\hat{\bar{p}}_{{\rm n},i} - \bar{y}_{\rm n}|$, where $\bar{y}_{\rm n}$ is the mean of the responses of the new observations. Once $\hat{I}$ is obtained, the classified mixed logistic model predictor (CMLMP) of $p_{{\rm n},k}$, for any $1 \le k \le n_{\rm new}$, is obtained as

$$\hat{p}_{{\rm n},k} = \left. \frac{\int \text{logit}^{-1}(x'_{{\rm n},k}\hat{\beta} + \hat{\sigma}\xi)\exp\{\psi(X_i, y_i, \hat{\beta}, \hat{\sigma}\xi)\}\phi(\xi)d\xi}{\int \exp\{\psi(X_i, y_i, \hat{\beta}, \hat{\sigma}\xi)\}\phi(\xi)d\xi} \right|_{i=\hat{I}}. \quad (7.44)$$

Alternatively, the mixed effect $p_{{\rm n},k}$ can be predicted as

$$\check{p}_{{\rm n},k} = \text{logit}^{-1}(x'_{{\rm n},k}\hat{\beta} + \hat{\alpha}_{\hat{I}}),$$

where $\hat{\alpha}_i$ is the EBP of α_i given by

$$\tilde{\alpha}_i = \sigma \frac{\text{E}[\xi \exp\{y_{i\cdot}\sigma\xi - \sum_{j=1}^{n_i} \log(1 + e^{x'_{ij}\beta + \sigma\xi})\}]}{\text{E}[\exp\{y_{i\cdot}\sigma\xi - \sum_{j=1}^{n_i} \log(1 + e^{x'_{ij}\beta + \sigma\xi})\}]}, \quad (7.45)$$

with β, σ replaced by $\hat{\beta}, \hat{\sigma}$, respectively.

7.5.4.1 Estimating uncertainty of CMLMP

[221] also developed a second-order unbiased estimator of MSPE for CMLMP. That is, their estimator was such that,

$$\text{E}(\widehat{\text{MSPE}}) = \text{MSPE} + o(m^{-1}). \quad (7.46)$$

If $\hat{\theta}$ is the CMLMP estimator of the mixed effect $\hat{\theta} = logit^{-1}(x'_{new}\beta + \alpha_I)$ with I being the true index for the new observations. Let $\psi = (\beta, \sigma^2, I)$. Then we can write,

$$\text{MSPE} = \text{E}(\hat{\theta} - \theta)^2 = \text{E}[\text{E}(\hat{\theta} - \theta)^2 | Y], \quad (7.47)$$

where $Y = (y, y_{new})$. The interior conditional expectation is a function of Y and ψ, namely,

$$\begin{aligned} \text{E}\{(\hat{\theta} - \theta)^2 | Y\} &= \hat{\theta}^2 - 2\hat{\theta}\text{E}(\theta|Y) + \text{E}(\theta^2|Y) & (7.48) \\ &= \hat{\theta}^2 - 2\hat{\theta}a_1(Y, \psi) + a_2(Y, \psi), & (7.49) \end{aligned}$$

where $a_s(Y, \psi) = \text{E}(\theta^2|Y)$.

Ref. [221] replaced $\hat{\psi}$ for ψ (a consistent estimator), thus producing a first-order unbiased estimator,

$$\mathrm{E}\{a(Y,\hat{\psi}) - a(Y,\psi)\} = O(m^{-1}). \tag{7.50}$$

Now, $\mathrm{E}(a(Y,\psi))$ and $\mathrm{E}(a(Y,\hat{\psi}))$ are functions of ψ. So defined the following: $b(\psi) = \mathrm{E}\{a(Y,\psi)\}$, $c(\psi) = \mathrm{E}\{a(Y,\hat{\psi})\}$ and $d(\psi) = b(\psi) - c(\psi)$. Then, first-order unbiasedness means that $d(\psi) = O(m^{-1})$. So if one again plugs in $\hat{\psi}$ for ψ, then, we have

$$d(\hat{\psi}) - d(\psi) = o(m^{-1}). \tag{7.51}$$

So a second order unbiased estimator can be written as,

$$\widehat{\mathrm{MSPE}} = a(Y,\hat{\psi}) + b(\hat{\psi}) - c(\hat{\psi}), \tag{7.52}$$

which [221] showed is such that,

$$\mathrm{E}(\widehat{\mathrm{MSPE}}) = \mathrm{MSPE} + o(m^{-1}). \tag{7.53}$$

The first term $a(Y,\hat{\psi})$ is typically $O(1)$ but the remaining difference $d(\hat{\psi})$ is evaluated using a Monte Carlo method similar to Sumca.

To do so, let P_ψ be the distribution of Y with ψ the true parameter vector. Given ψ, Y can be generated under $P_{\hat{\psi}}$. This is done K times and a Monte Carlo estimate of $d(\hat{\psi})$ generated, namely,

$$d(\hat{\psi}) \approx \frac{1}{K}\sum_{k=1}^{K} a(Y_{[k]}, \hat{psi}) - a(Y_{[k],\hat{\psi}_{[k]}}). \tag{7.54}$$

[221] showed that this Monte Carlo assisted MSPE estimator is second-order unbiased provided that $m/K \to 0$. The authors detailed how to specify the function $a(Y,\psi)$ for the matched CMLMP case. This is done by combining the training data and new data corresponding to I and expanding the responses. Then similar quantities to (7.42) are defined for $s \in 1, 2$ where evaluation is done by numerical integration or a Monte Carlo method.

7.6 Concluding remarks

We offer some remarks not just regarding the topics discussed in this section but for the entire monograph as well.

Although the E-MS was shown to have a good performance in the simulation study reported in the previous section, which focused on regression variable selection, its application to SAE has so far not been reported. A "straightforward" extension does not seem difficult, although the special interest of SAE needs to be taken into consideration, as discussed in Section 6.2 (last paragraph).

Beyond this section, there have been, for the most part, two types of robustness considerations, not just in SAE but in Statistics in general. One is robustness against outliers; the other is robustness against model misspecification. A recent work is under way by [165], who proposed a double-robust method for SAE under an area-level model consideration. The proposed SAE method features both types of robustness as mentioned above.

The new era of Big Data has brought new opportunities, and challenges, to SAE. The traditional setting of SAE problems is such that the sample sizes are small for, at least, some areas. In the new era, some may find that all of a sudden, the sample sizes are no longer so small due to the increased capability, and means, of data collection. For example, some survey data may be collected via the internet, rather than by mail-in questionnaires or telephone interviews. This brings an important issue regarding the quality of the data collection. For example, internet data collection may not be done statistically, or scientifically. It is true that we now have larger sample sizes, but do we still have good data? In SAE, we learned how to borrow strength from good data; we have not (yet) learned how to borrow strength from "bad data", or something between good and bad.

In terms of robustness, it would be useful to develop methods that are robust against bad data quality, and yet still be able to utilize useful information from the data. Clearly, some features of the data quality need to be incorporated into the robust method, if it can be developed.

Finally, there has been substantial and increasing interest in protecting privacy associated with the data. For example, the United States Census Bureau has made new differential privacy (DP) changes in its 2020 decennial enumeration data to protect confidentiality. See, for example, [2], [92]. On the other hand, DP presents a new challenge to SAE methods, whose purpose is to borrow strength by utilizing information about subpopulations.

In particular, the U.S. Census Bureau developed a DP algorithm called TopDown. See, for example, [138]. The description below is based on our understanding of the TopDown algorithm, where we adopt the notation from the above reference. To run the TopdDown algorithm, we first create the DP contaminated histogram, H, at the state level to mimic the true histogram H. Note that H is a $m \times 1$ vector with respect to the age groups. To introduce the DP noise, we generate the noisy measurements via

$$M = Q(H) + \text{Noise}. \tag{7.55}$$

where $Q(\cdot)$ is the identity function and Gaussian noise was used in our simulation. To observe more variability in the DP contaminated histogram, we considered the variance of Gaussian noise to be proportional to population size in the corresponding histogram H with the proportion $q \in (0, 1)$.

The $\tilde{H}$ can be obtained by a two-step optimization procedure. In the first step, a non-negative fractional histogram, H^*, is obtained by minimizing the L_2 norm between $Q(H^*)$ and M. We constrain H^* so that $\text{sum}(H^*) = N$.

In the second step, the non-negative integer histogram, $\tilde{H}$, is obtained by minimizing the L_1 norm between H^* and $\tilde{H}$, with the following constraints:

1. $\tilde{H}[x] > 0$ for all cells x;
2. $\left|\tilde{H}[x] - H^*[x]\right| \leq 1$ for all cells x;
3. $\sum_x \tilde{H}[x] = \sum_x H^*[x]$.

The optimizations can be performed by Linear programming.

Consider, for example, generating the DP contaminated population, called the Fuzzy population, for different age-groups at the county level. We created the DP contaminated histogram, H_i, with respect to the age groups for each county i, $i = 1, \ldots, n$ (n is the total number of counties in the U.S.). We apply a similar procedure as above at the state level. The noisy measurements for each state, $M_i = Q(H_i) + \text{Noise}_i$, are generated with proportional Gaussian variance at q. The H_i^* is obtained by minimizing the sum of L_2 norms between $Q(H_i^*)$ and M_i over i. We constrain H_i^* such that $\sum_{i=1}^n H_i^* = \tilde{H}$, so the county level histograms are consistent with the state level DP contaminated histogram. We add a new constraint at the county level: $0.5H_i[x] \leq H_i^* \leq 1.5H_i[x]$ for all i and all cell elements x, so that the difference H_i^* and H_i would not be too large for some age groups for small counties. The $\tilde{H}_i$ is obtained by minimizing the L_1 norm between H_i^* and $\tilde{H}_i$, with similar constraints as in the state level optimizations, for all i.

Table 7.7 presents a sample portion of the data generated by the (actual) TopDown algorithm. Column 2010 Pops are the actual population sizes pub-

TABLE 7.7
Sample comparison of county-level population data and fuzzy (DP-contaminated) population data by age-groups.

Age-Group	2010 Pops	2010 Fuzzy	Diff.	$\delta = 0.05$	Diff	$\delta = 0.2$	Diff
00-04 years	97652	97676	24	97635	-17	97549	-103
05-09 years	94546	94547	1	94575	29	94499	-47
10-14 years	91070	91058	-12	91053	-17	90978	-92
15-19 years	100394	100256	-138	100494	100	100172	-222
20-24 years	107049	106977	-72	106937	-112	106701	-348
25-29 years	113597	113643	46	113685	88	113684	87
30-34 years	114607	114651	44	114641	34	114223	-384
35-39 years	115275	115271	-4	115275	0	115194	-81
40-44 years	112216	112201	-15	112188	-28	112290	74
45-49 years	114111	114155	44	114174	63	114136	25
50-54 years	108506	108457	-49	108567	61	108682	176
55-59 years	94648	94659	11	94601	-47	94658	10
60-64 years	78854	78861	7	78769	-85	78869	15
65-69 years	52663	52719	56	52567	-96	52617	-46
70-74 years	37774	37733	-41	37750	-24	37688	-86
75-79 years	29185	29267	82	29231	46	29142	-43
80-84 years	23391	23386	-5	23354	-37	23438	47
85+ years	24733	24753	20	24660	-73	24863	130

lished by the 2010 U.S. Census; column 2010 Fuzzy are the corresponding DP contaminated population sizes published by the U.S. Census (with the unknown value of q), called fuzzy population; the next column is the difference between the fuzzy and actual population sizes. The last four columns are the fuzzy population sizes created with known q values: $q = 0.05$ and $q = 0.2$, and the corresponding differences.

As can be seen, the difference between the fuzzy and actual population sizes is quite small, even for $q = 0.2$. A question is whether the difference would affect inferential results for SAE. It seems unlikely, although one cannot make such a conclusion without a careful and formal study.

In a way, this is related to the robustness of the existing SAE methods to the contamination of DP errors. Clearly, research work is needed in this area to meet this new challenge from modern data science.

Bibliography

[1] Basu A., I. R. Harris, Hjort N. L., and M. C. Jones. Robust and efficient estimation by minimising a density power divergence. *Biometrika*, 85(3):549–559, 1998.

[2] J. M. Abowd. The U.S. Census Bureau adopts differential privacy. In *Proceedings of the 24th ACM SIGKDD International Conference on Knowledge Discovery and Data Mining.* KDD, London, UK, 2018.

[3] R. J. Adler. *An Introduction to Continuity, Extrema, and Related Topics for General Gaussian Processes.* Institute of Mathematical Statistics, Hayward, CA, 1990.

[4] H. Akaike. Information theory as an extension of the maximum likelihood principle. In *Second International Symposium on Information Theory, B. N. Petrov and F. Csaki eds.* Akademiai Kiado, Budapest, 1973.

[5] A. Azzalini and A. Capitanio. *The Skew-normal and Related Families.* Cambridge Univ. Press, New York, 2014.

[6] R. Bandyopadhyay. Benchmarking the observed best predictor, ph. d. dissertation, dept. of statist., univ. of calif., davis, ca, 2017.

[7] R. Bandyopadhyay and J. Jiang. Benchmarking the observed best predictor. *Calcutta Statist. Assoc. Bull.*, 75:116–140, 2023.

[8] G. E. Battese, R. M. Harter, and W. A. Fuller. An error-components model for prediction of county crop areas using survey and satellite data,. *J. Amer. Statist. Assoc.*, 80(401):28–36, 1988.

[9] A. E. Beaton and J. W. Tukey. The ®tting of power series, meaning polynomials, illustrated on band spectroscopic data. *Technometrics*, 16:147–185, 1974.

[10] J.-F. Beaumont, D. Haziza, and A. Ruiz. A unified approach to robust estimation in finite sampling. *Biometrika*, 100:555–569, 2013.

[11] W. R. Bell and E. T. Huang. Using the t-distribution to Deal with Outliers in Small Area Estimation. In *Proceedings of Statistics Canada Symposium: Methodological Issues in Measuring Population Health*, 2006.

[12] W.R. Bell, W. Basel, and J.J. Maples. *An overview of the U.S. Census Bureau's small area income and poverty estimates program. In Analysis of Poverty Data by Small Area Estimation, edited by M. Pratesi.* Wiley, New York, NY, first edition, 2016.

[13] R. Beran. Minimum Hellinger distance estimates for parametric models. *Ann. Statist.*, 5(3):445–463, 1977.

[14] R. W. Bohannon. Comfortable and maximum walking speed of adults aged 20-79 years: Reference values and determinants. *Age and Ageing*, 26:15–19, 1997.

[15] H. D. Bondell, A. Krishna, and S. K. Ghosh. Joint variable selection for fixed and random effects in linear mixed-effects models. *Biometrics*, 66:1069–1077, 2010.

[16] J. G. Booth and J. P. Hobert. Maximum generalized linear mixed model likelihood with an automated Monte Carlo EM algorithm. *J. Roy. Statist. Soc. B*, 61:265–285, 1999.

[17] L. Breiman. Random forests. *Machine Learning*, 22:5–32, 2001.

[18] K. W. Broman and T. P. Speed. A model selection approach for the identification of quantitative trait loci in experimental crosses. *J. Roy. Statist. Soc. B*, 64:641–656, 2002.

[19] F. B. Butar and P. Lahiri. On measures of uncertainty of empirical bayes small area estimators. *J. Statist. Planning Inference*, 112(1-2):63–76, 2003.

[20] S. Cai and J.N.K. Rao. Selection of auxiliary variables for three-fold linking modles in small area estimation: a simple and effective method. *Stats*, 5:128–138, 2022.

[21] S. Cai, J.N.K. Rao, L. Dumitrescu, and G. Chatrchi. Effective transformation-based variable selection under two-fold subarea models in small area estimation. *Statistics In Transition*, 21:68–83, 2020.

[22] J. A. Calvin and J. Sedransk. Bayesian and frequentist predictive inference for the patterns of care studies. *J. Amer. Ststist. Assoc.*, 86:36–48, 1991.

[23] J. R. Carpenter and M. G. Kenward. *Multiple Imputation and Its Application.* Wiley, New York, 2013.

[24] A. Chakraborty, G. K. Datta, and A. Mandal. Robust hierarchical Bayes small area estimation for nested error regression model. *Int. Stat. Rev.*, 87:158–176, 2019.

[25] R. Chambers and H. Chandra. A random effect block bootstrap for clustered data. *J. Comput. Graph. Stat.*, 22:452–470, 2013.

[26] R. Chambers, H. Chandra, N. Salvati, and N. Tzavidis. Outlier robust small area estimation, invited presentation. (2009), isi 2009, south africa.

[27] R. Chambers, H. Chandra, and N. Tzavidis. On bias-robust means squared error estimation for pseudo-linear small area estimation. *Surv. Methodol.*, 27:153–170, 2011.

[28] R. Chambers, N. Salvati, and N. Tzavidis. Semiparametric small area estimation for binary outcomes with application to unemployment estimation for local authorities in the uk. *J. Roy. Statist. Soc. A*, 179:453–479, 2014.

[29] R. Chambers and N. Tzavidis. M-quantile models for small area estimation. *Biometrika*, 93:255–268, 2006.

[30] R. L. Chambers. Outlier robust finite population estimation. *J. Amer. Statist. Assoc.*, 81:1063–1069, 1986.

[31] R. L. Chambers, H. Chandra, N. Salvati, and N. Tzavidis. Outlier robust small area estimation. *J. Roy. Statist. Soc. B*, 76:47–69, 2014.

[32] S. Chatterjee, P. Lahiri, and H. Li. Parametric bootstrap approximation to the distribution of eblup, and related prediction intervals in linear mixed models. *Ann. Statist.*, 36(3):1221–1245, 2008.

[33] J. Chen. Consistency of the MLE under mixture models. *Statist. Sci.*, 32(1):47–63, 2017.

[34] S. Chen, J. Jiang, and T. Nguyen. Observed best prediction for small area counts. *J. Survey Statist. Methodol.*, 3(1):136–161, 2015.

[35] Y. Chen, P. Lahiri, and N. Salvati. Effects of model misspecification on small area estimation, manuscript under review for publication.

[36] G. Claeskens and J. D. Hart. Goodness-of-fit tests in mixed models (with discussion). *TEST*, 18:213–239, 2009.

[37] D. R. Cox. Prediction intervals and empirical Bayes confidence intervals. In *Perspectives in Probability and Statistics, Papers in Honor of M. S. Bartlett, J. Gani, Ed.*, pages 47–55. Applied Probability Trust, University of Sheffield, Sheffield, 1975.

[38] C. Dao and J. Jiang. A modified Pearson's χ^2 test with application to generalized linear mixed model diagnostics. *Ann. Math. Sci. Appl.*, 1:195–215, 2016.

[39] K. Das, J. Jiang, and J. N. K. Rao. Mean squared error of empirical predictor. *Ann. Statist.*, 32:818–840, 2004.

[40] G. Datta. Model-based approach to small area estimation. In *Handbook of Statistics, Sample Surveys: Inference and Analysis, Rao, C. R. and Pfeffermann, D. Eds.*, volume 32, pages 251–288. Elsevier, New York, 2009.

[41] G. Datta and M. Ghosh. Bayesian prediction in linear models: Applications to small area estimation. *Ann. Statist.*, 19:1748–1770, 1991.

[42] G. Datta, J.N.K. Rao, and D.D. Smith. On measuring the variability of small area estimators under a basic area level model. *Biometrika*, 92:183–196, 2005.

[43] G. S. Datta, T. Kubokawa, J. N. K. Rao, and I. Molina. Estimation of mean squared error of model-based small area estimators. *TEST*, 20:367–388, 2011.

[44] G. S. Datta and P. Lahiri. Robust hierarchical Bayes estimation of small area characteristics in the presence of covariates and outliers. *J. Multivar. Anal.*, 54:310–328, 1995.

[45] G. S. Datta and P. Lahiri. A unified measure of uncertainty of estimated best linear unbiased predictors in small area estimation problems. *Statistica Sinica*, 10(2):613–627, 2000.

[46] G. S. Datta and P. Lahiri. Discussion on "Scales of evidence for model selection: Fisher versus Jeffreys" by B. Efron and A. Gous. In *Model Selection, P. Lahiri ed., IMS Lec. Ser.*, pages 208–256. Institute of Mathematical Statistics, 2001.

[47] G. S. Datta, P. Lahiri, and T. Maiti. Empirical bayes estimation of median income of four-person families by state using time series and cross-sectional data. *J. Stat. Plan. Inference*, 102(1):83–97, 2002.

[48] A. P. Dempster, N. M. Laird, and D. B. Rubin. Maximum likelihood from incomplete data via the EM algorithm (with discussion). *J. Roy. Statist. Soc. B*, 39:1–38, 1977.

[49] A. P. Dempster and L. M. Ryan. Weighted normal plots. *J. Amer. Ststist. Assoc.*, 80:845–850, 1985.

[50] P. J. Diggle, P. Heagerty, K. Y. Liang, and S. L. Zeger. *Analysis of Longitudinal Data.* Oxford Univ. Press, second edition, 2002.

[51] B. Efron. Bootstrap method: Another look at the jackknife. *Ann. Statist.*, 7(1):1–26, 1979.

[52] B. Efron. Tweedie's formula and selection bias. *J. Amer. Statist. Assoc.*, 106(496):1602–1614, 2011.

[53] B. Efron and D. V. Hinkley. Assessing the accuracy of the maximum likelihood estimator: observed versus expected Fisher information. *Biometrika*, 65:457–487, 1978.

[54] B. Efron and R. J. Tibshirani. *An Introduction to the Bootstrap*. Chapman & Hall/CRC, 1993.

[55] E. Fabrizi and P. Lahiri. A new approximation to the Bayes information criterion in finite population sampling. *STATISTICA*, 73:289–301, 2004.

[56] E. Fabrizi, N. Salvati, and C. Trivisano. Robust Bayesian small area estimation based on quantile regression. *Comput. Stat. Data Anal.*, 145:106900, 2020.

[57] J. Fan and R. Li. Variable selection via nonconcave penalized likelihood and its oracle properties. *J. Amer. Statist. Assoc.*, 96:1348–1360, 2001.

[58] J. Fan and Q. Yao. *Nonlinear Time Series: Nonparametric and Parametric Methods*. Springer, New York, 2003.

[59] R. E. Fay and R. A. Herriot. Estimates of income for small places: an application of James-Stein procedures to census data. *J. Amer. Statist. Assoc.*, 74(366a):269–277, 1979.

[60] W. H. Fellner. Robust estimation of variance components. *Technometrics*, 28:51–60, 1986.

[61] M. R. Ferrante and C. Trivisano. mall area estimation of the number of firms' recruits by using multivariate models for count data. *Surv. Methodol.*, 36:171–180, 2010.

[62] D. Firth and K. E. Bennett. Robust models in probability sampling. *J. Roy. Statist. Soc. B*, 60:3–21, 1998.

[63] R. A. Fisher. On the interpretation of chi-square from contingency tables, and the calculation of P. *J. Roy. Statist. Soc. B*, 85:87–94, 1922.

[64] J. Friedman. Multivariate adaptive regression splines (with discussion). *Ann. Statist.*, 19:1–67, 1991.

[65] J. Friedman. Greedy function approximation: A gradient boosting machine. *Ann. Statist.*, 29:1189–1232, 2001.

[66] W. A. Fuller. *Sampling Statistics*. Wiley, New York, 2009.

[67] N. Ganesh. Simultaneous credible intervals for small area estimation problems. *J. Multivariate Anal.*, 100(8):1610–1621, 2009.

[68] R. I. Garcia, J. G. Ibrahim, and H. Zhu. Variable selection for regression models with missing data. *Stat. Sin.*, 20:149–165, 2010.

[69] A. Gelman, J. B. Carlin, H. S. Stern, and D. B. Rubin. *Bayesian Data Analysis.* Chapman & Hall, New York, 1995.

[70] G. Gershunskaya and P. Lahiri. Robust empirical best small area finite population mean estimation using a mixture model. *Calcutta Statist. Assoc. Bull.*, 69(2):183–204, 2018.

[71] M. Ghosh, T. Ghosh, and M. Y. Hirose. Poisson counts, square root transformation and small area estimation: Square root transformation. *Sankhya B*, pages 1–23, 2021.

[72] M. Ghosh and P. Lahiri. Robust empirical bayes estimation of means from stratified samples. *J. Amer. Statist. Assoc.*, 82(400):1153–1162, 1987.

[73] M. Ghosh, T. Maiti, and A. Roy. Influence functions and robust Bayes and empirical Bayes small area estimation. *Biometrika*, 95(3):573–585, 2008.

[74] M. Ghosh and G. Meeden. Empirical bayes estimation in finite population sampling. *J. Amer. Statist. Assoc.*, 81:1058–1062, 1986.

[75] M. Ghosh, J. Myung, and F. Moura. R package robustsae, https://cran.r-project.org/web/packages/robustsae/index.html, 2016.

[76] M. Ghosh, N. Nangia, and D. Kim. Estimation of median income of four-person families: A Bayesian time series approach. *J. Amer. Statist. Assoc.*, 91:1423–1431, 1996.

[77] M. Ghosh and J. N. K. Rao. Small area estimation: An appraisal (with discussion). *Statist. Sci.*, 9(1):55–93, 1994.

[78] W. González-Mantiega, M. J. Lombardía, I. Molina, D. Morales, and L. Santamaría. Bootstrap mean squared error of a small-area EBLUP. *J. Statist. Comput. Simul.*, 78(5):443–462, 2008.

[79] I. Goodfellow, Y. Benjio, and A. Courville. *Deep Learning.* MIT Press, Cambridge, MA, first edition, 2016.

[80] P. J. Green and B.W. Silverman. *Nonparametric Regression and Generalized Linear Models: A roughness penalty approach.* Chapman and Hall/CRC, 1993.

[81] W. J. Guan, Z. Y. Ni, Y. Hu, W. H. Liang, C. Q. Ou, J. X. He, ..., and B. Du. Clinical characteristics of 2019 novel coronavirus infection in China. *New England J. Med.*, 382:1708–1720, 2020.

[82] W. Guo. Functional mixed effects models. *Biometrics*, 58:121–128, 2002.

[83] N. Hajarisman. Two-level hierarchical bayesian poisson models for small area estimation of infant mortality rates, ph. d. dissertation, dept. of math. natural sci., bogor agricultural univ., indonesia, 2013.

[84] J. Hajek. Comments on paper by D. Basu. In *Foundations of Statistical Inference (eds V. P. Godambe and D. A. Sprott)*, page 236. Holt, Rinehart and Winston, Toronto, 1971.

[85] A. Hajjem, F. Bellavance, and D. Larocque. Mixed-effects random forest for clustered data. *J. Statist. Comput. Simulation*, 84:1313–1328, 2014.

[86] A. R. Hall. *Generalized method of moments, Advanced Texts in Econometrics.* Oxford Univ. Press, 2005.

[87] C. A. Hall and W. W. Meyer. Optimal error bounds for Cubic spline interpolation. *J. Approx. Theory*, 16:8105–122, 1976.

[88] P. Hall and T. Maiti. Nonparametric estimation of mean-squared prediction error in nested-error regression models. *Ann. Statist.*, 34(4):1733–1750, 2006.

[89] P. Hall and T. Maiti. On parametric bootstrap methods for small area prediction. *J. Roy. Statist. Soc. B*, 68(2):221–238, 2006.

[90] B. Han. Conditional Akaike information criterion in the Fay-Herriot model. *Stat. Methodol.*, 11:53–67, 2013.

[91] T. J. Hastie and R. J. Tibshirani. *Generalized Additive Models.* Chapman & Hall, CRC, 1990.

[92] M. B. Hawes. Implementing differential privacy: Seven lessons from the 2020 United States Census. In *Harvard Data Science Review.* MIT Press, Cambridge, MA, 2018.

[93] D. Hedeker, R. D. Gibbons, and B. R. Flay. Random-effects regression models for clustered data with an example from smoking prevention research. *J. Consulting Clinical Psych.*, 62:757–765, 1994.

[94] C. R. Henderson. Estimation of genetic parameters (abstract). *Ann. Math. Statist.*, 21:309–310, 1950.

[95] N. C. Henderson, R. Varadhan, and T. A. Louis. Improved small domain estimation via compromise regression weights. *J. Amer. Statist. Assoc.*, 118(544):2793–2809, 2023.

[96] S. E. Hosseini, D. Shahsavani, M. R. Rabiei, M. Arashi, and H. Baghishani. Small area estimation using a semiparametric spatial model with application in insurance. *Symmetry*, 14:2194, 2022.

[97] K. Hu, J. Choi, A. Sim, and J. Jiang. Best predictive generalized linear mixed model with predictive lasso for high-speed network data analysis. *Int. J. Stat. Prob.*, 4:132–148, 2015.

[98] G.B. Huang, Q.Y. Zhu, and C.K. Siew. Extreme learning machine: Theory and applications. *Neurocomputing*, 70:489–501, 2006.

[99] P. J. Huber. Robust regression: Asymptotics, conjectures, and monte carlo. *Ann. Statist.*, 1(5):799–821, 1973.

[100] P. J. Huber and E. M. Ronchetti. *Robust Statistics.* Wiley, second edition, 2009.

[101] J. G. Ibrahim, H. Zhu, R. I. Carcia, and R. Guo. Fixed and random effects selection in mixed effects models. *Biometrics*, 67:495–503, 2011.

[102] J. G. Ibrahim, H. Zhu, and N. Tang. Model selection criteria for missing-data problems using the EM algorithm. *J. Amer. Statist. Assoc.*, 103:1648–1658, 2008.

[103] Cisco Systems Inc. Netflow services and applications, white paper.

[104] J. Jiang. REML estimation: Asymptotic behavior and related topics. *Ann. Statist.*, 24:255–286, 1996.

[105] J. Jiang. Consistent estimators in generalized linear mixed models. *J. Amer. Statist. Assoc.*, 93:720–729, 1998.

[106] J. Jiang. Goodness-of-fit tests for mixed model diagnostics. *Ann. Ststist.*, 29:1137–1164, 2001.

[107] J. Jiang. Mixed-effects models with random cluster sizes. *Stat. Probab. Lett.*, 53:201–206, 2001.

[108] J. Jiang. Empirical methods of moments and its applications. *J. Stat. Plan. Inference*, 115(1):69–84, 2003.

[109] J. Jiang. *Linear and Generalized Linear Models and Their Applications.* Springer, New York, first edition, 2007.

[110] J. Jiang. The subset argument and consistency of MLE in GLMM: Answer to an open problem and beyond. *Ann. Statist.*, 41:177–195, 2013.

[111] J. Jiang. The fence methods. *Advances in Statistics*, 2014:1–14, 2014.

[112] J. Jiang. *Robust Mixed Model Analysis.* World Scientific, Singapore, 2019.

[113] J. Jiang. *Large Sample Techniques for Statistics.* Springer, second edition, 2022.

[114] J. Jiang, H. Jia, and H. Chen. Maximum posterior estimation of random effects in generalized linear mixed models. *Stat. Sin.*, 11:97–120, 2001.

[115] J. Jiang and P. Lahiri. Empirical best prediction for small area inference with binary data. *Ann. Inst. Statist. Math.*, 53(2):217–243, June 2001.

[116] J. Jiang and P. Lahiri. Mixed model prediction and small area estimation (with discussion). *TEST*, 15(1):1–96, 2006.

[117] J. Jiang, P. Lahiri, and T. Nguyen. A unified monte-carlo jackknife for small area estimation after model selection. *Ann. Math. Sci. Appl.*, 3(2):405–438, 2018.

[118] J. Jiang, P. Lahiri, and S. Wan. A unified jackknife theory for empirical best prediction with m-estimation. *Ann. Statist.*, 30(6):1782–1801, 2002.

[119] J. Jiang, P. Lahiri, and C. Wu. A generalization of the Pearson's χ^2 goodness-of-fit test with estimated cell frequencies. *Sankhya A*, 63:260–0276, 2001.

[120] J. Jiang, C. Li, D. Paul, C. Yang, and H. Zhao. On high-dimensional misspecified mixed model analysis in genome-wide association study. *Ann. Statist.*, 44:2127–2160, 2016.

[121] J. Jiang and T. Nguyen. Small area estimation via heteroscedastic nested-error regression. *Can. J. Stat.*, 40(3):588–603, 2012.

[122] J. Jiang and T. Nguyen. *The Fence Methods.* World Scientific, Singapore, 2016.

[123] J. Jiang and T. Nguyen. *Linear and Generalized Linear Mixed Models and Their Applications.* Springer, New Work, second edition, 2021.

[124] J. Jiang, T. Nguyen, and J. S. Rao. A simplified adaptive fence procedure. *Stat. Probab. Lett.*, 79:625–629, 2009.

[125] J. Jiang, T. Nguyen, and J. S. Rao. Fence method for nonparametric small area estimation. *Surv. Methodol.*, 36:3–11, 2010.

[126] J. Jiang, T. Nguyen, and J. S. Rao. Best predictive small area estimation. *J. Amer. Statist. Assoc.*, 106(494):732–745, 2011.

[127] J. Jiang, T. Nguyen, and J. S. Rao. Invisible fence methods and the identification of differentially expressed gene sets. *Statist. Interface*, 4:403–415, 2011.

[128] J. Jiang, T. Nguyen, and J. S. Rao. Observed best prediction via nested-error regression with potentially misspecified mean and variance. *Surv. Methodol.*, 41:37–55, 2015.

[129] J. Jiang, T. Nguyen, and J. S. Rao. The E-MS algorithm: model selection with incomplete data. *J. Amer. Statist. Assoc.*, 110:1136–1147, 2015.

[130] J. Jiang and J. S. Rao. Robust Small Area Estimation: An Overview. *Annual Review of Statistics and Its Applications*, 7(1):337–360, 2020.

[131] J. Jiang, J. S. Rao, J. Fan, and T. Nguyen. Classified mixed model prediction. *J. Amer. Statist. Assoc.*, 113:269–279, 2018.

[132] J. Jiang, J. S. Rao, Z. Gu, and T. Nguyen. Fence methods for mixed model selection. *Ann. Statist.*, 36:1669–1692, 2008.

[133] J. Jiang and M. Torabi. Sumca: Simple, unified, monte-carlo assisted approach to second-order unbiased mean-squared prediction error estimation. *J. Roy. Statist. Soc. B*, 82(2):467–485, 2020.

[134] J. Jiang and M. Torabi. Goodness-of-fit test with a robustness feature. *TEST*, 31:76–100, 2022.

[135] X. Jin, B. P. Carlin, and S. Banerjee. Generalized hierarchical multivariate CAR models for area data. *Biometrics*, 61:950–961, 2005.

[136] V. D. Jiongo, D. Haziza, and P. Duchesne. Controlling the bias of robust small-area estimators. *Biometrika*, 100:843–858, 2013.

[137] R. N. Kackar and D. A. Harville. Approximations for standard errors of estimators of fixed and random effects in mixed linear models. *J. Amer. Statist. Assoc.*, 79(388):853–962, 1984.

[138] D. Kifer. Consistency with external knowledge: The topdown algorithm, presentation slides, 2020.

[139] R. Koenker. *Quantile Regression.* Cambridge Univ. Press, first edition, 2005.

[140] R. Koenker and G. Bassett. Regression quantiles. *Econometrica*, 46:33–50, 1978.

[141] H. Kozumi and G. Kobayashi. Gibbs sampling methods for Bayesian quantile regression. *J. Statist. Comput. Simul.*, 81(11):1565–1578, 2011.

[142] P. Krennmair and T. Schmid. Flexible domain prediction using mixed effecxts random forests, technical report. (2022), freie universitat berlin.

[143] P. Krennmair, N. Wurz, and T. Schmid. Tree-based machine learning in small area estimation. *The Survey Statistician*, 86:22–31, 2022.

[144] A.K. Kreutzmann, S. Pannier, N. Rojas-Perilla, T. Schmid, M. Temple, and N. Tzavidis. The R package emdi for estimating and mapping regionally disaggregated indicators. *J. Stat. Softw.*, 91:1–33, 2019.

[145] A. K. Kreutzmann, S. Pannier, N. Rojas-Perilla, T. Schmid, M. Templ, and N. Tzavidis, The R package emdi for estimating and mapping regionally disaggregated indicators https://cran.r-project.org/web/packages/emdi/emdi.pdf, 2018.

[146] T. Kubokawa, S. Sugasawa, M. Ghosh, and S. Chaudhuri. Prediction in heteroscedastic nested error regression models with random dispersions. *Stat. Sin.*, 26(2):465–492, 2016.

[147] D. Kurisu, T. Ishihara, and S. Sugasawa. Adaptively robust small area estimation: Balancing robustness and efficiency of empirical Bayes confidence intervals. In *arXiv:2018.11551v2.* Cornel Univ. Press, 2022.

[148] M. H. Kutner, C. J. Nachtsheim, J. Neter, and W. Li. *Applied Linear Statistical Models.* McGraw-Hill, New York, fifth edition, 2005.

[149] P. Lahiri. Model Selection. In *P. Lahiri ed., IMS Lec. Ser.* Institute of Mathematical Statistics, 2001.

[150] P. Lahiri and J. N. K. Rao. Robust estimation of mean squared error of small area estimators. *J. Amer. Statist. Assoc.*, 90(430):758–766, 1995.

[151] P. Lahiri and J. Suntornchost. Variable selection for linear mixed models with applications to small area estimation. *Sankhya B*, 77:312–320, 2015.

[152] N. Lange and L. Ryan. Assessing normality in random effects models. *Ann. Ststist.*, 17:624–642, 1989.

[153] E. L. Lehmann and G. Casella. *Theory of Point Estimation.* Springer, New Work, second edition, 1998.

[154] R. Lehtonen and A. Veijanen. Logistic generalised regression estimators. *Surv. Methodol.*, 24:51–56, 1998.

[155] R. Lehtonen and A. Veijanen. Design-based methods of estimation for domains and small area. In *Sample Surveys: Inference and Analysis, D. Pfeffermann and C. R. Rao, Eds.).*, volume 29B, pages 219–249. Handbook of Statistics, North-Holland, Amsterdam, 2009.

[156] K. Lele, S. R. Nadeem and B. Schmuland. Estimability and likelihood inference for generalized linear mixed models using data cloning. *J. Amer. Statist. Assoc.*, 105:1617–1625, 2010.

[157] S. R. Lele, B. Dennis, and F. Lutscher. Data cloning: easy maximum likelihood estimation for complex ecological models using Bayesian Markov chain Monte Carlo methods. *Ecol. Lett.*, 10:551–563, 2007.

[158] Y. Li and P. Lahiri. A simple adaptation of variable selection software for regression models to select variables in nested error regression models. *Sankhya B*, 81:302–317, 2019.

[159] R. Little and D. Rubin. *Statistical Analysis with Missing Data.* Wiley, Hoboken, NJ, second edition, 2014.

[160] J. S. Liu. *Monte Carlo Strategies in Scientific Computing.* Springer, New York, 2004.

[161] X. Liu, H. Ma, and J. Jiang. Assessing uncertainty in small area estimation involving model misspecification: A one-bring-one route. *SCIENCE CHINA Mathematics*, 52:779–808, 2022.

[162] X. Liu, H. Ma, and J. Jiang. That Prasad-Rao is robust: Estimation of mean squared prediction error of observed best predictor under potential model misspecification. *Stat. Sin.*, 32:2217–2240, 2022.

[163] S. L. Lohr. *Sampling: Design and Analysis.* Brooks/Cole, Cengage Learning, second edition, 2010.

[164] M. J. Lombardía and S. Sperlich. Semiparametric inference in generalized mixed effects models. *J. Roy. Statist. Soc. B*, 70(5):913–930, 2008.

[165] H. Ma and J. Jiang. Double robust small area estimation, technical report. (2024), dept. of statist., univ. of calif., davis, ca.

[166] T. Maiti, H. Ren, and S. Sinha. Prediction error of small area predictors shrinking both means and variance. *Scand. J. Statist.*, 41(3):775–790, 2014.

[167] T. Maiti, S. Sinha, and P.-S. Zhong. Functional mixed effects model for small area estimation. *Scand. J. Stat.*, 43:886–903, 2016.

[168] D. Malec and P. Müller. A Bayesian semi-parametric model for small area estimation. In *Pushing the Limits of Contemporary Statistics: Contributiona in Honor of Jayanta K. Ghosh, IMS Collections*, volume 3, pages 223–236. Institute of Mathematical Statistics, 2008.

[169] D. Malec, J. Sedransk, C. L. Moriarity, and F. B. LeClere. Small area inference for binary variables in the National Health Interview Survey. *J. Amer. Statist. Assoc.*, 92:815–826, 1997.

[170] J. T. Massey, T. F. Moore, V. L. Parsons, and W. Tadros. Design and estimation for the National Health Interview Survey, 1985–94. In *National Center for Health Statistics, Vital and Health Statistics*, volume 2, page 110. U.S. Department of Health and Human Services, 1989.

[171] N. S. Matloff. Use of regression functions for improved estimation of means. *Biometrika*, 68(3):685–689, 1981.

[172] P. May, K.S. McConville, G.G. Moisen, J. Bruening, and R. Dubayah. A spatially varying model for small area estimates of biomass density across the contiguous United States. *Remote Sensing and Environment*, 286:113420, 2023.

[173] P. McCullagh and J. A. Nelder. *Generalized Linear Models.* Chapman and Hall, London, second edition, 1989.

[174] G. Mendez and S. Lohr. Estimating residual variance in random forest regression. *Computational Statistics and Data Analysis*, 55:2937–2950, 2011.

[175] J. Meza and P. Lahiri. A note on the Cp statistic under the nested error regression model. *Surv. Methodol.*, 31:105–109, 2005.

[176] L. Mohadjer, J.N.K. Rao, B. Liu, T. Krenzke, and W. Van De Kerckhove. Hierarchical Bayes small area estimates of adult literacy using unmatched sampling and linking models. *J. Indian Soc. Agricultural Statist.*, 66:55–63, 2012.

[177] G. Molenberghs and M. G. Kenward. *Missing data in Clinical Studies.* Wiley, New York, 2007.

[178] I. Molina and Y. Marhuenda. R package sae, https://cran.r-project.org/web/packages/sae/sae.pdf, 2018.

[179] I. Molina and J. N. K. Rao. Small area estimation of poverty indicators. *Canadian J. Statist.*, 38(3):369–385, 2010.

[180] D. S. Moore. Chi-square tests. In *Studies in Statistics (R. V. Hogg ed.), Studies in Math.*, volume 19. Math. Assoc. Amer., 1978.

[181] C. N. Morris and C. L. Christiansen. Hierarchical models for ranking and for identifying extremes with applications. In *Bayes Statistics*, volume 5. Oxford Univ. Press., 1995.

[182] S. Müller, J. L. Scealy, and A. H. Welsh. Model selection in linear mixed models. *Statist. Sci.*, 28:135–167, 2013.

[183] R. Münnich, J. P. Burgard, and M. Vogt. Small area estimation for population counts in the German Census 2011. In *Section on Survey Research Methods, JSM 2009, Washington, D.C.* American Statistical Association, 2009.

[184] T. Nguyen and J. Jiang. Estimating the mean squared prediction error of the observed best predictor associated with small area counts: A computationally oriented approach. *Can. J. Stat.*, page in press, 2023.

[185] T. Nguyen, J. Jiang, and J.S. Rao. Assessing uncertainty for classified mixed model prediction. *J. Statist. Comp. Simulation*, 92:249–261, 2022.

[186] R. Nishii. Asymptotic properties of criteria for selection of variables in multiple regression. *Ann. Statist.*, 12:758–765, 1984.

[187] X. Niu, Z. Pang, and Z. Wang. Robust small area estimation for unit level model with density power divergence. *PLoS ONE*, 18(11):e0288639, 2023.

[188] J. D. Opsomer, F. J. Breidt, G. Claeskens, G. Kauermann, and M. G. Ranalli. Nonparametric small area estimation using penalized spline regression. *J. Roy. Statist. Soc. B.*, 70(1):265–286, 2008.

[189] Z. Pang, B. Lin, and J. Jiang. Regularisation parameter selection via bootstrapping. *Austral. New Zealand J. Statist.*, 58:335–356, 2016.

[190] P. Parker. Nonlinear Fay-Herriot models for small area estimation using random weight neural networks. *J. Off. Stat.*, 40:317–332, 2024.

[191] D. Pfeffermann. Small area estimation—new developments and directions. *Int Stat Rev*, 70(1):125–143, 2002.

[192] D. Pfeffermann. New important developments in small area estimation. *Statist. Sci.*, 28(1):40–68, 2013.

[193] D. Pierce. The asymptotic effect of substituting estimators for parameters in certain types of statistics. *Ann. Ststist.*, 10:475–478, 1982.

[194] J. Plass, T. Augustin, M. Cattaneo, and G. Schollmeyer. Statistical modelling under epistemic data imprecision: Some results on estimating multinomial distributions and logistic regression for coarse categorical data. In *Proc. ISIPTA'15, T. Augustin, S. Doria, E. Miranda, and E. Quaeghebeur eds*, pages 247–256. Institute of Mathematical Statistics, 2015.

[195] J. Plass, A. Omar, and T. Augustin. Towards a cautious modelling of missing data in small area estimation. *Proc. Mach. Learn. Res.*, 62:253–264, 2017.

[196] N. G. N. Prasad and J. N. K. Rao. The estimation of mean squared errors of small area estimators. *J. Amer. Statist. Assoc.*, 85(409):163–171, 1990.

[197] J. O. Ramsay and B. W. Silverman. *Functional Data Analysis.* Springer, New York, second edition, 2005.

[198] J. N. K. Rao. *Small Area Estimation.* Wiley, New York, first edition, 2003.

[199] J. N. K. Rao and I. Molina. *Small Area Estimation.* Wiley, New York, second edition, 2015.

[200] J. N. K. Rao, S. K. Sinha, and L. Dumitrescu. Robust small area estimation under semi-parametric mixed models. *Can. J. Stat.*, 42:126–141, 2014.

[201] J. N. K. Rao and M. Yu. Small area estimation by combining time series and cross-sectional data. *Can. J. Stat.*, 22(4):511–528, 1994.

[202] B. J. Reich and L. B. Smith. Bayesian quantile regression for censored data. *Biometrics*, 69:651–660, 2013.

[203] K. Reluga, M. J. Lombardía, and S. Sperlich. Bootstrap-based statistical inference for linear mixed effects under misspecifications. *Comput. Stat. Data Anal.*, 195:107956, 2024.

[204] W. Ren, J. Li, A. Erciulescu, T. Krenzke, and L. Mohadjer. A variable selection method for small area estimation modeling of the proficiency of adult completency. *Stats*, 5:689–713, 2022.

[205] J. A. Rice. *Mathematical Statistics and data analysis.* Duxbury Press, Belmont, CA, second edition, 1995.

[206] G. K. Robinson. That BLUP is a good thing: The estimation of random effects (with discussion). *Statist. Sci.*, 6:15–51, 1991.

[207] D. B. Rubin. Inference and missing data. *Biometrika*, 63:581–592, 1976.

[208] G. Schwartz. Estimating the dimension of a model. *Ann. Statist.*, 6:461–464, 1978.

[209] S. R. Searle, G. Casella, and C. E. McCulloch. *Variance Components.* Wiley, New York, 1992.

[210] S. G. Self and K. Y. Liang. Asymptotic properties of maximum likelihood estimators and likelihood ratio tests under nonstandard conditions. *J. Amer. Statist. Assoc.*, 82:605–610, 1987.

[211] J. Shao and D. Tu. *The Jackknife and Bootstrap.* Springer, New York, second edition, 1995.

[212] R. Shibata. Approximate efficiency of a selection procedure for the number of regression variables. *Biometrika*, 71:43–49, 1984.

[213] G. Shmueli. To explain or to predict. *Statist. Sci.*, 25:289–310, 2010.

[214] F. Shokoohi and M. Torabi. Semi-parametric small-area estimation by combining time-series and cross-sectional data methods. *Aust. N. Z. J. Stat.*, 60:323–342, 2018.

[215] S. K. Sinha and J. N. K. Rao. Robust small area estimation. *Canadian J. Stat.*, 37:381–309, 2009.

[216] D. J. Spiegelhalter, N. G. Best, B. P. Carlin, and van der Linde. Bayesian measures of model complexity and fit (with discussion). *J. Roy. Statist. Soc. B*, 64:583–639, 2002.

[217] S. Sugasawa. Robust empirical Bayes small area estimation with density power divergence. *Biometrika*, 107(2):467–480, 2020.

[218] S. Sugasawa, Y. Kawakubo, and G. S. Datta. Observed best selective prediction in small area estimation. *J. Multivar. Anal.*, 173:383–392, 2019.

[219] S. Sugasawa and T. Kubokawa. Heteroscedastic nested error regression models with variance functions. *Stat. Sin.*, 27(3):1101–1123, 2017.

[220] S. Sugasawa and T. Kubokawa. *Mixed Effects Models and Small Area Estimation.* Springer, Singapore, 2023.

[221] H. Sun, T. Nguyen, Y. Luan, and J. Jiang. Classifed mixed logistic model prediction. *J. Multivar. Anal.*, 168:63–74, 2018.

[222] M. Sverchkov and D. Pfeffermann. Prediction of finite population totals based on the sample distribution. *Surv. Methodol.*, 30:79—-92, 2004.

[223] M. Sverchkov and D. Pfeffermann. Small area estimation under informative sampling and not missing at random non-responses. *J. R. Statist. Soc. A*, 181:981–1008, 2018.

[224] R. J. Tibshirani. Regression shrinkage and selection via the Lasso. *J. Roy. Statist. Soc. B*, 58:385–395, 1996.

[225] M. Torabi. Spatial generalized linear mixed models with multivariate CAR models for area data. *Spatial Statistics*, 10:12–26, 2014.

[226] M. Torabi. Spatial generalized linear mixed models in small area estimation. *Can. J. Stat.*, 47:426–437, 2019.

[227] M. Torabi and J. Jiang. Estimation of mean squared prediction error of empirically spatial predictor of small area means under a linear mixed model. *J. Stat. Plan. Inference*, 208:82–93, 2020.

[228] F. Vaida and S. Blanchard. Conditional Akaike information for mixed-effects models. *Biometrika*, 92:351–370, 2005.

[229] S. van Buuren, J. P. L. Brand, C. G. M. Groothuis-Oudshoorn, and D. B. Rubin. Fully conditional specification in multivariate imputation. *J. Statist. Comput. Simulation*, 76:1049–1064, 2005.

[230] M. Viljanen, L. Meijerink, L. Zwakhals, and J. van de Kassteele. A machine learning approach to small area estimation: predicting the health, housing and well-being of the populatio of the Netherlands. https://doi.org.10.1186/s12942-022-00304-5. *Int. J. Health Geogr.*, 21:4, 2022.

[231] G. Wahba. Improper priors, spline smoothing and the problem of guarding against model errors in regression. *J. Roy. Statist. Soc. B*, 40:364–372, 1978.

[232] M. Wand. Smoothing and mixed models. *Comput. Statist.*, 18:223–249, 2003.

[233] J. Wang, W. A. Fuller, and Y. Qu. Small area estimation under a restriction. *Surv. Methodol.*, 34:29–36, 2008.

[234] Y. Wang, N. A. Ponce, P. Wang, J. D. Opsomer, and H. Yu. Generating health estimates by zip code: A semiparametric small area estimation approach using the california health interview survey. *Am. J. Public Health*, 105:2534–2540, 2015.

[235] A. H. Welsh and E. Ronchetti. Bias-calibrated estimation from sample surveys containing outliers. *J. Roy. Statist. Soc. B*, 60:413–428, 1998.

[236] H. White. Maximum Likelihood Estimation of Misspecified Models. *Econometrica*, 50(1):1–25, 1982.

[237] S. N. Wood. *Generalized Additive Models.* Chapman and Hall/CRC, second edition, 2017.

[238] G. Yan and J. Sedransk. Bayesian diagnostic techniques for detecting hierarchical structure. *Bayesian Anal.*, 2:735–760, 2007.

[239] G. Yan and J. Sedransk. A note on Bayesian residuals as a hierarchical model diagnostic technique. *Stat Papers*, 51:1–10, 2010.

[240] Z. Yang, Y. Luan and J. Jiang (2022), The asymptotic distribution of robust maximum likelihood estimator with Huber function for the mixed spatial autoregressive model with outliers, *Communications in Statistics – Theory and Methods* 52, 6311–6340.

[241] J. Ye. On measuring and correcting the effects of data mining and model selection. *J. Amer. Statist. Assoc.*, 93:120–131, 1998.

[242] Y. You and J. N. K. Rao. A pseudo-empirical best linear unbiased prediction approach to small area estimation using survey weights. *Can. J. Stat.*, 30:431–439, 2002.

[243] J. Zhang and J. Jiang. Modified observed best prediction strategies for small area estimation with unit-level data, technical report, dept. of statist., univ. of calif., davis, ca.

[244] H. Zheng and R. J. Little. Penalized spline model-based estimation of the finite population total from probability-proportional-to-size samples. *J. Off. Stat.*, 19:99–117, 2003.

[245] H. Zou. The adaptive Lasso and its oracle properties. *J. Amer. Statist. Assoc.*, 101:1418–1429, 2006.

Index

FSC
www.fsc.org
MIX
Papier | Fördert
gute Waldnutzung
FSC® C083411

Kolibri 360 GmbH
Ferdinand-Jühlke-Straße 7
99095 Erfurt, Deutschland
produktsicherheit@kolibri360.de